Paul Shaw
11/84

SPREAD SPECTRUM COMMUNICATIONS

Volume I

ELECTRICAL ENGINEERING
COMMUNICATIONS AND SIGNAL PROCESSING

Raymond L. Pickholtz, Series Editor

Anton Meijer and Paul Peeters
Computer Network Architectures

Marvin K. Simon, Jim K. Omura, Robert A. Scholtz, and Barry K. Levitt
Spread Spectrum Communications, Volume I

William W Wu
Elements of Digital Satellite Communication, Volume I

Also of interest:

Victor B. Lawrence, Joseph L. Lo Cicero, and Laurence B. Milstein, editors
IEEE Communication Society's Tutorials in Modern Communications

Wushow Chou, Editor-in-Chief
Journal of Telecommunication Networks

SPREAD SPECTRUM COMMUNICATIONS

Volume I

Marvin K. Simon
Jet Propulsion Laboratory

Jim K. Omura
University of California

Robert A. Scholtz
University of Southern California

Barry K. Levitt
Jet Propulsion Laboratory

COMPUTER SCIENCE PRESS

Computer Science Press
11 Taft Court
Rockville, Maryland 20850

1 2 3 4 5 6   Printing   Year   89 88 87 86 85

Library of Congress Cataloging in Publication Data
Main entry under title:

Spread spectrum communications (V.I)
     Bibliography: p.
     Includes index.
     1. Spread spectrum communications—Collected works.
I. Simon, Marvin.
TK5102.5.S6662   1985   621.38′0413   84-4959
ISBN 0-88175-012-3 (Vol. I)
ISBN 0-88175-017-4 (Set)

VOLUME I
CONTENTS

Contents to Other Volumes

VOLUME II
CONTENTS

Preface

VOLUME III
CONTENTS

PREFACE

Not more than a decade ago, the discipline of spread-spectrum (SS) communications was primarily cloaked in secrecy. Indeed, most of the information available on the subject at that time could be found only in documents of a classified nature.

Today the picture is noticeably changed. The open literature abounds with publications on SS communications, special issues of the *IEEE Transactions on Communications* have been devoted to the subject, and the formation of an annual conference on military communications, MILCOM, now offers a public forum for presentation of unclassified (as well as classified) papers dealing with SS applications in military systems. On a less formal note, many tutorial and survey papers have recently appeared in the open literature, in addition to which presentations on a similar level have taken place at major communications conferences. Finally, as further evidence we cite the publication of several books dealing either with SS communications directly or as part of the more general electronic countermeasures (ECM) and electronic counter-counter measures (ECCM) problem. References to all these forms of public documentation are given in Section 1.7 of Chapter 1, Volume I.

The reasons behind this proliferation can be traced to many sources. While it is undoubtedly true that the primary application of SS communications is still in the development of enemy jam-resistant communication systems for the military, a large part of which takes place within the confines of classified programs, the emergence of other applications in both the military and civilian sectors is playing a role of ever-increasing importance. For example, to minimize mutual interference, the flux density of transmissions from radio transmitters often must be maintained at acceptably low radiation levels. A convenient way of meeting these requirements is by spreading the power spectrum of the signal before transmission and despreading it after reception. This is the non-hostile equivalent of the military low-probability-of-intercept (LPI) signal design.

Another instance where SS techniques are particularly useful in a non-anti-jam application is in the area of multiple-access communications wherein many users desire to share a single communication channel. Here the assignment of a unique SS sequence to each user allows him or her to simultaneously transmit over the common channel with a minimum of mutual interference. This often simplifies the network control requirements to coordinate users of the available channel capacity.

Still another example is the requirement for extremely accurate position location using several satellites in synchronous and asynchronous orbits. Here, satellites transmitting pseudorandom noise sequences modulated onto the transmitted carrier signal provide the means for accomplishing the required range and distance determination at any point on the earth.

Finally, SS techniques offer the advantage of improved reliability of transmission in frequency-selective fading and multipath environments. Here the improvement stems from the fact that spreading the information bandwidth of the transmitted signal over a wide range of frequencies reduces its vulnerability to interference located in a narrow frequency band and often provides some diversity gain at the receiver.

At the heart of all these potential applications lies the increasing use of digital forms of modulation for transmitting information, which itself is driven by the tremendous advances that have been made over the last decade in microelectronics. No doubt this trend will continue, and thus it should not be surprising that more and more applications for spread-spectrum techniques will continue to surface. Indeed the state-of-the-art is advancing so rapidly (e.g., witness the recent improvements in frequency synthesizers boosting frequency hop rates from the Khops/sec to the Mhops/sec ranges over SS bandwidths in excess of a GHz) that today's primarily theoretical concepts in a particular situation will be realized in practice tomorrow.

Unclassified research and developments in spread-spectrum communications have reached a point of maturity necessary to justify a textbook on SS communications that goes far beyond the level of those available on today's market. Such is the purpose of *Spread Spectrum Communications*. Contained within the fourteen chapters of its three volumes is an in-depth treatment of SS communications that should appeal to the specialist already familiar with the subject as well as the neophyte with little or no background in the area. The book is organized into five parts within which the various chapters are for the most part self-contained. The exception to this is that Chapter 3, Volume I dealing with basic concepts and system models is a basis for many of the other chapters that follow it. As would be expected, the more traditional portions of the subject are treated in the first two parts, while the latter three parts deal with the more specialized aspects. Thus the authors envision that an introductory one-semester course in SS communications to be taught on a graduate level in a university might cover all or parts of Chapters 1, 3, 4, 5 of Volume I, Chapters 1 and 2 of Volume II, and Chapters 1 and 2 of Volume III.

In composing the technical material presented in *Spread Spectrum Communications*, the authors have intentionally avoided referring by name to specific modern SS systems that employ techniques such as those discussed in many of the chapters. Such a choice was motivated by the desire to offer a unified approach to the subject that stresses fundamental principles rather than specific applications. Nevertheless, the reader should feed confident that the broad experience of the four authors ensures that the material is practically significant as well as academically inspiring.

In writing a book of this magnitude, we acknowledge many whose efforts should not go unnoticed either by virtue of a direct or indirect contribution. Credit is due to Paul Green for originally suggesting the research that uncovered the material in Chapter 2, Volume I, and Bob Price for tireless sleuthing which led to much of the remarkable information presented there. Chapter 5, Volume I benefitted significantly from the comments of Lloyd Welch, whose innovative research is responsible for some of the elegant sequence designs presented there. Per Kullstam helped clarify the material on DS/BPSK analysis in Chapter 1, Volume II. Paul Crepeau contributed substantially to the work on list detectors. Last, but by no means least, the authors would like to thank James Springett, Gaylord Huth, and Richard Iwasaki for their contribution to much of the material presented in Chapter 4, Volume III.

Several colleagues of the authors have aided in the production of a useful book by virtue of critical reading and/or proofing. In this regard, the efforts of Paul Crepeau, Larry Hatch, Vijay Kumar, Sang Moon, Wei-Chung Peng, and Reginaldo Polazzo, Jr. are greatly appreciated.

It is often said that a book cannot be judged by its cover. The authors of *Spread Spectrum Communications* are proud to take exception to this commonly quoted cliche. For the permission to use the historically significant noise-wheel cover design (see Chapter 2, Volume I, Section 2.2.5), we gratefully acknowledge the International Telephone and Telegraph Corp.

Marvin K. Simon
Jim K. Omura
Robert A. Scholtz
Barry K. Levitt

To

Sidney, Belle, Anita, Brette, and Jeffrey Simon
Shomatsu and Shizuko Omura
Lolly, Michael, and Paul Scholtz
Beverly Kaye

for a variety of reasons known only to the authors

Part 1

INTRODUCTION TO SPREAD-SPECTRUM COMMUNICATION

Chapter 1

A SPREAD-SPECTRUM OVERVIEW

Over thirty years have passed since the terms spread-spectrum (SS) and noise modulation and correlation (NOMAC) were first used to describe a class of signalling techniques possessing several desirable attributes for communication and navigation applications, especially in an interference environment. What are these techniques? How are they classified? What are those useful properties? How well do they work? Preliminary answers are forthcoming in this introductory chapter.

We will motivate the study of spread-spectrum systems by analyzing a simple game, played on a finite-dimensional signal space by a communications system and a jammer, in which the signal-to-interference energy ratio in the communication receiver's data detection circuitry serves as a payoff function. The reader is hereby forewarned that signal-to-interference ratio calculations alone cannot illustrate many effects which, in subtle ways, degrade more realistic performance ratios, e.g., bit-error-rate in coded digital SS systems. However, the tutorial value of the following simple energy calculations soon will be evident.

1.1 A BASIS FOR A JAMMING GAME

The following abstract scenario will be used to illustrate the need for spectrum spreading in a jamming environment, to determine fundamental design characteristics, and to quantify one measure of SS system performance. Consider a synchronous digital communication complex in which the communicator has K transmitters available with which to convey information to a cooperating communicator who possesses K matching receivers (see Figure 1.1). Assume for simplicity that the communication signal space has been "divided equally" among the K transmitters. Hence, with a bandwidth W_{ss} available for communicating an information symbol in a T_s second interval $(0, T_s)$, the resultant transmitted-signal function space of dimension approximately $2T_s W_{ss}$ is divided so that each transmitter

1

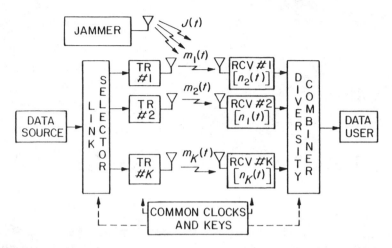

Figure 1.1. The scenario for a game between a jammer and a communication system complex.

has a D-dimensional subspace, $D = 2T_sW_{ss}/K$, in which to synthesize its output signal. Denote an orthonormal basis for the total signal space by $\psi_k(t)$, $k = 1, 2, \ldots, 2T_sW_{ss}$, i.e.,

$$\int_0^{T_s}\psi_j(t)\psi_k^*(t)\,dt = \begin{cases} 1, & j = k \\ 0, & j \neq k \end{cases} \tag{1.1}$$

where the basis functions may be complex valued, and ( )* denotes conjugation. Then the signal emitted by the k-th transmitter is of the form

$$m_k(t) = \sum_{j \in \mathcal{N}_k} a_j\psi_j(t), \tag{1.2}$$

where

$$\mathcal{N}_k = \{\, j\colon (k-1)D < j \leq kD \,\} \tag{1.3}$$

and $\{a_j\}$ is a data-dependent set of coefficients. We will refer to the above as an *orthogonal communication system complex of multiplicity K*.

Of course, real systems generally radiate real signals. The reader may wish to view $m_k(t)$ as the modulation on the radiated signal $\mathrm{Re}\{m_k(t)\exp(j\omega_ct + \theta)\}$. Without loss of generality, we can dispense with the shift to RF during this initial discussion.

In a simplified jamming situation, the signal $z_i(t)$ observed at the i-th receiver in the receiving complex might be

$$z_i(t) = \sum_{k=1}^{K} m_k(t) + J(t) + n_i(t), \tag{1.4}$$

where $n_i(t)$ represents internally generated noise in the i-th receiver, $J(t)$ is an externally generated jamming signal, and the K-term sum represents the total output signal of the transmitter complex. One signal processing strategy for the i-th receiver is to project the received signal onto the set of basis functions for the i-th transmitter's signal space, thereby calculating

$$z_j = \int_0^{T_s} z_i(t)\psi_j^*(t)\,dt \quad \text{for all } j \in \mathcal{N}_i. \tag{1.5}$$

In the absence of jamming and receiver noise, the properties of the orthonormal basis insure that $z_j = a_j$, and thus, the i-th receiver correctly discovers the data dependent set of coefficients $\{a_j\}$, $j \in \mathcal{N}_i$, used by the i-th transmitter.

Both the jamming and receiver noise signals can be expanded in terms of the orthonormal basis as

$$J(t) = \sum_{j=1}^{2T_sW_{ss}} J_j\psi_j(t) + J_0(t), \tag{1.6}$$

$$n_i(t) = \sum_{j=1}^{2T_sW_{ss}} n_{ij}\psi_j(t) + n_{0i}(t), \tag{1.7}$$

where $J_0(t)$ represents that portion of the jamming signal orthogonal to all of the $2T_sW_{ss}$ basis functions used in producing the composite signal. The receiver noise component $n_{0i}(t)$ likewise is orthogonal to all possible transmitted signals. These representations indicate that, in general, the projection (1.5) of $z_i(t)$ onto $\psi_j(t)$ in the i-th receiver produces

$$z_j = a_j + J_j + n_{ij} \quad \text{for all } j \in \mathcal{N}_i. \tag{1.8}$$

The everpresent thermal noise random variable n_{ij}, assumed complex Gaussian, independent, and identically distributed for different values of i and/or j, represents the relatively benign receiver perturbations in the absence of jamming. The jamming signal coefficients J_j are less easily classified, and from the jammer's point of view, hopefully are unpredictable by the receiver.

The total energy E_J in the jamming signal $J(t)$ over the time interval $(0, T_s)$ is given by

$$E_J = \int_0^{T_s} |J(t)|^2\,dt = \sum_{j=1}^{2T_sW_{ss}} |J_j|^2 + \int_0^{T_s} |J_0(t)|^2\,dt \tag{1.9}$$

Obviously, the energy term involving $J_0(t)$ serves no useful jamming purpose, and henceforth, will be assumed zero. (In keeping with this conservative aspect of communication system design, we also assume that the jammer has full knowledge of timing and of the set $\{\psi_j(t)\}$ of basis functions.) The sum in (1.9) can be partitioned into K parts, the i-th part representing the energy E_{Ji} used to jam the i-th receiver. Thus,

$$E_J = \sum_{i=1}^{K} E_{Ji}, \quad E_{Ji} = \sum_{j \in \mathcal{N}_i} |J_j|^2. \tag{1.10}$$

A similar partition holds for the total transmitted signal energy E_s, namely

$$E_S = \sum_{i=1}^{K} E_{Si}, \quad E_{Si} = \sum_{j \in \mathcal{N}_i} |a_j|^2, \tag{1.11}$$

E_{Si} being the energy used by the i-th transmitter. The additive partitions (1.10), (1.11) are a direct result of the orthogonality requirement placed on the signals produced by the transmitter complex.

The above signal representations and calculations have been made under the assumption that the channel is ideal, causing no attenuation, delay, or distortion in conveying the composite transmitted signal to the receiver complex, and that synchronous clocks are available at the transmitter and receiver for determining the time interval $(0, T_s)$ of operation. Hence, important considerations have been suppressed in this initial discussion, so that we may focus on one major issue facing both the communication system designer and the jammer designer, namely their allocations of transmitter energy and jammer energy over the K orthogonal communication links.

1.2 ENERGY ALLOCATION STRATEGIES

Within the framework of an orthogonal communication system complex of multiplicity K, let's consider the communicator and jammer to use the following strategies for allocating their available energies, E_S and E_J respectively, to the K links.

Communicators' strategy: Randomly select K_S links, $K_S \leq K$, for equal energy allocations, each receiving E_S/K_S units. The remaining links are not utilized.

Jammer's strategy: Randomly select K_J receivers for equal doses of jamming energy, each receiving E_J/K_J units. The remaining channels are not jammed.

The quantity K_S is referred to as the *diversity factor* of the communication system complex. When K_S exceeds unity, the receiver must employ a diversity combining algorithm to convert the outputs of the K_S chosen links into a single output for the system user. The performance measure to be employed here, in determining the effectiveness of these strategies, will not depend on specifying a particular diversity combining algorithm.

The randomness required of these strategies should be interpreted as meaning that the corresponding adversary has no logical method for predicting the choice of strategy, and must consider all strategies equally likely. Furthermore, random selection of communication links by the transmitter should not affect communication quality since all available links are assumed to have equal attributes. (Examples of link collections with non-uniform attributes will be considered in Volume II, Chapter 2.)

The receiving complex, having knowledge of the strategy selected for communication, will collect all E_S units of transmitted energy in the K_S receivers remaining in operation. However, the amount of jamming energy collected by those same K_S receivers is a random variable whose value is determined by which of the $\binom{K}{K_J}$ jamming strategies is selected ($\binom{\vdots}{\vdots}$ denotes a binomial coefficient). Under the equally likely strategy assumption, the probability that the jammer strategy will include exactly N of the K_S receivers in use, is given by

$$\Pr(N) = \begin{cases} \dfrac{\dbinom{K_S}{N}\dbinom{K - K_S}{K_J - N}}{\dbinom{K}{K_J}}, & N_{\min} \leq N \leq N_{\max} \\ \\ 0, & \text{otherwise} \end{cases} \qquad (1.12)$$

where

$$N_{\min} = \max(0, K_J + K_S - K) \qquad (1.13)$$

$$N_{\max} = \min(K_S, K_J). \qquad (1.14)$$

Using (1.12)–(1.14), it is possible to compute the expected total effective jamming energy $E_{J\text{eff}}$ sensed by the K_S receivers, namely

$$E_{J\text{eff}} = \frac{E_J}{K_J}\mathbf{E}\{N\}, \qquad (1.15)$$

$\mathbf{E}$ being the expected value operator. Despite the complicated form of $\Pr(N)$, it can be verified that

$$\mathbf{E}\{N\} = \frac{K_J K_S}{K}, \qquad (1.16)$$

and hence, that

$$E_{J\text{eff}} = \frac{E_J K_S}{K}. \qquad (1.17)$$

More generally, it can be verified that when the communicators use the strategy described above, (1.17) is the average total effective jamming energy for any arbitrary distribution of jamming energy.

This idealized situation leads one to conclude, based on (1.17), that the receiver can minimize the jammer's effectiveness energy-wise by not using diversity, i.e., by using $K_S = 1$. Furthermore, the multiplicity K of the orthogonal communication system complex should be made as large as possible to reduce $E_{J\text{eff}}$, i.e., the complex should be designed to use all of the available bandwidth. The energy-optimal communication strategy ($K_S = 1$) using a single one of the K available communication links, is called a *pure spread-spectrum strategy*. This strategy, with its accompanying threat to use any of K orthogonal links, increases the total signal-to-jamming ratio from E_S/E_J at each receiving antenna's terminals to KE_S/E_J at the output of the designated receiver, and therefore qualifies as an *anti-jam* (AJ) modulation technique.

The improvement $E_J/E_{J\text{eff}}$ in signal-to-jamming ratio will be called the *energy gain EG* of the signalling strategy played on the orthogonal communication system complex.

$$EG = \frac{E_J}{E_{J\text{eff}}} = \frac{K}{K_S} = \frac{2T_s W_{ss}}{K_S D}. \tag{1.18}$$

Hence, the energy gain for a pure SS strategy is the multiplicity factor of the complex. In this fundamental form (1.18), the energy gain is the ratio of the signal space dimension $2T_s W_{ss}$ perceived by the jammer for potential communication use to the total dimension $K_S D$ of the K_S links' D-dimensional signal spaces which the receiver must observe. For a fixed $T_s W_{ss}$ product, this definition of energy gain makes no distinction between diversity and SS strategies using the same signal space dimension $K_S D$. The reader may recognize the fact that the quantity called the multiplicity factor, or energy gain in this chapter, is sometimes referred to as the processing gain of the SS system. This nomenclature is by no means universally accepted, and we will instead identify the term *processing gain PG* with the ratio W_{ss}/R_b, where R_b is the data rate in bits/second. It is easily verified from (1.18) that processing gain and energy gain are identical when $R_b = K_S D/2T_s$, e.g., for binary orthogonal signalling ($D = 2$) with no diversity ($K_S = 1$).

Two key assumptions were made in showing that the pure SS strategy is best: (1) The channel is ideal and propagates all signals equally well, and (2) the proper performance measure is the total effective jamming energy. If either of the above assumptions is not acceptable, then the jammer's strategy may influence the performance measure, and the optimum diversity factor K_S may be greater than one. Indeed, in later chapters it is shown that the use of bit-error rate (BER) as a performance measure implies that the optimum diversity factor can exceed unity.

Let's summarize the requirements characteristic of a digital spread-spectrum communication system in a jamming environment:

1. The bandwidth (or equivalently the link's signal-space dimension D) required to transmit the data in the absence of jamming is much less than the bandwidth W_{ss} (or equivalently the system's signal space dimension $2T_s W_{ss}$) available for use.
2. The receiver uses inner product operations (or their equivalent) to confine its operation to the link's D-dimensional signal space, to demodulate the signal, and thereby to reject orthogonal jamming waveform components.
3. The waveforms used for communication are randomly or pseudorandomly selected, and equally likely to be anywhere in the available

bandwidth (or equivalently, anywhere in the system's $2T_sW_{ss}$ dimensional signal space).

The term *pseudorandom* is used specifically to mean random in appearance but reproducible by deterministic means.

We will now review a sampling of the wide variety of communication system designs which possess SS characteristics.

1.3 SPREAD-SPECTRUM SYSTEM CONFIGURATIONS AND COMPONENTS

A pure spread-spectrum strategy, employing only a single link at any time, can be mechanized more efficiently than the system with potential diversity factor K, shown in Figure 1.1. In an SS system, the K transmitter-receiver pairs of Figure 1.1 are replaced by a single wideband communication link having the capability to synthesize and detect all of the waveforms potentially generated by the orthogonal communication system complex. The pure SS strategy of randomly selecting a link for communication is replaced with an equivalent approach, namely, selecting a D-dimensional subspace for waveform synthesis out of the system's $2T_sW_{ss}$-dimensional signal space. This random selection process must be independently repeated each time a symbol is transmitted. Independent selections are necessary to avoid exposing the communication link to the threat that the jammer will predict the signal set to be used, will confine his jamming energy to that set, and hence, will reduce the apparent multiplicity and energy gain to unity.

Three system configurations are shown in Figure 1.2, which illustrate basic techniques that the designer may use to insure that transmitter and receiver operate synchronously with the same apparently random set of signals. The portions of the SS system which are charged with the responsibility of maintaining the unpredictable nature of the transmission are double-boxed in Figure 1.2. The *modus operandi* of these systems is as follows:

1. *Transmitted reference* (TR) systems accomplish SS operation by transmitting two versions of a wideband, unpredictable carrier, one ($x(t)$) modulated by data and the other ($r(t)$) unmodulated (Figure 1.2(a)). These signals, being separately recovered by the receiver (e.g., one may be displaced in frequency from the other), are the inputs to a correlation detector which recovers the data modulation. The wideband carrier in a TR-SS system may be a truly random, wideband noise source, unknown by transmitter and receiver until the instant it is generated for use in communication.

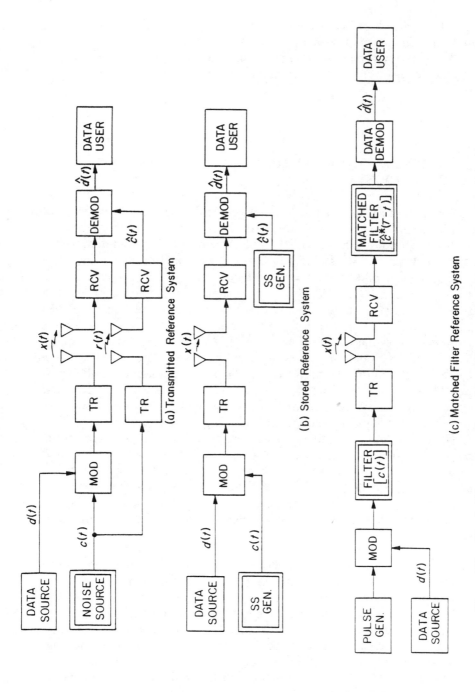

Figure 1.2. Simple SS system configurations. (The notation $\hat{z}(t)$ is used to denote an estimate of $z(t)$.)

2. *Stored reference* (SR) systems require independent generation at transmitter and receiver of pseudorandom wideband waveforms which are
identical in their essential characteristics (Figure 1.2(b)). The receiver's
SS waveform generator is adjusted automatically to keep its output in
close synchronism with the arriving SS waveform. Data detection, then,
is accomplished by cross-correlation. The waveform generators are initialized prior to use by setting certain parameters in the generating
algorithm, thereby denying the jammer knowledge of the waveform set
being used (even if the jammer has succeeded in determining the generator's structure).
3. *Matched filter* (MF) systems generate a wideband transmitted signal by
pulsing a filter having a long, wideband, pseudorandomly controlled
impulse response (Figure 1.2(c)). Signal detection at the receiver employs
an identically pseudorandom, synchronously controlled, matched filter
which performs the correlation computation. Matched filter systems
differ from SR systems primarily in the manner in which the inner-product detection process is mechanized, and hence, have externally observed
properties similar to those of SR systems.

Certainly, a pure TR system has several fundamental weaknesses including:
(1) The system is easily spoofed since a jammer can in principle transmit a
pair of waveforms which are accepted by the receiver, (2) relatively poor
performance occurs at low signal levels because noise and interference are
present on both signals which are cross-correlated in the receiver, (3) the
data is easily determined by any listener who has access to both transmitted
signals, and (4) the TR system's two channels may require extra bandwidth
and may be difficult to match. Some of the problems associated with TR
systems may be mitigated by randomly changing parameters of one of the
communication links (e.g., by protecting one of the TR wideband links with
an SR-like technique). Historical examples of SR-protected TR systems will
be given in the next chapter.

Spread-spectrum waveform generators for SR systems employing the
following general modulation formats have been built. The output of an SS
waveform generator is given the generic name $c(t)$ and is a (possibly
complex-valued) baseband representation of the SS waveform.

1. *Recorded modulation*: The waveform $w(t)$ of duration T_p is recorded, and
if necessary, extended periodically to give

$$c(t) = \sum_n w(t - nT_p).\tag{1.19}$$

The utility of this type of signal is limited by the problem of distributing
recordings to transmitter and receiver so that reuse of a waveform is not
necessary.
2. *Frequency hopping* (FH): Assuming that $p(t)$ is a basic pulse shape of
duration T_h (usually called the *hop time*), frequency-hopping modulation

has the form

$$c(t) = \sum_n \exp[j(2\pi f_n t + \phi_n)] p(t - nT_h). \tag{1.20}$$

In all likelihood, the complex baseband signal $c(t)$ never physically appears in the transmitter or receiver. Instead, the pseudorandomly generated sequence $\{f_n\}$ of frequency shifts will drive a frequency synthesizer to produce a real-valued IF or RF carrier-modulated version of $c(t)$. The sequence $\{\phi_n\}$ of random phases is a by-product of the modulation process.

3. *Time hopping* (TH): Assuming that the pulse waveform $p(t)$ has duration at most T_s/M_T, a typical time hopping waveform might be

$$c(t) = \sum_n p\left(t - \left(n + \frac{a_n}{M_T}\right)T_s\right). \tag{1.21}$$

In this example, time has been segmented into T_s second intervals, with each interval containing a single pulse pseudorandomly located at one of M_T locations within the interval.

4. *Direct sequence* (DS) *modulation*: Spread-spectrum designers call the waveform

$$c(t) = \sum_n c_n p(t - nT_c) \tag{1.22}$$

direct sequence modulation. Here, the output sequence $\{c_n\}$ of a pseudorandom number generator is linearly modulated onto a sequence of pulses, each pulse having a duration T_c called the *chip time*.

5. *Hybrid modulations*: Each of the above techniques possesses certain advantages and disadvantages, depending on the system design objectives (AJ protection is just one facet of the design problem). Potentially, a blend of modulation techniques may provide better performance at the cost of some complexity. For example, the choice

$$c(t) = \prod_i c^{(i)}(t) \tag{1.23}$$

may capture the advantages of the individual wideband waveforms $c^{(i)}(t)$ and mitigate their individual disadvantages.

Three schemes seem to be prevalent for combining the data signal $d(t)$ with the SS modulation waveform $c(t)$ to produce the transmitted SS signal $x(t)$.

1. *Multiplicative modulation*: Used in many modern systems, the transmitted signal for multiplicative modulation is of the form

$$x(t) = \text{Re}\{d(t)c(t)e^{j(\omega_c t + \phi_T)}\}. \tag{1.24}$$

Mechanization simplicity usually suggests certain combinations of data and SS formats, e.g., binary phase-shift-keyed (BPSK) data on a DS signal, or multiple frequency-shift-keyed (MFSK) data on a FH signal. These modulation schemes are the ones of primary interest in this book.

2. *Delay modulation*: Suggested for use in several early systems, and a natural for mechanization with TH-SS modulation, this technique transmits the signal

$$x(t) = \text{Re}\{c(t - d(t))e^{j(\omega_c t + \phi_T)}\}. \tag{1.25}$$

3. *Independent (switching) modulation*: Techniques (1) and (2) are susceptible to a jamming strategy in which the jammer forwards the transmitted signal to the receiver with no significant additional delay (a severe geometric constraint on the location of the jammer with respect to the transmitter and receiver), but with modified modulation. This repeater strategy, which if implementable, clearly reduces the multiplicity factor K of the SS system to unity, can be nullified by using a transmitted signal of the form

$$x(t) = \text{Re}\{c^{(d(t))}(t)e^{j(\omega_c t + \phi_T)}\}. \tag{1.26}$$

Here the data signal, quantized to M levels, determines which of M distinct SS modulations $c^{(d)}(t)$, $d = 1, \ldots, 2, \ldots, M$, is transmitted. The key assumption here is that even though the jammer can observe the above waveform, it cannot reliably produce an alternate waveform $c^{(j(t))}(t)$, $j(t) \neq d(t)$, acceptable to the receiver as alternate data modulation. The cost of independent data modulation is a clearly increased hardware complexity.

The data demodulation process in a digital SS system must compute inner products in the process of demodulation. That is, the receiver must mechanize calculations of the form

$$(m_T, m_R) = \int_0^{T_S} m_T(t) m_R^*(t) \, dt \tag{1.27}$$

where in general $m_T(t)$ and $m_R(t)$ represent complex baseband signals and (m_T, m_R) is their inner product. However, one or both of these complex baseband signals usually appears in modulated form as a real IF or RF signal

$$x_i(t) = \text{Re}\{m_i(t) e^{j(\omega_i t + \phi_i)}\}, \ i = T, R. \tag{1.28}$$

The inner product (1.27) can be recovered from the modulated signal(s) in several ways, as illustrated in Figure 1.3. For example, the receiver can first demodulate the signal $x_T(t)$ to recover the real and imaginary parts of $m_T(t)$ and then proceed with straightforward correlation or matched filtering operations using baseband signals. On the other hand, as indicated in Figures 1.3(c) and 1.3(d), there are alternative ways to compute the inner product, which do not require that both signals be shifted to baseband first.

In all cases, the heart of the SS receiver is its synchronization circuitry, and the heartbeats are the clock pulses which control almost all steps in

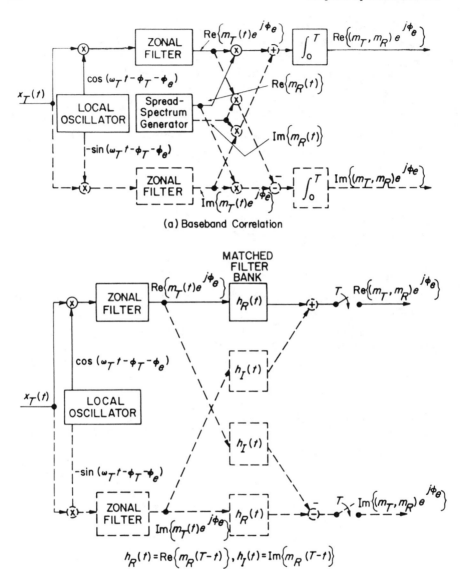

(a) Baseband Correlation

(b) Baseband Matched Filtering

Figure 1.3. Examples of correlation-computing block diagrams. The dashed portions of the diagrams can be eliminated when the modulations $m_T(t)$ and $m_R(t)$ are real and, in addition, the local oscillator is phase-coherent, i.e., $\phi_e \approx 0$. Solid line processing is often called the "in-phase" channel, while the dashed line processing is called the "quadrature" channel.

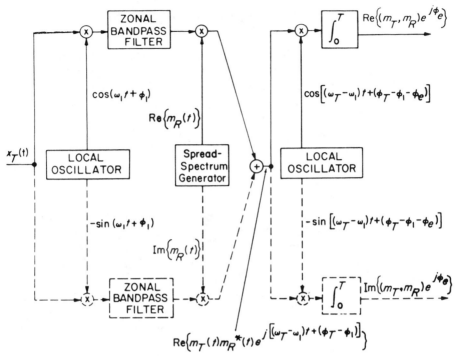

(c) IF Multiplication, Baseband Integration

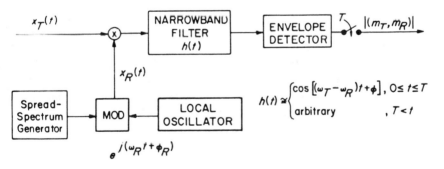

(d) IF Multiplication and Integration
(Bandpass Correlator)

Figure 1.3. *Continued.*

forming the desired inner product. Recovery of $\mathrm{Re}\{(m_T, m_R)\}$ and $\mathrm{Im}\{(m_T, m_R)\}$ requires three levels of synchronization.

1. *Correlation interval synchronization*: Correlators require pulses to indicate when the interval of integration is to begin and when it is to end. In the bandpass correlator of Figure 1.3(d), interval sync not only provides the timing for the sampling operation, but also initializes the narrowband filter's state to zero at the beginning of each correlation interval. Typically, in DS systems these signals correspond to the data symbol clock pulses. In FH systems in which the data symbol time exceeds the hop time, the interval sync pulses must indicate the duration of a single frequency, since correlation operations spanning random phase transitions are not generally useful.
2. *SS generator synchronization*: Timing signals are required to control the epoch of the system's SS waveform generator's output and the rate at which that output is produced. Direct sequence systems employ a clock ticking at the chip rate $1/T_c$ for this purpose, while FH systems have a similar clock operating at the hopping rate $1/T_h$.
3. *Carrier synchronization*: Ideal reduction of the SS signal to baseband in the receiver is possible if a local oscillator (or oscillator network) is available whose output is in frequency and phase synchronism with the received signal's carrier (i.e., $\phi_e = 0$ in Figure 1.3). The above level of carrier sync is often available in DS systems, but usually only frequency synchronism is attained in FH systems.

In some SS systems, the above synchronization signals are derived from a single clock; in others, the carrier local oscillator is independent of the clock signals which control its modulation. Automatic control circuitry generally is included to align the receiver's clocks for proper demodulation of the incoming signal, although some systems have been built in which ultrastable clocks are initially aligned and then are allowed to drift in a free-running mode until communication is concluded. Proper operation of the correlation computing circuits generally requires control of the symbol clock epoch to within a small fraction of the correlation interval's duration T. Similarly, it is necessary to adjust the SS generator clock's ticks to within a small fraction of the reciprocal of the SS modulation's short-term bandwidth, i.e., the bandwidth of the energy spectrum of the SS reference waveform within a correlation interval. Section 1.5.3 will indicate that the SS generator clock error for DS and FH systems must be a small fraction of T_c and T_h, respectively, to maintain correlator operation at nearly maximum output signal levels, as required.

Frequency synchronous operation of correlation detectors requires that the phase drift between the incoming carrier (excluding SS modulation) and the receiver's local oscillator, over a correlation interval, be a fraction of radian, i.e., the quantity ϕ_e in Figure 1.3 may be assumed nearly constant during the correlation computation. (Phase synchronism of the local oscillator requires, in addition, that ϕ_e be near zero.) The bandpass correlator is a frequency synchronous device requiring that the input to its narrowband filter be centered in its passband to an error tolerance of a fraction of the reciprocal of the correlation time.

Output threshold crossing techniques, similar to those used in radar detection, are an alternative to MF output sampling in Figure 1.3(b), and may have higher tolerance to synchronization errors than SR/DS systems. However, any realized tolerance to synchronization errors implies a potential weakness to repeater jamming.

1.4 ENERGY GAIN CALCULATIONS FOR TYPICAL SYSTEMS

The following examples illustrate the energy gain calculation for basic SS systems. In each case, the SS system is viewed (in the terminology of Section 1.2) as replacing an orthogonal communication system complex, and its multiplicity factor K determined, thereby evaluating the energy gain of the system via (1.18).

Example 1.1. Let's examine a SS system using binary (± 1) DS spreading modulation as in (1.22), multiplicative data modulation (1.24), and single-channel phase-synchronous detection (solid line portion of Figure 1.3(a)) over the data symbol duration T_s of N_c chip times, i.e., $T_s = N_c T_c$. Hence, the pseudorandom quantities which are known to the receiver, but unknown to the jammer, are the DS pulse modulation sequence $c_1, c_2, \ldots, c_{N_c}$ and the carrier phase ϕ_T. The inner product of two such waveforms with different pseudorandom variables and data modulations is given by

$$\int_0^{T_s} \text{Re}\left\{ dc(t) e^{j(\omega_c t + \phi_T)} \right\} \text{Re}\left\{ d'c'(t) e^{j(\omega_c t + \phi_T')} \right\} dt$$

$$= \frac{dd' \cos\left(\phi_T - \phi_T'\right)}{2} \int_0^{T_s} c(t) c'(t)\, dt$$

$$= \frac{dd' \cos\left(\phi_T - \phi_T'\right)}{2} \sum_{n=1}^{N_c} c_n c_n' \int_0^{T_c} |p(t)|^2\, dt. \qquad (1.29)$$

For any particular choice of $[c_1, \ldots, c_{N_c}]$ and ϕ_T, and regardless of the values of the data modulation d and d', constants over $(0, T_s)$, the two signals are orthogonal if either $\phi_T - \phi_T' = \pi/2$ or $[c_1', \ldots, c_{N_c}']$ is orthogonal to $[c_1, \ldots, c_{N_c}]$. Since the set of real N_c-tuples forms an N_c-dimensional space (and furthermore an orthogonal basis of vectors with ± 1 entries can be found when N_c is a multiple of 4), and since carrier phase differences of magnitude $\pi/2$ cause orthogonality, the jammer is forced to view his waveform selection problem as being defined for an orthogonal communication system complex with multiplicity factor K given by

$$K = 2N_c = 2T_s/T_c. \tag{1.30}$$

Example 1.2. Suppose that two independently hopped SS waveforms, $c^{(0)}(t)$ and $c^{(1)}(t)$, of the form (1.20) are employed in a switching modulation scheme as in (1.26) to transmit binary data. The data symbols' duration T_s spans M_D hop times. Frequency-synchronous correlation computations in the receiver are then carried out over individual hop times, and the correlator's sync clock produces a pulse every T_h seconds. The signal parameters, known to the receiver but not to the jammer, are the two pseudorandom hopping sequences $f_{01}, f_{02}, \ldots, f_{0M_D}$ and $f_{11}, f_{12}, \ldots, f_{1M_D}$ which represent the two possible data symbols. Two similar FH waveforms have inner product

$$\int_0^{T_h} \cos\big(2\pi(f_c + f_{d1})t + \phi_1\big)\cos\big(2\pi(f_c + f'_{d'1})t + \phi'_1\big)\,dt$$

$$= \frac{1}{2}\int_0^{T_h}\cos\big(2\pi(f_{d1} - f'_{d'1})t + (\phi_1 - \phi'_1)\big)\,dt$$

$$= \frac{1}{4\pi(f_{d1} - f'_{d'1})}\Big[\cos(\phi_1 - \phi'_1)\sin\big(2\pi(f_{d1} - f'_{d'1})T_h\big)$$

$$+ \sin(\phi_1 - \phi'_1)\big(\cos(2\pi(f_{d1} - f'_{d'1})T_h) - 1\big)\Big].$$

$$\tag{1.31}$$

Orthogonality between two such waveforms is guaranteed, regardless of the values of the phases (ϕ_1, ϕ'_1) and data symbols (d, d'), provided that

$$f_{d1} - f'_{d'1} = \frac{k}{T_h} \tag{1.32}$$

where k is any non-zero integer, for all d, d' in $\{0, 1\}$. We assume that two such orthogonal frequency waveforms are used by the communication system as f_{0n} and f_{1n}, with a different pair for each n, $1 \leq n \leq M_D$.

If the transmitter and receiver are capable of producing and observing M_F distinct orthogonal tones (assume M_F is even for convenience), it is clear that during each hop the jammer must contemplate combatting a pure SS strategy on an orthogonal communication system complex of multiplicity $M_F/2$. Therefore, during each hop time a single link in the orthogonal system complex requires four dedicated orthonormal basis functions (e.g., sines and cosines at two distinct frequencies), and uses $4M_D$ such functions over M_D hops. By the same reasoning the number of basis functions available to the entire complex is $2M_FM_D$, and hence, the energy gain of (1.18) is given by

$$EG = \frac{2M_FM_D}{4M_D} = \frac{M_F}{2}. \tag{1.33}$$

Example 1.3. One possible hybrid SS communication system employs TH, FH, and DS modulations to produce the wideband waveform

$$c^{(d(t))}(t) = \sum_n e^{j(2\pi f_n t + \phi_n)} \sum_{m=1}^{N_c} c_{nmd_n} p\left(t - mT_c - \left(n + \frac{b_n}{M_T}\right)T_s\right)$$

$$(1.34)$$

in which $p(t)$ is a unit-amplitude rectangular pulse of chip time duration T_c. A total of N_c pulses, modulated by the sequence $\{c_{nmd_n}\}$, are concatenated to produce a DS waveform of duration T_h, which in turn is frequency-hopped to one of M_F frequencies and time-hopped to one of M_T time intervals within the symbol time T_s. Hence,

$$T_s = M_T T_h = M_T N_c T_c. \tag{1.35}$$

Modulation by the n-th M-ary data symbol d_n is accomplished by switching the DS modulation to the d_n-th of M orthogonal vectors $c_{n1d_n}, \ldots, c_{nN_c d_n}$. Two such hybrid SS signals have inner product given by

$$\int_0^{T_s} \sum_{m=1}^{N_c} c_{0md_0} p\left(t - mT_c - \frac{b_0}{M_T}T_s\right)\cos(2\pi(f_c + f_0)t + \phi_0)$$

$$\times \sum_{m'=1}^{N_c} c'_{0m'd'_0} p\left(t - m'T_c - \frac{b'_0}{M_T}T_s\right)\cos(2\pi(f_c + f'_0)t + \phi'_0)\, dt$$

$$= \delta_{b_0 - b'_0} \cdot \frac{1}{4\pi(f_0 - f'_0)}\left[\cos(\phi_0 - \phi'_0)\sin(2\pi(f_0 - f'_0)T_c)\right.$$

$$\left. + \sin(\phi_0 - \phi'_0)(\cos(2\pi(f_0 - f'_0)T_c) - 1)\right]\sum_{m=1}^{N_c} c_{0md_0}c'_{0md'_0}.$$

$$(1.36)$$

Here δ_z is the Kronecker delta function, which is one if $z = 0$ and is zero otherwise. Orthogonality of these waveforms can be achieved, regardless of the data values (d, d') and random phase values (ϕ_0, ϕ'_0), if one or more of the following conditions holds:

1.
$$\sum_{m=1}^{N_c} c_{0md}c'_{0md} = 0, \text{ for all } d, d'. \tag{1.37}$$

2.
$$b_0 \neq b'_0. \tag{1.38}$$

3.
$$f_0 - f'_0 = \frac{k}{T_c}, \text{ for } k \text{ integer}, k \neq 0. \tag{1.39}$$

The signal variables, known *a priori* to the receiver, but not the jammer, are the pseudorandom DS chip values $\{c_{0md}: d = 1, 2, \ldots, M, m = 1, 2, \ldots, N_c\}$, the hop frequency f_0, and the time interval index b_0. Hence, the receiver must observe the signal in a $2M$-dimensional space whose basis over the interval $(0, T_s)$ consists of $\text{Re}\{c^{(d)}(t)\exp(j2\pi f_c t)\}$ evaluated for each of the M values of d, with the random hop phase ϕ_0 set at 0 and at $\pi/2$. The jammer, however, must choose his waveform to jam a signal space of dimension $2N_c M_F M_T$, whose basis consists of the sines and cosines of N_c orthogonal DS modulations hopped over M_F orthogonal tones and M_T disjoint time intervals. Therefore, the nominal energy gain EG of this system is

$$EG = \frac{N_c M_F M_T}{M}.\tag{1.40}$$

The nominal minimum bandwidth W_{ss} required to implement the orthogonality requirements $(1 - 3)$ is determined by the minimum hop frequency spacing $1/T_c$ to be M_F/T_c. Using this fact and substituting (1.35) into (1.40) indicates that the energy gain for this hybrid SS system in a closely packed design is $T_s W_{ss}/M$.

1.5 THE ADVANTAGES OF SPECTRUM SPREADING

We have seen the advantages of making a jammer counteract an ensemble of orthogonal communication systems. The bandwidth increase which must accompany this SS strategy has further advantages which we will outline here.

1.5.1 Low Probability of Intercept (LPI)

Spectrum spreading complicates the signal detection problem for a surveillance receiver in two ways: (1) a larger frequency band must be monitored, and (2) the power density of the signal to be detected is lowered in the spectrum-spreading process. The signal may have further desirable attributes based in part on LPI, such as low probability of position fix (LPPF) which includes both intercept and direction finding (DFing) in its evaluation, or low probability of signal exploitation (LPSE) which may append additional effects, e.g., source identification.

For now let's simply evaluate the power spectral density (PSD) of an SS signal to determine its general properties. Consider a signal of the form

$$x(t) = \text{Re}\{m_T(t)e^{j(2\pi f_c t + \phi_T)}\},\tag{1.41}$$

where $m_T(t)$ represents the total modulation as a result of data and spectrum-spreading effects, and ϕ_T is a random phase variable uniformly distributed on $(-\pi, \pi)$. The PSD $S_x(f)$ of the waveform $x(t)$ is defined as

$$S_x(f) \triangleq \lim_{T \to \infty} \frac{1}{2T}\text{E}\{|\mathbf{F}_{2T}\{x(t)\}|^2\}.\tag{1.42}$$

Here $\mathbf{F}_{2T}$ is the time-limited Fourier transform

$$\mathbf{F}_{2T}\{x(t)\} \triangleq \int_{-T}^{T} x(t) e^{-j2\pi ft}\, dt, \qquad (1.43)$$

and hence, $\mathbf{E}\{|\mathbf{F}_{2T}\{x(t)\}|^2\}$ is the average energy spectral density of a $2T$ second signal segment. Conversion of this energy density to a power density is accomplished by division by $2T$, and $S_x(f)$ then is simply the limiting form as T becomes large. Averaging over the random phase ϕ_T leads to the relation

$$S_x(f) = \tfrac{1}{4}\left[S_m(f - f_c) + S_m(-f - f_c)\right], \qquad (1.44)$$

where $S_m(f)$ is the PSD of the modulation $m_T(t)$. We will now evaluate $S_m(f)$ for two basic SS modulation designs.

1. *DS/BPSK modulation*: The waveform corresponding to a DS signal antipodally modulated by binary data is

$$m_T(t) = \sum_n c_n d_{\lfloor n/N_c \rfloor}\, p(t - nT_c), \qquad (1.45)$$

in which the data value $d_{\lfloor n/N_c \rfloor}$ has the opportunity to change every N_c chip times ($\lfloor x \rfloor$ denotes the integer part of x), and $p(t)$ is a pulse shape which is non-zero only in the interval $(0, T_c)$. The calculation of $S_m(f)$ is simplified by letting the limit parameter T grow in multiples of the symbol time $N_c T_c$. Hence,

$$S_m(f) = \lim_{K \to \infty} \frac{1}{2KN_cT_c} \mathbf{E}\left\{|\mathbf{F}_{2KN_cT_c}\{m_T(t)\}|^2\right\}. \qquad (1.46)$$

By converting the sum index n in (1.45) to $kN_c + m$ where k ranges from $-K$ to $K-1$ and m ranges between 0 and $N_c - 1$, it is possible to show that

$$\mathbf{F}_{2KN_cT_c}\{m_T(t)\} = P(f) \sum_{k=-K}^{K-1} d_k e^{-j2\pi f k N_c T_c} \sum_{m=0}^{N_c-1} c_{kN_c+m} e^{-j2\pi f m T_c}, \qquad (1.47)$$

where $P(f)$ is the Fourier transform of the chip pulse $p(t)$. Inserting this transform into (1.46) and ensemble averaging over the data sequence $\{d_k\}$ which we assume to be composed of independent binary random variables, each equally likely to be $+1$ or -1, gives

$$S_m(f) = \lim_{K \to \infty} \frac{1}{2K} \sum_{k=-K}^{K-1} \frac{|P(f)|^2}{T_c}$$

$$\times \frac{1}{N_c} \mathbf{E}\left\{\left|\sum_{m=0}^{N_c-1} c_{kN_c+m} e^{-j2\pi f m T_c}\right|^2\right\}. \qquad (1.48)$$

We consider two possible assumptions regarding the nature of the direct sequence $\{c_n\}$.

a. *Random DS Modulation*: If $\{c_n\}$ is a sequence of independent, identically distributed random variables, each equally likely to be $+1$ or -1, then the ensemble average in (1.48) simplifies to N_c, and

$$S_m(f) = \frac{|P(f)|^2}{T_c}. \tag{1.49}$$

This PSD is sketched in Figure 1.4 for several possible pulse shapes. Note that $1/T_c$ is a rough measure of the widths of these spectra.

b. *Periodic DS Modulation*: When $\{c_n\}$ is generated by a finite state device acting as a pseudorandom number generator, then $\{c_n\}$ must be periodic with some period N, i.e.,

$$c_{n+N} = c_n \quad \text{for all } n. \tag{1.50}$$

Hence, nothing random remains in (1.48) and the expected value operator can be dropped. Furthermore, the value of the sum on m in (1.48) must also be a periodic function of K, with period P_c corre-

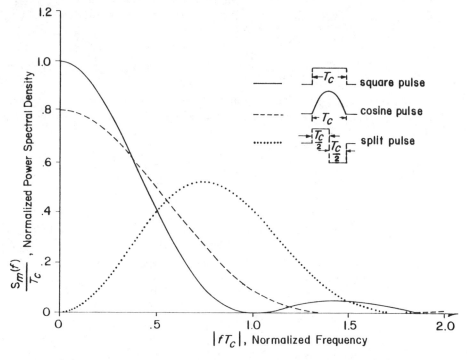

Figure 1.4. The power spectral density of random DS/BPSK modulation for three different chip pulse shapes.

sponding to the smallest value of k such that kN_c is a multiple of N. Hence,

$$P_c = \frac{\text{lcm}(N, N_c)}{N_c}, \tag{1.51}$$

where $\text{lcm}(\cdot, \cdot)$ denotes the least common integer multiple of its arguments, and

$$S_m(f) = \frac{|P(f)|^2}{T_c}$$

$$\times \frac{1}{P_c} \sum_{k=0}^{P_c-1} \frac{1}{N_c} \left| \sum_{m=0}^{N_c-1} c_{kN_c+m} e^{-j2\pi f m T_c} \right|^2. \tag{1.52}$$

When $P_c = N$, then (1.52) can be simplified still further by taking the k-sum inside the two sums of the squared absolute value, to give,

$$S_m(f) = \frac{|P(f)|^2}{NN_cT_c} \sum_{m=0}^{N_c-1} \sum_{m'=0}^{N_c-1} R_c(m - m') e^{-j2\pi f(m-m')T_c}, \tag{1.53}$$

where $R_c(\cdot)$ is the periodic autocorrelation function of the sequence $\{c_n\}$, i.e.,

$$R_c(k) \triangleq \sum_{m=0}^{N-1} c_{m+k}c_m = \sum_{m=0}^{N-1} c_{mN_c+k}c_{mN_c}. \tag{1.54}$$

The latter sum in (1.54) is the periodic correlation function of the sequence $\{c_{mN_c}\}$ which is called the *decimation* of $\{c_m\}$ by N_c. Such decimations preserve the periodic autocorrelation properties of the sequence when the greatest common divisor of N and N_c is unity. Summing like terms in (1.53) and using the symmetry of $R_c(k)$ gives

$$S_m(f) = \frac{|P(f)|^2}{T_c}$$

$$\times \frac{1}{N} \left[R_c(0) + 2 \sum_{k=1}^{N_c-1} R_c(k)\left(1 - \frac{k}{N_c}\right)\cos(2\pi f k T_c) \right]. \tag{1.55}$$

Since $\{c_n\}$ is composed of ± 1 elements, then $R_c(0) = N$, and this PSD is identical to that of random DS modulation provided that $R_c(k) = 0$ for $k = 1, 2, \ldots, N_c - 1$. This latter condition is one reasonable objective for pseudorandom number generator design.

2. *FH/FSK modulation*: The use of M-ary data, frequency-shift-keyed onto an FH signal at a rate of M_D hops per data symbol, creates the modulation

$$m_T(t) = \sum_n e^{j[2\pi(f_n + d_{\lfloor n/M_D \rfloor})t + \phi_n]} p(t - nT_h), \tag{1.56}$$

where $p(t)$ is a hop pulse waveform which is non-zero only in the interval $(0, T_h)$, $\{f_n\}$ is the sequence of frequency hop values, $\{d_k\}$ is the data sequence, and $\{\phi_n\}$ is a sequence of independent, uniformly distributed, random phase variables.

In calculating $S_m(f)$ using the approach in (1.42) we will take the limit as the integer $K = T/T_h$ becomes large. Hence,

$$S_m(f) = \lim_{K \to \infty} \frac{1}{2KT_h} \mathbf{E}\{|\mathbf{F}_{2KT_h}\{m_T(t)\}|^2\}. \tag{1.57}$$

The required Fourier transform is given by

$$\mathbf{F}_{2KT_h}\{m_T(t)\} = \sum_{n=-K}^{K-1} e^{j\phi_n}\mathbf{F}\{e^{j2\pi(f_n+d_{\lfloor n/M_D\rfloor})t}p(t - nT_h)\} \tag{1.58}$$

where $\mathbf{F}$ denotes the ordinary Fourier transform. Evaluating the transform in (1.58), substituting in (1.57), and averaging over the random phase sequence, gives

$$S_m(f) = \lim_{K \to \infty} \frac{1}{2KT_h} \sum_{n=-K}^{K-1} \mathbf{E}\{|P(f - f_n - d_{\lfloor n/M_D\rfloor})|^2\}, \tag{1.59}$$

where $P(f)$ is the transform of the hop pulse. Assuming that the data sequence elements are independent, identically distributed random variables chosen from the set $\mathscr{D}$ with probability function $P_D(\cdot)$, gives

$$S_m(f) = \sum_{d \in \mathscr{D}} P_D(d) \lim_{K \to \infty} \frac{1}{2KT_h} \sum_{n=-K}^{K-1} \mathbf{E}\{|P(f - f_n - d)|^2\}. \tag{1.60}$$

We now consider two models for the hopping sequence.

a. *Random Hopping.* When $\{f_n\}$ is a sequence of identically distributed random variables with values selected from a frequency set $\mathscr{F}$ according to a probability distribution function $P_F(\cdot)$, then

$$S_m(f) = \frac{1}{T_h} \sum_{d \in \mathscr{D}} \sum_{f' \in \mathscr{F}} P_D(d)P_F(f')|P(f - f' - d)|^2. \tag{1.61}$$

b. *Periodic Hopping.* If $\{f_n\}$ is sequence of frequencies with period N, generated by a finite state machine, and if $N_{f'}$ denotes the number of occurrences of the frequency value f' in one sequence period, then

$$S_m(f) = \frac{1}{T_h} \sum_{d \in \mathscr{D}} \sum_{f' \in \mathscr{F}} P_D(d)\frac{N_{f'}}{N}|P(f - f' - d)|^2. \tag{1.62}$$

Notice that the PSD of the deterministically generated sequence will match that of the randomly generated sequence with $P_F(f) = N_f/N$. Furthermore, the number of hops M_D spanned by a data symbol is not a factor in the spectral density of the resultant signal in either case.

A rough measure of the hop pulse transform's width is $1/T_h$. Since integer multiples of $1/T_h$ are required for orthogonal randomly phased tones of duration T_h, the assumption that the radiated tones $d + f$, $f \in \mathscr{F}$ and $d \in \mathscr{D}$, are orthogonal leads to the conclusion that the minimum required bandwidth is approximately MM_F/T_h, $|\mathscr{F}| = M_F$, $|\mathscr{D}| = M$.

The PSDs which we have just calculated correspond to asymptotically long-term ensemble-averaged, time-normalized, energy spectra. Attempts at interception of these signals in reality will be made over short time intervals, usually much shorter than the periods of the pseudorandom number generators driving the SS waveform generator. Hence, signals with comparable PSDs may have differing short-term energy density characteristics, and hence, different LPI capabilities.

1.5.2 Independent Interference Rejection and Multiple-Access Operation

We have already discussed the energy gain achievable against a jammer whose radiated signal is generated without knowledge of the key parameters used in generating the SS transmitter's modulation. We refer to this type of interference as *independent*, the connotation also applying to in-band interference from other friendly communication systems.

The ability of a SS system to reject independent interference is the basis for the multiple-access capability of SS systems, so called because several SS systems can operate in the same frequency band, each rejecting the interference produced by the others by a factor approximately equal to its energy gain. This asynchronous form of spectrum sharing is often called *spread-spectrum multiple-access* (SSMA) or *code-division multiple-access* (CDMA).

As an illustration of SSMA operation, consider a transmitted signal $x_T(t)$,

$$x_T(t) = \text{Re}\left\{ \sqrt{P_T} m_T(t) e^{j(2\pi f_c t + \phi_T)} \right\}, \tag{1.63}$$

and an interfering signal $x_I(t)$,

$$x_I(t) = \text{Re}\left\{ \sqrt{P_I} m_I(t - t_I) e^{j(2\pi(f_c + \Delta)t + \phi_I)} \right\}, \tag{1.64}$$

impinging on a receiver which is frequency synchronous with the transmitted signal and which computes the inner product of the received signal with a reference modulation $m_R(t)$. We assume without loss of generality that $m_T(t)$, $m_I(t)$, and $m_R(t)$ are unit power waveforms, i.e., the time-averaged and ensemble-averaged value of $|m_j(t)|^2$ is one for $j = T, I, R$. The output of the receiver correlator is the sum of two terms, corresponding to the desired inner product $\sqrt{P_T} v_T$ and the interference inner product $\sqrt{P_I} v_I$. These terms for the k-th correlation interval $((k - 1)T, kT)$, normalized to unit input power, are

$$v_T(k) = (m_T, m_R) = \int_{(k-1)T}^{kT} m_T(t) m_R^*(t) \, dt \cdot e^{j\phi_e}, \tag{1.65}$$

$$v_I(k) = (m_I, m_R) = \int_{(k-t)T}^{kT} m_I(t - t_I) m_R^*(t) e^{j2\pi\Delta t} \, dt \, e^{j(\phi_I - \phi_T + \phi_e)}, \tag{1.66}$$

where ϕ_e is the phase tracking error in the receiver (see Figure 1.3), and Δ is the carrier frequency offset between the interference and the transmitted signal. The average signal-to-interference energy at the input to the correlator is simply

$$\text{SIR}_{\text{in}} \triangleq \frac{\left\langle \mathbf{E}\left\{ P_T \int_{(k-1)T}^{kT} |m_T(t)|^2 \, dt \right\} \right\rangle}{\left\langle \mathbf{E}\left\{ P_I \int_{(k-1)T}^{kT} |m_I(t)|^2 \, dt \right\} \right\rangle} = \frac{P_T}{P_I}, \tag{1.67}$$

and the frequency-synchronous correlator's output signal-to-interference measurement ratio is

$$\text{SIR}_{\text{out(freq sync)}} = \frac{\left\langle \mathbf{E}\{ |v_T(k)|^2 \} \right\rangle}{\left\langle \mathbf{E}\{ |v_I(k)|^2 \} \right\rangle} \cdot \text{SIR}_{\text{in}}. \tag{1.68}$$

Here $\langle \cdot \rangle$ denotes discrete time averaging over the parameter k.

The above calculation is based on the supposition that outputs from both the in-phase and quadrature channels of the correlator are necessary, i.e., both the real and imaginary parts of $v_T(k)$ are necessary in the reception process. If instead reception is phase-synchronous, then $v_T(k)$ is real, and on the average only half of the interference power contributes to the disturbance of $\text{Re}\{ v_T(k) \}$. Hence, for phase-synchronous detection,

$$\text{SIR}_{\text{out}(\phi \text{ sync})} = 2 \cdot \text{SIR}_{\text{out(freq sync)}}. \tag{1.69}$$

The key to further analysis is the evaluation of the interference level at the correlator output.

$$\mathbf{E}\{ |v_I(k)|^2 \} = \mathbf{E}\left\{ \iint_{(k-1)T}^{kT} m_I(t - t_I) m_I^*(t' - t_I) e^{j2\pi\Delta(t-t')} \right.$$

$$\left. \times m_R^*(t) m_R(t') \, dt \, dt' \right\}. \tag{1.70}$$

The variable t_I denotes an arbitrary time shift of the interference modulation relative to the reference modulation, this shift being inserted to model the fact that the interference is assumed asynchronous with respect to all clocks generating the transmitted signal. Stored reference modulations (e.g., $m_I(t)$ if the interference is SSMA in nature) are generally *cyclo-stationary*, i.e., they may be made stationary by inserting a time-shift random variable (e.g., t_I) which is uniformly distributed on a finite interval. Hence, we assume that t_I is uniformly distributed on a finite interval which makes $m_I(t - t_I)$ a wide-sense stationary random process. This assumption and the further assumption that any random variables in $m_I(t)$ are independent of any in $m_R(t)$, effectively allow us to treat asynchronous multiple-access interference and jamming identically in average power calculations.

Averaging (1.70) over t_I gives

$$\mathbf{E}\{|v_I(k)|^2\} = \int\int_{(k-1)T}^{kT} R_I(t-t')e^{j2\pi\Delta(t-t')}\mathbf{E}\{m_R^*(t)m_R(t')\}\, dt\, dt'$$

$$= \int_{-\infty}^{\infty} S_I(f-\Delta)\mathbf{E}\{|M_{Rk}(f)|^2\}\, df, \tag{1.71}$$

where $R_I(\cdot)$ and $S_I(\cdot)$ are the autocorrelation function and PSD respectively of the unit power process $m_I(t-t_I)$. That is,

$$R_I(\tau) = \mathbf{E}\{m_I(t+\tau-t_I)m_I^*(t-t_I)\} = \int_{\infty}^{\infty} S_I(f)e^{j2\pi f\tau}\, df, \tag{1.72}$$

and $M_{Rk}(f)$ is the Fourier transform of the reference modulation $m_R(t)$ in the k-th correlation interval.

$$M_{Rk}(f) = \int_{(k-1)T}^{kT} m_R(t)e^{-j2\pi ft}\, dt. \tag{1.73}$$

Certainly when the correlation time T is not a multiple of the reference modulation generator's period (and this must be the case to avoid problems with simple repeater jammers), then $|M_{Rk}(f)|^2$ will vary with k.

Using the fact that the modulation forms are normalized to unit energy, and combining (1.65)–(1.68) and (1.71), demonstrates that the improvement in signal-to-interference ratio achieved in the correlation calculation is

$$\frac{\text{SIR}_{\text{out(freq sync)}}}{\text{SIR}_{\text{in}}} = \frac{\left\langle \mathbf{E}\left\{ \left| \int_{(k-1)T}^{kT} m_T(t)m_R^*(t)\, dt \right|^2 \right\} \right\rangle}{\int_{-\infty}^{\infty} S_I(f-\Delta)\langle \mathbf{E}\{|M_{Rk}(f)|^2\}\rangle\, df}. \tag{1.74}$$

Again we emphasize that (1.74) applies both to asynchronous multiple-access interference and independent jamming.

Example 1.4. Suppose that a DS/BPSK communication system employs binary antipodal modulation and phase coherent reception over a symbol time T_s. Hence, in the k-th correlation interval, $m_T(t)$ in (1.45) is equal to $c(t)$ or $-c(t)$, and reception is performed by correlating the received signal with $c(t)$ as given in (1.22). Then,

$$M_{Rk}(f) = \mathbf{F}\left\{ \sum_{n=(k-1)N_c}^{kN_c-1} c_n p(t-nT_c) \right\}$$

$$= P(f) \sum_{n=(k-1)N_c}^{kN_c-1} c_n e^{-j2\pi fnT_c}, \tag{1.75}$$

where again N_c denotes the number of chips per data bit, $\mathbf{F}$ denotes the Fourier transform operation, and $P(f)$ is the Fourier transform of the chip pulse $p(t)$. Denoting the period of $\{c_n\}$ by N and time-averaging the squared magnitude of $M_{Rk}(f)$ over k gives

$$\langle |M_{Rk}(f)|^2 \rangle = \frac{1}{P_c} \sum_{k=0}^{P_c-1} |M_{Rk}(f)|^2$$

$$= T_s S_m(f), \qquad (1.76)$$

where P_c is defined in (1.51) and $S_m(f)$ is given by (1.52). This result can be verified analytically, despite the fact that $S_m(f)$ is the PSD of the data-modulated waveform $m_T(t)$ while (1.76) represents an average of short-term energy spectra of $m_R(t)$. However, this result is reasonable because averaging over the zero-mean independent data symbols, which leads to the result (1.52), breaks up the PSD calculation into a time average of short-term energy spectra.

The SIR improvement ratio for a DS/BPSK system is determined by substituting (1.76) into (1.74) and using the fact that $m_T(t) = dm_R(t)$ in a correlation interval.

$$\frac{\text{SIR}_{\text{out}(\phi\,\text{sync})}}{\text{SIR}_{\text{in}}} = 2\left[\frac{1}{T_s}\int_{-\infty}^{\infty} S_I(f - \Delta)S_m(f)\,df\right]^{-1}. \qquad (1.77)$$

Since both $S_I(f)$ and $S_m(f)$ have fixed areas and are non-negative, the worst-case interference PSD $S_I(f - \Delta)$ in (1.77) is a Dirac delta function located at the frequency which maximizes $S_m(f)$. Therefore,

$$\frac{\text{SIR}_{\text{out}(\text{freq sync})}}{\text{SIR}_{\text{in}}} \geq 2\left[\frac{1}{T_s}\max_f S_m(f)\right]^{-1}. \qquad (1.78)$$

A well-designed spreading sequence $\{c_n\}$ should result in a PSD $S_m(f)$ close to that of a purely random sequence, namely $|P(f)|^2/T_c$. In keeping with our power normalizations, let's assume that

$$p(t) = \begin{cases} 1, & 0 \leq t < T_c \\ 0, & \text{otherwise} \end{cases} \quad |P(f)| = \left|\frac{\sin(\pi f T_c)}{\pi f}\right|, \qquad (1.79)$$

and hence, $\max_f |P(f)| = T_c$. Therefore, a well-designed spreading sequence has $\max_f S_m(f) = T_c$, and from (1.78)

$$\frac{\text{SIR}_{\text{out}(\phi\,\text{sync})}}{\text{SIR}_{\text{in}}} \geq \frac{2T_s}{T_c} = 2N_c. \qquad (1.80)$$

This lower bound on signal-to-interference ratio improvement is the energy gain indicated in Example 1.1.

When the interference is spectrally similar to the transmitted signal, e.g., in the SSMA case, and both signals possess the spectra of a purely random binary sequence modulated on rectangular chip pulses (1.79), then

$$\frac{\text{SIR}_{\text{out}(\phi\,\text{sync})}}{\text{SIR}_{\text{in}}} = 2\left[\frac{1}{T_s}\int_{-\infty}^{\infty} S_m^2(f)\,df\right]^{-1}$$

$$= 2\left[\frac{1}{T_s T_c^2}\int_{-\infty}^{\infty}|P(f)|^4\,df\right]^{-1} = 3N_c. \qquad (1.81)$$

The above example illustrates the following points concerning SIR improvement calculations based on long-term energy averages:

1. The jammer can minimize the signal-to-interference ratio by transmitting a single tone at the frequency of the transmitter spectrum's peak. However, this is not the best jamming strategy if the receiver has an additional notch filtering capability, or if the designer is interested in the more significant bit-error-rate (BER) design criterion.
2. The above SIR improvement calculation is valid even if $m_I(t) = m_T(t)$, randomization by time shift t_I being enough to make $m_T(t - t_I)$ and $m_T(t)$ quite distinct on the average. This suggests that several systems could use identical SS waveforms, provided the probability that they arrive in near synchronism at a receiver (for any reason, natural or jam-motivated) is virtually zero.
3. Cross-correlation functions and cross-spectra between asynchronous interference and desired signal are not a specific factor in the signal-to-interference ratio improvement based on long-term averages.

Clearly the use of long-term averages in a SIR-based figure-of-merit has led to these simple results.

Similar results can be achieved for other forms of SS modulation. The DS example was particularly simple because the receiver used only one correlator. Corresponding analyses of SS systems using higher dimensional signal sets, e.g., FH/FSK, must consider the total signal energy and total interference energy collected in a set of correlators.

1.5.3 High-Resolution Time-of Arrival (TOA) Measurements

Not all interference waveforms satisfy the independence and randomly asynchronous assumptions used in Section 1.5.2 to reaffirm the energy gain capability of SS systems. Here are some examples which are illustrated pictorially in Figure 1.5.

1. *Multipath*: Additional propagation paths from transmitter to receiver may produce undesirable interference in a correlator synchronized to a signal arriving via a specified path. For example, a single additional path may produce an interfering signal of the form (1.64) with

$$m_I(t) = m_T(t), \qquad (1.82)$$

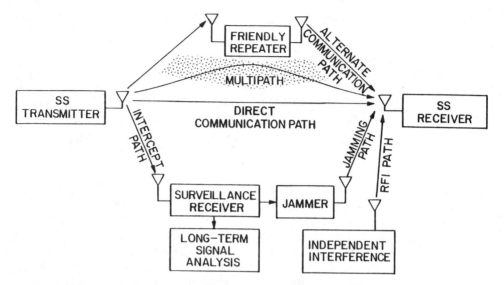

Figure 1.5. A scenario for SS link operation.

at a fixed delay t_I corresponding to the incremental propagation delay between the interference path and the communication signal path.

2. *Repeater Jamming*: This is a form of artificial multipath, in which the jammer attempts to receive the SS signal, somehow alter the data modulation, and then broadcast the result. Hence, if the modulation is multiplicatively changed, then the signal retransmitted by the jammer may be of the form (1.64) with

$$m_I(t) = d'(t)m_T(t). \qquad (1.83)$$

In this case, t_I corresponds to the additional propagation delay encountered over the propagation path through the surveillance/jamming system.

In both of these cases, the time shift parameter t_I and frequency offset Δ of (1.64) are nearly constant or slowly varying over restricted ranges. The interference's incremental delay t_I is a positive quantity when a direct natural propagation path is used for communication. On the other hand, communication via a friendly repeater or an indirect path may result in a negative value for t_I.

The average response of a correlation detector to the above types of signal-related interference can be determined by evaluating (1.66) or (1.71), e.g.,

$$\left\langle \mathbf{E}\{|v_I(k)|^2\} \right\rangle = \left\langle \mathbf{E}\left\{\left|\int_{(k-1)T}^{kT} m_I(t - t_I)m_R^*(t)e^{j2\pi\Delta t}\,dt\right|^2\right\} \right\rangle. \qquad (1.84)$$

Here it is assumed that the receiver's correlator reference signal $m_R(t)$ is synchronized to the signal arriving via the communication path, and that t_I and Δ are the incremental time and frequency shifts incurred by the interference signal during the k-th correlation interval. At this point, we must depart from the analysis of the previous section, because t_I is fixed within a limited range in this interference scenario and cannot be used as an averaging variable.

The shape of $\langle \mathbf{E}\{|v_I(k)|^2\}\rangle$ as a function of t_I and Δ will indicate the time and frequency *resolution* capability of the SS signal structure, i.e., the ability of the receiver's correlation detector to discriminate against versions of the transmitted signal which do not arrive in synchronism with the receiver's clocks. The mean-squared value of the integral in (1.84) is the *cross-ambiguity function* of the waveforms $m_I(t)$ and $m_R(t)$ at offsets t_I and Δ, and hence, we are embarking on a study of the time- and ensemble-averaged ambiguity function provided by an SS waveform/modulation system. The theory of auto-ambiguity functions (i.e., $m_I(t) = m_R(t)$) states that the time-width of the function's central and largest peak at $t_I = 0$ and $\Delta = 0$ is inversely proportional to the rms bandwidth of the signal upon which the correlator acts. Hence, one might expect under certain conditions that SS receivers are especially sensitive to synchronization errors, and possess high time-of-arrival (TOA) resolution capabilities.

The evaluation of (1.84) for repeater modulation of the form (1.83), under the assumption that the added modulation $d'(t)$ is an independent stationary random process, yields

$$\langle \mathbf{E}\{|v_{RJ}(k)|^2\}\rangle = \Big\langle \mathbf{E}\Big\{ \iint_{(k-1)T}^{kT} R_{d'}(t - t')e^{j2\pi\Delta(t-t')}$$

$$\times m_T(t - t_I)m_R^*(t)m_T^*(t' - t_I)m_R(t')\, dt\, dt' \Big\rangle$$

$$= \int_{-\infty}^{\infty} S_{d'}(f)\langle \mathbf{E}\{|v_M(k)|^2\}\rangle df \qquad (1.85)$$

where $R_{d'}(\cdot)$ and $S_{d'}(\cdot)$ are the autocorrelation function and PSD, respectively, of $d'(t)$; $v_{RJ}(k)$ and $v_M(k)$ are the values of $v_I(k)$ for repeater jamming (1.83) and multipath (1.84) respectively, and

$$|v_M(k)|^2 = \left| \int_{(k-1)T}^{kT} m_T(t - t_I)m_R^*(t)e^{j2\pi(f+\Delta)t}\, dt \right|^2 . \qquad (1.86)$$

Henceforth, for simplicity we set

$$\nu \triangleq f + \Delta. \qquad (1.87)$$

Equation (1.85) indicates that the effect of the jammer's added modulation $d'(t)$ is to average an equivalent multipath interference measure $\langle \mathbf{E}\{|v_M(k)|^2\}\rangle$ at offsets t_I and $f + \Delta$ over f (weighted by $S_{d'}(f)$) to determine the repeater jamming measure $\langle \mathbf{E}\{|v_{RJ}(k)|^2\}\rangle$.

We will now evaluate the multipath interference measure for two basic SS waveform designs.

1. *DS/BPSK modulation*: In this case $m_T(t)$ is given by (1.45) and $m_R(t)$ is simply the SS code $c(t)$ in (1.22). Breaking up the integral over the data symbol interval of duration $T = T_s$ in (1.86) into a sum of integrals over the chip intervals of $m_R(t)$, gives

$$|v_M(k)|^2 = \left| \sum_{n=(k-1)N_c}^{kN_c-1} \sum_{m=-\infty}^{\infty} d_{\lfloor m/N_c \rfloor} c_m c_n \right.$$

$$\left. \times \int_{nT_c}^{(n+1)T_c} p(t - t_I - mT_c) p^*(t - nT_c) \cdot e^{j2\pi\nu t} dt \right|^2. \quad (1.88)$$

The integral in (1.88) can be simplified by using the fact that the pulse shape $p(t)$ is non-zero only in $(0, T_c)$ and by defining the quantities N_I (and integer) and τ_I to satisfy

$$-t_I = N_I T_c + \tau_I, \qquad 0 \leq \tau_I < T_c. \quad (1.89)$$

Then

$$\int_{nT_c}^{(n+1)T_c} p(t - t_I - mT_c) p^*(t - nT_c) e^{j2\pi\nu t} dt$$

$$= \begin{cases} X_p(\tau_I, \nu; T_c) e^{j2\pi\nu n T_c}, \\ \quad \text{for } m = N_I + n \\ X_p(\tau_I - T_c, \nu; T_c) e^{j2\pi\nu n T_c}, \\ \quad \text{for } m = N_I + n + 1 \\ 0, \text{ otherwise} \end{cases} \quad (1.90)$$

where $X_p(\tau, \nu; T)$ is the ambiguity function of the pulse waveform $p(t)$ generally defined as

$$X_p(\tau, \nu; T) = \int_0^T p(t + \tau) p^*(t) e^{j2\pi\nu t} dt \quad (1.91)$$

for pulse waveforms $p(t)$ non-zero only in the time interval $(0, T)$. The limited values of m which produce non-zero results in (1.90) further simplify (1.88) to

$$|v_M(k)|^2 = \left| \sum_{n=(k-1)N_c}^{kN_c-1} e^{j2\pi\nu n T_c} \left[d_{\lfloor (N_I+n)/N_c \rfloor} c_{N_I+n} c_n X_p(\tau_I, \nu; T_c) \right. \right.$$

$$\left. \left. + d_{\lfloor (N_I+n+1)/N_c \rfloor} c_{N_I+n+1} c_n X_p(\tau_I - T_c, \nu; T_c) \right] \right|^2. \quad (1.92)$$

Equation (1.92) is a convenient starting point for time and ensemble averaging operations.

We will carry out the ensemble-averaging process under the assumption that the spectrum-spreading sequence $\{c_n\}$ is composed of independent, identically distributed random variables, equally likely to be $+1$ or -1, and that the independent data sequence $\{d_k\}$ likewise, is composed solely of $+1$'s and -1's. The expansion of (1.92) then requires the moments

$$
\mathbf{E}\left\{d_{\lfloor(N+m)/N_c\rfloor}d_{\lfloor(N+n)/N_c\rfloor}c_{N+m}c_mc_{N+n}c_n\right\} = \begin{cases} 1 & \text{if } N = 0 \\ 1 & \text{if } N \neq 0, n = m \\ 0 & \text{otherwise} \end{cases}
$$

$$(1.93)$$

$$
\mathbf{E}\left\{d_{\lfloor(N+m)/N_c\rfloor}d_{\lfloor(N+n+1)/N_c\rfloor}c_{N+m}c_mc_{N+n+1}c_n\right\} = 0 \text{ for all } N, m, n,
$$

$$(1.94)$$

and the fact that

$$
\left|\sum_{n=(k-1)N_c}^{kN_c-1} e^{j2\pi\nu nT_c}\right|^2 = \left|\frac{\sin(\pi\nu T_s)}{\sin(\pi\nu T_c)}\right|^2.
$$

$$(1.95)$$

Expanding the squared sum in (1.92) as a double sum and simplifying, gives

$$
\mathbf{E}\left\{|v_M(k)|^2\right\} = \begin{cases} \left|\dfrac{\sin(\pi\nu T_s)}{\sin(\pi\nu T_c)}\right|^2 |X_p(\tau_I,\nu;T_c)|^2 + N_c|X_p(\tau_I - T_c,\nu;T_c)|^2, \\ \qquad\qquad \text{for } N_I = 0, \\[6pt] \left|\dfrac{\sin(\pi\nu T_s)}{\sin(\pi\nu T_c)}\right|^2 |X_p(\tau_I - T_c,\nu;T_c)|^2 + N_c|X_p(\tau_I,\nu;T_c)|^2, \\ \qquad\qquad \text{for } N_I = -1, \\[6pt] N_c\left[|X_p(\tau_I,\nu;T_c)|^2 + |X_p(\tau_I - T_c,\nu;T_c)|^2\right], \\ \qquad\qquad \text{otherwise.} \end{cases}
$$

$$(1.96)$$

Because of the manner in which we modelled $\{c_n\}$ and $\{d_k\}$, the above mean-squared value of $v_M(k)$ is not a function of k, and hence, no further averaging is necessary.

When the chip pulse $p(t)$ is unit amplitude and rectangular in shape, with duration T, then

$$
|X_p(\tau,\nu;T_c)|^2 = \begin{cases} \left|\dfrac{\sin[\pi\nu(T - |\tau|)]}{\pi\nu}\right|^2, & |\tau| < T \\ 0, & \text{otherwise.} \end{cases}
$$

$$(1.97)$$

With the aid of (1.89) and (1.97), the average ambiguity surface (1.96) can be evaluated as a function of t_I and ν, and the results are displayed in Figure 1.6 for the case $N_c = 100$.

When there is no time or frequency mismatch, i.e., when $t_I = 0$ and $\nu = 0$, then the normalized detector output (1.96) is $N_c^2 T_c^2$. Alternatively, when $|t_I| \geq T_c$, then (1.96) is upper-bounded by $N_c T_c^2$. Hence, multipath and repeater jamming are reduced by a factor $1/N_c$ when the interference's incremental delay exceeds a chip time. This is a firm basis for stating that the TOA resolution of a high energy gain DS/BPSK system is approximately one chip time.

2. *FH/FSK modulation*: When FH signalling of the form (1.56) is employed, a correlator set to detect the data frequency d_R correlates the received signal with the reference

$$m_R(t) = \sum_n e^{j[2\pi(f_n + d_R)t + \phi_n']} p(t - nT_h). \tag{1.98}$$

where $p(t)$ is a hop pulse of duration T_h which is non-zero only on $(0, T_h)$, and $\{\phi_n'\}$ is a sequence of uniformly distributed phase variables. The correlator resets every hop interval, i.e., $T = T_h$, and the squared

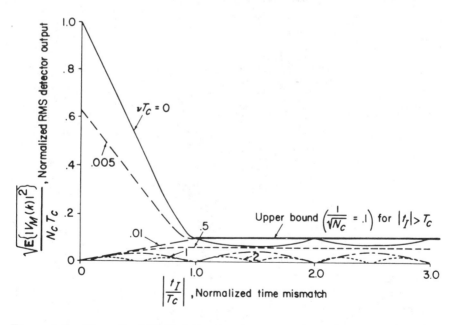

Figure 1.6. Normalized RMS detector output as a function of normalized time mismatch t_I/T_c for several values of normalized frequency mismatch νT_c. Random DS modulation is assumed, with an energy gain of $N_c = 100$ and square chip pulses.

k-th correlator output (1.86) is

$$|v_M(k)|^2 = \left| \int_{(k-1)T_h}^{kT_h} \sum_m e^{j[2\pi(f_m + d \ \lfloor m/M_D \rfloor)(t-t_I) + \phi_m]} p(t - t_I - mT_h) \right.$$

$$\left. \times e^{-j[2\pi(f_{k-1} + d_R)t + \phi'_{k-1}]} p^*(t - (k-1)T_h) e^{j2\pi\nu t} \, dt \right|^2. \quad (1.99)$$

Only two terms in the sum over m contribute to the value of $|v_M(k)|^2$ for a given k. Defining the integer N_I and τ_I to satisfy

$$-t_I = N_I T_h + \tau_I, \qquad 0 \leq \tau_I < T_h, \qquad (1.100)$$

and using techniques similar to those seen earlier, result in

$$|v_M(k)|^2 = |e^{j\theta} X_p\big(\tau_I, f_{N_I + k - 1} + d_{\lfloor (N_I + k - 1)/M_D \rfloor} - f_{k-1} - d_R + \nu; T_h\big)$$

$$+ e^{j\theta'} X_p\big(\tau_I - T_h, f_{N_I + k} + d_{\lfloor (N_I + k)/M_D \rfloor}$$

$$-f_{k-1} - d_R + \nu; T_h\big)|^2, \qquad (1.101)$$

where θ and θ' are independent, uniformly distributed phase variables, and $X_p(\cdot, \cdot; \cdot)$ is defined by (1.91).

We continue on the assumption that $\{f_n\}$ is a sequence of random variables, each chosen from a set $\mathscr{F}$ of possible frequencies according to the probability distribution $P_F(\cdot)$. Averaging over θ, θ', and $\{f_n\}$ in (1.101), produces a variety of expressions depending on the values of N and the data-induced frequency shifts. For simplicity, we assume that all the involved data variables are equal. Ensemble-averaging (1.101) produces

$$\mathbf{E}\{|v_M(k)|^2\} = \begin{cases} |X_p(\tau_I, \nu; T_h)|^2 + |X_p(\tau_I - T_h, \nu; T_h)|^2, & N_I = 0 \\[2mm] \overline{|X_p(\tau_I - T_h, \nu; T_h)|^2} + \overline{|X_p(\tau_I, \nu; T_h)|^2}, & N_I = -1 \\[2mm] \overline{|X_p(\tau_I, \nu; T_h)|^2} + \overline{|X_p(\tau_I - T_h, \nu; T_h)|^2}, & \text{otherwise} \end{cases}$$

$$(1.102)$$

for $d_R = d_{\lfloor (N_I + k - 1)/M_D \rfloor} = d_{\lfloor (N_I + k)/M_D \rfloor}$, where

$$\overline{|X_p(\tau, \nu; T_h)|^2} \triangleq \sum_{f' \in \mathscr{F}} \sum_{f'' \in \mathscr{F}} P_F(f') P_F(f'') |X_p(\tau, f' + f'' + \nu; T_h)|^2. \qquad (1.103)$$

We will use (1.102) as a basis for evaluating the TOA resolution capability of FH signals.

Before proceeding, we will develop a useful upper bound on (1.103). For this purpose we assume that $\mathscr{F}$ is composed of M_F frequencies, equally likely and uniformly spaced $1/T_h$ Hz apart (the minimum spacing required for orthogonality over a hop time). For this calculation's purposes, this assumption is conservative in the sense that the hop frequencies in an operational system may be further apart to support FSK data and maintain the orthogonality of all possible waveforms. Assuming that the pulse shape is rectangular, the corresponding ambiguity function (1.97) can be overbounded by

$$|X_p(\tau, \nu; T_h)|^2 \le \begin{cases} T_h^2, & \nu < \dfrac{1}{\pi T_h}, \\[2ex] \left(\dfrac{1}{\pi \nu}\right)^2, & \nu \ge \dfrac{1}{\pi T_h}. \end{cases} \qquad (1.104)$$

Now define the integer k_0 such that

$$\left| \frac{k_0}{T_h} + \nu \right| = \min_k \left| \frac{k}{T_h} + \nu \right| \qquad (1.105)$$

or equivalently

$$\nu = -\frac{k_0}{T_h} + \frac{\delta}{T_h}, \qquad |\delta| \le \frac{1}{2}. \qquad (1.106)$$

Noting that there are $M_F - k_0$ frequency pairs (f', f''), each from $\mathcal{F}$, for which $f' - f'' = k_0/T_h$, it can be shown that

$$\overline{|X_p(\tau, \nu; T_h)|^2} \le \frac{1}{M_F^2} \left[(M_F - k_0)T_h^2 + \sum_{\substack{f' \in \mathcal{F} \\ f'-f'' \ne k_0/T_h}} \sum_{f'' \in \mathcal{F}} \frac{1}{\pi^2(f' - f'' + \nu)^2} \right]$$

$$\le \frac{1}{M_F^2} \left[(M_F - k_0)T_h^2 \right.$$

$$\left. + \sum_{\substack{k_1=1 \\ k_1-k_2 \ne k_0}}^{M_F} \sum_{k_2=1}^{M_F} \frac{T_h^2}{\pi^2(|k_1 - k_2 - k_0| - 1/2)^2} \right].$$

$$(1.107)$$

The last inequality in (1.107) follows from (1.105) and (1.106), and worst case choice of δ. The weakest bound (1.107) occurs when ν is close to zero and $k_0 = 0$, and hence, considering that case and letting $m = k_1 -$

k_2, gives

$$\overline{|X_p(\tau, \nu; T_h)|^2} \leq \frac{1}{M_F^2}\left[M_F T_h^2 + \frac{T_h^2}{\pi^2} \sum_{m=-M_F+1}^{M_F-1} (M_F - |m|)\left(\frac{2}{2|m|-1}\right)^2 \right]$$

$$< \frac{T_h^2}{M_F}\left[1 + \frac{8}{\pi^2} \sum_{m=1}^{\infty} \frac{1}{(2m-1)^2}\right] = \frac{2T_h^2}{M_F}. \tag{1.108}$$

The resolution capability of FH/SS communication system can now be estimated by noting from (1.102) and (1.108) that

$$\frac{E\{|v_M(k)|^2\}|_{t_l=0, \nu=0}}{E\{|v_M(k)|^2\}|_{|t_l|\geq T_h}} > \frac{M_F}{2}. \tag{1.109}$$

This is the basis for stating that a high energy gain FH/FSK communication system has TOA resolution on the order of T_h. That is, (1.109) indicates that frequency-synchronous correlation detectors should be able to reject the desired FH/FSK signal, if it arrives out of time synchronism by more than T_h seconds.

1.6 DESIGN ISSUES

Based on the analyses presented in this chapter, spread-spectrum techniques promise an attractive approach to the design of communication systems which must operate in an interference environment. However, major design issues have been obscured thus far by the apparent simplicity of the concept.

1. How does the receiver acquire and retain synchronization with the received signal's clocks, especially in the presence of interference?
2. Can the appearance of randomness in modulation selection be achieved deterministically by a stored-reference SS system?
3. In what ways does the communication vs. jamming game change when the payoff function is bit-error rate?
4. What are the effects of imperfect system operation/modelling (e.g., synchronization errors, non-uniform channel characteristics across the communication band) on performance estimates?
5. How should data detectors be designed when the nature of the interference is not known *a priori*?

These are a sample of the questions which must be answered before a realistic system design can be achieved.

Several truths have been demonstrated by this introductory treatment. First, excess bandwidth is required to employ randomized signalling strategies against interference. That is, spreading the spectrum is necessary in this gaming approach to interference rejection. Secondly, if either side, communicator or jammer, fails to completely randomize its signalling strategy, the opponent in principle may observe this fact, adapt to this failing, and take advantage of the situation. Barring this event, the energy gain of the SS system on the average will be achieved by the communicator unless the jammer can avoid complying with a rule of the game.

1.7 REFERENCES

In most cases the results of this chapter and the issues raised herein will be fully discussed in greater detail in later chapters. The references listed here are general references to background material, tutorials, and paper collections.

1.7.1 Books on Communication Theory

[1] W. B. Davenport, Jr., and W. L. Root, *Random Signals and Noise*. New York: McGraw-Hill, 1958.

[2] D. Middleton, *An Introduction to Statistical Communication Theory*. New York: McGraw-Hill, 1960.

[3] S. W. Golomb, ed., *Digital Communications with Space Applications*. Englewood Cliffs, NJ: Prentice-Hall, 1964.

[4] J. M. Wozencraft and I. M. Jacobs, *Principles of Communication Engineering*. New York: John Wiley, 1965.

[5] M. Schwartz, W. R. Bennett, and S. Stein, *Communication Systems and Techniques*. New York: McGraw-Hill, 1966.

[6] A. J. Viterbi, *Principles of Coherent Communication*. New York: McGraw-Hill, 1966.

[7] S. Stein and J. J. Jones, *Modern Communication Principles*. New York: McGraw-Hill, 1967.

[8] H. L. Van Trees, *Detection, Estimation, and Modulation Theory* (3 vols.). New York: McGraw-Hill, 1968.

[9] G. L. Turin, *Notes on Digital Communication*. New York: Van Nostrand Reinhold, 1969.

[10] J. B. Thomas, *An Introduction to Statistical Communication Theory*. New York: John Wiley, 1969.

[11] J. J. Stiffler, *Theory of Synchronous Communication*. Englewood Cliffs, NJ: Prentice-Hall, 1971.

[12] W. C. Lindsey, *Synchronization Systems in Communication and Control*. Englewood Cliffs, NJ: Prentice-Hall, 1972.

[13] W. C. Lindsey and M. K. Simon, *Telecommunication Systems Engineering*. Englewood Cliffs, NJ: Prentice-Hall, 1973.

[14] R. E. Ziemer and W. H. Tranter, *Principles of Communications*. Boston, MA: Houghton Mifflin, 1976.

[15] J. J. Spilker, Jr., *Digital Communications by Satellite*. Englewood Cliffs, NJ: Prentice-Hall, 1977.
[16] S. Haykin, *Communication Systems*. New York: John Wiley, 1978.
[17] A. J. Viterbi and J. K. Omura, *Principles of Digital Communication and Coding*. New York: McGraw-Hill, 1979.
[18] J. G. Proakis, *Digital Communication*. New York: McGraw-Hill, 1983.

1.7.2 Books on Resolution and Ambiguity Functions

[19] C. E. Cook and M. Bernfeld, *Radar Signals*. New York: Academic Press, 1967.
[20] A. W. Rihaczek, *Principles of High Resolution Radar*. New York: McGraw-Hill, 1969.
[21] G. W. Deley, "Waveform Design," in *Radar Handbook*, M. Skolnik, ed., New York: McGraw-Hill, 1970.

1.7.3 Recent Books and Proceedings on Spread-Spectrum Communications

[22] *Proceedings of the 1973 Symposium on Spread Spectrum Communications*. Naval Electronics Laboratory Center, San Diego, CA, March 13–16, 1973.
[23] *Spread Spectrum Communications*. Lecture Series No. 58, Advisory Group for Aerospace Research and Development, North Atlantic Treaty Organization, July 1973 (AD 766914).
[24] R. C. Dixon, ed., *Spread Spectrum Techniques*. New York: IEEE Press, 1976.
[25] R. C. Dixon, *Spread Spectrum Systems*. New York: John Wiley, 1976.
[26] L. A. Gerhardt and R. C. Dixon, eds., "Special Issue on Spread Spectrum Communications," *IEEE Trans. Commun.*, COM-25, August 1977.
[27] D. J. Torrieri, *Principles of Military Communication Systems*. Dedham, MA: Artech House, 1981.
[28] C. E. Cook, F. W. Ellersick, L. B. Milstein, and D. L. Schilling, eds., "Special Issue on Spread Spectrum Communications," *IEEE Trans. Commun.*, COM-30, May 1982.
[29] J. K. Holmes, *Coherent Spread Spectrum Systems*. New York: John Wiley, 1982.
[30] R. H. Pettit, *ECM and ECCM Techniques for Digital Communication Systems*. Belmont, CA: Lifelong Learning Publications, 1982.
[31] *MILCOM Conference Record*, 1982 IEEE Military Communications Conference, Boston, MA, October 17–20, 1982.
[32] *Proceedings of the 1983 Spread Spectrum Symposium*, Long Island, NY. Sponsored by the Long Island Chapter of the IEEE Commun. Soc., 807 Grundy Ave., Holbrook, NY.

1.7.4 Spread-Spectrum Tutorials and General Interest Papers

[33] R. A. Scholtz, "The spread spectrum concept," *IEEE Trans. Commun.*, COM-25, pp. 748–755, August 1977.
[34] M. P. Ristenbatt and J. L. Daws, Jr., "Performance criteria for spread spectrum

communications," *IEEE Trans. Commun.*, COM-25, pp. 756–763, August 1977.

[35] C. L. Cuccia, "Spread spectrum techniques are revolutionizing communications," *MSN*, pp. 37–49, Sept. 1977.

[36] J. Fawcette, "Mystic links revealed," *MSN*, pp. 81–94, Sept. 1977.

[37] W. F. Utlaut, "Spread spectrum—principles and possible application to spectrum utilization and allocation," *ITU Telecommunication J.*, vol. 45, pp. 20–32, Jan. 1978. Also see *IEEE Commun. Mag.*, Sept. 1978.

[38] "Spread Spectrum: An Annotated Bibliography," National Telecommunications and Information Administration, Boulder, CO, May 1978 (PB 283964).

[39] J. J. Spilker, "GPS signal structure and performance characteristics," *Navigation*, vol. 25, pp. 121–146, Summer 1978.

[40] W. M. Holmes, "NASA's tracking and data relay satellite system," *IEEE Commun. Mag.*, vol. 16, pp. 13–20, Sept. 1978.

[41] R. E. Kahn, S. A. Gronemeyer, J. Burchfield, and R. C. Kunzelman, "Advances in packet radio technology," *Proc. IEEE*, vol. 66, pp. 1468–1496, Nov. 1978.

[42] A. J. Viterbi, "Spread spectrum communications—myths and realities," *IEEE Commun. Mag.*, vol. 17, pp. 11–18, May 1979.

[43] P. W. Baier and M. Pandit, "Spread spectrum communication systems," *Advances in Electronics and Electron Physics*, vol. 53, pp. 209–267. Sept. 1980.

[44] R. L. Pickholtz, D. L. Schilling, and L. B. Milstein, "Theory of spread spectrum communications—a tutorial," *IEEE Trans. Commun.*, COM-30, pp. 855–884, May 1982.

[45] N. Krasner, "Optimal detection of digitally modulated signals," *IEEE Trans. Commun.*, COM-30, pp. 885–895, May 1982.

[46] C. E. Cook and H. S. Marsh, "An introduction to spread spectrum," *IEEE Commun. Mag.*, vol. 21, pp. 8–16, March 1983.

[47] M. Spellman, "A comparison between frequency hopping and direct sequence PN as antijam techniques," *IEEE Commun. Mag.*, vol. 21, pp. 26–33, July 1983.

[48] A. B. Glenn, "Low probability of intercept," *IEEE Commun. Mag.*, vol. 21, pp. 26–33, July 1983.

Chapter 2

THE HISTORICAL ORIGINS OF SPREAD-SPECTRUM COMMUNICATIONS[1]

"Whuh? Oh," said the missile expert. "I guess I was off base about the jamming. Suddenly it seems to me that's so obvious, it must have been tried and it doesn't work."

"Right, it doesn't. That's because the frequency and amplitude of the control pulses make like purest noise–they're genuinely random. So trying to jam them is like trying to jam FM with an AM signal. You hit it so seldom, you might as well not try."

"What do you mean, random? You can't control anything with random noise."

The captain thumbed over his shoulder at the Luanae Galaxy. "They can. There's a synchronous generator in the missiles that reproduces the same random noise, *peak by pulse. Once you do that, modulation's no problem. I don't know* how *they do it. They just do. The Luanae can't explain it; the planetoid developed it."*

England put his head down almost to the table. "The same random," he *whispered from the very edge of sanity.*

from "The Pod in the Barrier" by Theodore Sturgeon, in *Galaxy*, Sept. 1957; reprinted in *A Touch of Strange* (Doubleday, 1958).

Led by the Global Positioning System (GPS) and the Joint Tactical Information Distribution System (JTIDS), the spread-spectrum (SS) concept has

---

[1]Major portions of the material in this chapter have been adapted from three historical papers [1] © 1982 IEEE, [2] © 1983 IEEE, [3] © 1983 IEEE with the permission of the Institute of Electrical and Electronic Engineers.

39

emerged from its cloak of secrecy. And yet the history of this robust military communication technique remains largely unknown to many modern communication engineers. Was it a spark of genius or the orderly evolution of a family of electronic communication systems that gave birth to the spread-spectrum technique? Was it, as Frank Lehan said, an idea whose time had come? Was the spread-spectrum technique practiced in World War II, as Eugene Fubini declares? Was it invented in the 1920s as the U.S. Patent Office records suggest? Was Theodore Sturgeon's lucid description of a jam-proof guidance system precognition, extrasensory perception, or a security leak? Let's examine the evidence, circa 1900–1960, concerning the development of spread-spectrum communications.

2.1 EMERGING CONCEPTS

Before we can assess the ingenuity which went into the development of the first spread-spectrum systems, we must examine the development of communication theory and technology. When all the data has been reviewed, a case can be made for evolution and selection as the method by which progress has been made. A search for the origin of SS communications can become mired in the "definitionsmanship" (a word coined by Robert Price in [3]) attendant to classification by order, genus, and species. The following will provide a database from which the reader may draw conclusions.

2.1.1 Radar Innovations

From the 1920s through World War II, many systems incorporating some of the characteristics of spread-spectrum systems were studied. The birth of RADAR, i.e., RAdio Detection And Ranging, occurred in the mid-1920s when scientists used echo sounding to prove the existence of an ionized gas layer in the upper atmosphere. British scientists E. V. Appleton and M. A. F. Barnett performed this feat by transmitting a frequency modulated wave upward and listening for the return echo [4]. Applications of this concept to aircraft instrumentation were obvious and FM altimetry became a reality in the 1930s, with all major combatants in World War II making use of this technology [5]. Typically, linear-sawtooth or sinusoidal modulations were used in these early systems. The frequency modulation generally serves two purposes: 1) it ameliorates the problem of interference leakage of the transmitted signal directly into the receiver, and 2) it makes possible the measurement of propagation delay and, hence, range.

Historically, the development of pulsed radars has received more attention than that of continuous wave (CW) radars, since isolation of the transmitting and receiving systems is a lesser problem in this case. By the end of World War II, the Germans were developing a linear FM pulse compression (chirp) system called Kugelschale, and a pulse-to-pulse frequency-hopping radar called Reisslaus [6]. In the 1940s, Prof. E. Huttman was issued a German patent on a chirp pulse radar, while U.S. patents on this type of system were first filed by R. H. Dicke in 1945 and by S. Darlington in 1949 [7]. The mid-1940s also saw the formulation of the matched filter concept for maximum signal-to-noise ratio (SNR) pulse detection by North [8] and Van Vleck and Middleton [9]. This development indicated that the performance of optimum signal detection procedures in the presence of white noise depends only on the ratio of signal energy to noise power spectral density, thus, leaving the choice of waveform open to satisfy other design criteria (e.g., LPI or AJ). Resolution, accuracy, and ambiguity properties of pulse waveforms finally were placed on a sound theoretical basis by P. M. Woodward [10] in the early 1950s.

Spectrum spreading was a natural result of the Second World War battle for electronic supremacy, a war waged with jamming and anti-jamming tactics. On the Allied side by the end of the war, every heavy bomber, excluding Pathfinders, on the German front was equipped with at least two jammers developed by the Radio Research Laboratory (RRL) at Harvard [11]. The use of chaff was prevalent, the Allies consuming 2000 tons per month near the end. On the German side, it is estimated that at one time, as many as 90 percent of all available electronic engineers were involved in some way in a tremendous, but unsuccessful, AJ program. Undoubtedly Kugelschale and Reisslaus were products of this effort.

In a postwar RRL report [11], the following comment on AJ design is notable:

> In the end, it can be stated that the best anti-jamming is simply good
> engineering design and the spreading of the operating frequencies.

Certainly, spectrum spreading for jamming avoidance (AJ) and resolution, be it for location accuracy or signal discrimination (AJ), was a concept familiar to radar engineers by the end of the war.

In the late 1950s and early 1960s, the East German scientist F. H. Lange toured Europe and the United States collecting (unclassified) material for a book on correlation techniques. Published first in 1959 with its third edition being translated into English [12] a few years later, Lange's book contains some references all but unnoticed by researchers on this side of the Atlantic. The most intriguing of these is to the work of Gustav Guanella (Figure 2.1) of Brown, Boveri, and Company in Switzerland. Among Guanella's approximately 100 patents is one [13] filed in 1938, containing all the technical characteristics of an SR-SS radar! The radiated signal in Guanella's CW radar is "composed of a multiplicity of different frequencies the energies of which are small compared with the total energy" of the signal. His prime examples of such signals are acoustic and electrical noise, and an oscillator whose frequency is "wobbled at a high rate between a lower and upper limit."

Ranging is accomplished by adjusting an internal signal delay mechanism to match the external propagation delay experienced by the transmitted

Figure 2.1. Gustav Guanella, Swiss pioneer of noise-modulated radar and speech privacy systems. (Photo courtesy of I. Wigdorovits.)

signal (see Figure 2.2(c)). Delay matching errors are detected by cross correlating the internally delayed signal with a 90 degree phase-shifted (across the whole transmission band) version of the received signal. Thus, if the transmitted signal is of the form

$$\sum_n a_n \cos(\omega_n t + \phi_n),$$

the propagation delay is τ_p, and the internal delay is τ_i, then the measured error is proportional to

$$\sum_n a_n^2 \sin\left[\omega_n(\tau_p - \tau_i)\right].$$

This ensemble of phase locked loops, all rolled up into one neat package, possesses a tracking-loop S-curve which looks like the Hilbert transform of the transmitted signal's autocorrelation function. Undoubtedly, Guanella's patent contains possibly the earliest description of a delay-locked loop. Guanella used the same type of error-sensing concept in an earlier patent filed in 1936 [14]. Many of his inventions are cited as prior art in later patents.

In addition to accurate range measurement, the noise-radar patent indicates improved performance against interference. Guanella evidently did not pursue these intriguing claims, but instead turned his innovative talents to the field of speech scrambling.

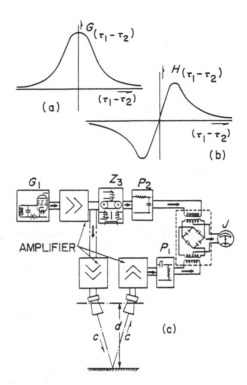

Figure 2.2. Guanella's noise radar patent [13] (redrawn). Part (c) shows a noise source (G_1), recording media (Z_3) for mechanizing the adjustable internal delay, filters (P_1 and P_2) whose design controls the measurement characteristic read on meter J. Parts (a) and (b) show two possible characteristics as functions of the time delay mismatch $\tau_1 - \tau_2$. The characteristic of part (b) was used later in the disclosure to describe delay-lock feedback control of the internal delay.

2.1.2 Developments in Communication Theory

In 1915, E. T. Whittaker concluded his search for a distinctive function among the set of functions, all of which take on the same specified values at regularly spaced points along the real line [16]. This "function of royal blood whose distinguished properties set it apart from its bourgeois brethren" [17] is given by

$$x(t) = \sum_n x(n/2W)\sin[\pi(2Wt - n)]/[\pi(2Wt - n)]$$

where $x(n/2W)$ represents the specified values and $x(t)$ is the cardinal function of the specified values, a function whose Fourier transform is strictly band limited in the frequency domain [15]—[18]. Based on this result, the sampling theory used in a communication context by Hartley [19], Nyquist [20], Kotelnikov [21], and Shannon [22] states that a function band limited to W Hz can be represented without loss of information by samples spaced $1/(2W)$ seconds apart. Generalizations [23], [24] of this result indicate that a set of approximately $2TW$ orthogonal functions of T seconds duration and occupying W Hz can be constructed. In SS theory, this provides the connection between the number of possible orthogonal signaling formats and system bandwidth. Although earlier, Nyquist [20] and later Gabor [25] both had argued using Fourier series that $2TW$ samples should be sufficient to represent a T-second segment of such a band-limited signal, it was Shannon who made full use of this classical tool.

Probabilistic modeling of information flow in communication and control systems was the brainchild of the preeminent mathematican Norbert Wiener of the Massachusetts Institute of Technology (M.I.T.). In 1930, Wiener published his celebrated paper "Generalized Harmonic Analysis" [26] developing the theory of spectral analysis for nonperiodic infinite-duration functions. When World War II began, Wiener was asked by the National Defense Research Committee (NDRC) to produce a theory for the optimal design of servomechanisms. Potential military applications for this theory existed in many gunfire control problems [27]. The resultant work [28], published initially in 1942 as a classified report and often referred to as the "Yellow Peril," laid the groundwork for modern continuous-parameter estimation theory. By 1947, Wiener's filter design techniques were in the open literature [29].

Claude E. Shannon, who had known Wiener while a graduate student at M.I.T., joined the Bell Telephone Laboratories (BTL) in 1941, where he began to establish a fundamental theory of communication within a statistical framework. Much of his work, motivated in good part by the urge to find basic cryptographic and cryptanalytic design principles [30], was classified well past the end of the Second World War. In a paper [22] first presented in 1947, Shannon invoked the cardinal expansion in formulating a capacity for delivering information (negentropy [31]) over channels perturbed solely by additive Gaussian noise. He showed that this channel capacity was maximized by selectively spreading the signaling spectrum so that wherever deployed within designated bandwidth confines–but only there–the sum of its power spectral density plus that of the independent noise should lie as uniformly low as possible, yet utilize all the average transmitter power available (see Figure 2.3). Moreover, this capacity was met by sending a set of noise-like waveforms and distinguishing between them at the receiver via a minimum-distance criterion akin to correlation-testing the observed signal against locally stored waveform replicas.

Even though Shannon's theory [22] did not apply directly to many jamming/interference situations, his result might be construed as the solution to a game theory problem in a jamming situation. When queried in 1982 by Robert Price [3], Shannon replied that there could well have been such a military application in the back of his mind. Regardless of his research motives, Shannon's remarkable concepts and results profoundly

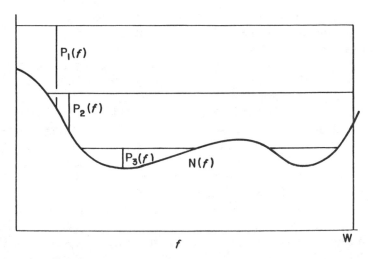

Figure 2.3. This "water pouring" illustration redrawn from Shannon's thought-provoking paper [22] depicts the communication-capacity-maximizing densities $(P_i(f))$ over the available bandwidth W, as a function of frequency f, for three different power levels. The optimal distribution may be viewed as the gravity-driven result of pouring a fluid, having a total volume proportional to the available transmitter power, into a rectangular vessel of width W, whose bottom profile in the f direction corresponds uniformly, and by the same proportion, to the power density $N(f)$ of the additively interfering Gaussian noise.

influenced communication engineers' thinking. Nearly a decade later, further applications of game theory to jamming situations were presented by William Root [32], [33], Nelson Blachman [34], [35], and Lloyd Welch [36], all then associated with SS system developments.

Driven by the intense interest in the theories of Wiener and Shannon, the Institute of Radio Engineers (IRE) formed the Professional Group on Information Theory, which commenced publishing in 1953 [37]. The first three chairmen of this Group were, in order, Nathan Marchand, William G. Tuller, and Louis deRosa. Marchand and deRosa, close friends, were at that time playing key roles in the development of SS systems; Tuller had independently but rather heuristically arrived at one of Shannon's capacity formulas.

2.1.3 Correlator Mechanization

One of the difficult problems which Guanella faced (by this account without any knowledge of Wiener's work) was to fabricate a device which will perform a weighted correlation computation on two inputs. Specifically, a means was needed for taking two inputs $x_1(t)$ and $x_2(t)$ and computing

$$y(t) = \int^t x_1(u)x_2(u)w(t-u)\,du$$

where $y(t)$ is the device output and $w(t)$ is the weighting function. The difficulty here is not with the weighting (i.e., filtering) operation, but with the prior multiplication of $x_1(t)$ by $x_2(t)$, and in particular with the range of inputs over which accurate multiplication can be accomplished. As shall be seen later, the ability to mechanize the correlation operation precisely is essential in building high-performance SS systems.

In 1942, Nathan Marchand, then a 26-year-old engineer working for ITT's Federal Telephone and Radio Corporation in New York, discussed his radio receiver invention with ITT engineer and patent attorney Paul Adams. Marchand had developed a converter for demodulating a received FM signal of known frequency wobbulation by mixing it with a time-aligned, heterodyned replica of the wobbulated signal to produce a signal of constant intermediate frequency (IF) which could then be narrow-band filtered. The receiver's anti-multipath attributes designed by Marchand and additional anti-interference features suggested by Adams appear in a 1947 patent [38]. Later during World War II, after studying Wiener's "Yellow Peril," Marchand was able to dub his converter a bandpass correlator.

However, Marchand was neither the first nor the last to propose the bandpass correlator, a similar device being contained in Purington's 1930 patent application [39].

At M.I.T. in 1947, Prof. Yuk Wing Lee commenced research into the implications of Wiener's theories and the new directions they inspired for engineering science. Soon thereafter, Lee was joined by Jerome Wiesner and Thomas Cheatham, and their collective efforts led to the development of the first high-performance electronic correlators. In August 1949, they applied for a patent [40] and in October they reported applications of correlation techniques to detection problems [41]. Continuing this work, Henry Singleton proceeded to innovate an all-digital correlator [42].

2.1.4 Protected Communications

It has been found [6], [43, p. 1] that secrecy, jamming, and anti-jamming had each begun to stir attention as early as 1901. (Note: For further details see [44, pp. 38—39], where also a recommendation is mentioned that "the homing-pigeon service should be discontinued as soon as some system of wireless telegraphy is adopted.") The first two of these three differing adversary-recourses were astutely exercised together in reporting, by "wireless," the America's Cup yacht race of that year. They were invoked by a thereby countermeasures-pioneering engineer who was competing against both Guglielmo Marconi and Lee De Forest for live coverage of this nautical event.

Even earlier, in 1899, Marconi had experimented with frequency-selective reception in response to worries about radio interference [6], [43]. The Navy, whose due concern had prompted this most basic advance in the radio art, reemphasized its apprehensions by warnings "... during 1904 on the dangers of relying on wireless in war" that "raised the questions of enemy intercept and interference which were to remain and grow into the major research and operational fields of radio intelligence and radio countermeasures" [45, p. 1243].

The earliest patent [46] presently construed by the U.S. Patent Office as being spread spectrum in character was filed in 1924 by Alfred N. Goldsmith, one of the three founders of the IRE. Goldsmith proposed to counteract the multipath-induced fading effects encountered in shortwave communication by

> radiating a certain range of wave frequencies which are modulated in accordance with the signal and actuating a receiver by means of energy collected on all the frequencies, preferably utilizing a wave which is continuously varied in wave frequency over a certain range of cycles recurring in a certain period.

Certainly, we can identify this as a form of FM-SS transmission. However, the envisioned data modulation was by amplitude (AM) with reception by a broadly tuned AM receiver. Hence, the correlation detector necessary to achieve the full benefits of SS operation was not inherent in Goldsmith's disclosure. For a World War II disclosure on an FM-SS chirp communication system with a more sophisticated receiver, claiming a primitive form of diversity reception for multipath signals and a capability against narrowband interference, see [47].

In 1935, Telefunken engineers Paul Kotowski and Kurt Dannehl applied for a German patent on a device for masking voice signals by combining them with an equally broad-band noise signal produced by a rotating generator [48]. The receiver in their system had a duplicate rotating generator, properly synchronized so that its locally produced noise replica could be used to uncover the voice signal. The U.S. version of this patent was issued in 1940, and was considered prior art in a later patent [49] on DS-SS communication systems. Certainly, the Kotowski-Dannehl patent exemplifies the transition from the use of key-stream generators for discrete data encryption [50] to pseudorandom signal storage for voice or continuous signal encryption. Several elements of the SS concept are present in this patent, the obvious missing notion being that of purposeful bandwidth expansion.

The Germans used Kotowski's concept as the starting point for developing a more sophisticated capability that was urgently needed in the early years of World War II. Gottfried Vogt, a Telefunken engineer under Kotowski, remembers testing a system for analog speech encryption in 1939. This employed a pair of irregularly slotted or sawtoothed disks turning at different speeds, for generating a noise-like signal at the transmitter, to be modulated/multiplied by the voice signal. The receiver's matching disks were synchronized by means of two transmitted tones, one above and one below the encrypted voice band. This system was used on a wire link from

Germany, through Yugoslavia and Greece, to a very- and/or ultra-high frequency (VHF/UHF) link across the Mediterranean to General Erwin Rommel's forces in Derna, Libya.

In January 1943, British troops captured, from General Rommel's forces in North Africa, a communication transceiver called the "Optiphone" (or "Photophone"). Developed in Germany in the mid-1930s (and covered by U.S. Patent No. 2010313 to the Zeiss works), this system could provide voice communications over a light-path up to four miles long under reasonably clear atmospheric conditions [51]. While it would be far-fetched to view such (prelaser) incoherent optical communications as SS in nature, the interesting aspect here is that this apparatus seems to have been a further example of Rommel's employment of relatively advanced technology in secure communications.

Dr. Richard Gunther, an employee of the German company Siemens and Halske during World War II, recalls another speech encryption system involving bandwidth expansion and noise injection. In a fashion similar to the Western Electric B1 Privacy System, the voice subbands were pseudo-randomly frequency scrambled to span 9 kHz and pure noise was added to fill in the gaps. The noise was later eliminated by receiver filtering in the speech restoration process. Tunis was the terminus of a link operated at 800 MHz and protected by this system.

In turning next to another battlefield of the Second World War, the following point of explanation is appropriate. Classical wideband FM systems have not been classified within the particular species described as "modern spread-spectrum communications" since this bandwidth expanding FM technique does not encompass all of the attributes described in Chapter 1. However, there is no doubt that wideband FM belongs to the genus of spectrum-spreading systems, and in fact fulfilled an urgent need for jam-resistant communications during a crisis in the war. This singular event took place at the late-1944 "Battle of the Bulge" which surged past the crossroads town of Bastogne, in the Belgian Ardennes, where extended American forces had been trapped and General George S. Patton's Third Army was storming to their rescue. Quoting at some length from [52, pp. 163–164]:

> On 26 December Patton succeeded in forcing a narrow corridor through the German tanks ringing Bastogne. It was only three hundred yards wide, but it opened the way to the American troops cut off there and punctured the German bulge, which began slowly to deflate under combined British and American pressure.
>
> Now came the first and only battle test of Jackal, the high-powered airborne radio jammer AN/ART-3 developed by the Signal Corps for the AAF [Army Air Forces]. The First and Third U.S. Armies had been reluctant to try Jackal jamming because a portion of the frequency band used by their tank radios overlapped into the German band to be jammed. Earlier tests in

England had indicated somewhat inconclusively that little or no interference would be caused, since American radio for armored forces was FM while similar German sets and Jackal were AM [amplitude modulation]. Now that nearly the whole of the German *Sixth Panzer Army* was in the Ardennes fighting, it seemed a good opportunity to test Jackal.

Accordingly, beginning on 29 December and continuing through 7 January [1945], Eighth Air Force B-24's based in England and bearing the Jackal jammers blaring full blast, flew in relays over the battle area, coinciding with a Third Army counterthrust in the vicinity of Bastogne. The first results seemed inconclusive, but, according to later reports from German prisoners, Jackal effectively blanketed German armored communications during these crucial days. Nor were the American tankmen inconvenienced or made voiceless by the overlap in frequencies. The jammer effectively filled the German AM receivers with a meaningless blare, while the American FM sets heard nothing but the voices of the operators.

This intriguing FM episode dramatized in real life (as twice again referred to in [52, p. 318, footnote 57; p. 324]) the simple FM/AM analogy later given in Theodore Sturgeon's science-fiction portrayal of SS communications, given at the beginning of this chapter. That FM radio (and FM radio-relay, too, each unique to the U.S. Army in this worldwide war, per [52, pp. 21, 107] was even available, "beyond anything either the enemy or the other allied nations possessed" [52, p. 631], was due in large measure to three men. Through experiments conducted in the mid-1930s with the close cooperation of its inventor, Edwin Armstrong, (the then) Col. Roger B. Colton and Maj. James D. O'Connell recognized the clear superiority of this modulation method and pushed hard to have it ready for communicators on the battlefield. One colonel exclaimed: "I feel that every soldier that lived through the war with an Armored Unit owes a debt that he does not even realize to General Colton" [52, p. 631]. Armstrong patriotically donated to every military service of the United States, free of any royalty payments, license to the use of all his inventions. (Note: Additional information on Maj. Armstrong's contributions to the Signal Corps is given in [53].)

The Allies' voice communication system that supplied the highest level of voice security for Roosevelt, Churchill, Truman, and others [3], [54], [55] in top secret conversations, was developed by Bell Telephone Laboratories at the beginning of the war. Officially called the X-System or Project X-61753 at Bell Labs, nicknamed the "Green Hornet," and code-named "Sigsaly" by the Signal Corps, this system contained several significant innovations related to the signal processing contained in its novel vocoding apparatus [55], [225]. (Other early vocoders are discussed in [56].)

The most striking security feature of the X-System, was its use of non-repeating, prerecorded keys which were combined modulo 6 with elaborately generated speech samples, to provide the complete security of a one-time-pad crypto-system, at a data rate of 1500 bps. Copies of each key were pressed on phonograph discs by Muzak, and the best studio-quality turntables for playing these records were incorporated into the thirty racks of equipment making up a Sigsaly radio station. The weight of the million dollar station, i.e., the operational equipment, a 30 kw power source, air conditioning, a complete set of spare parts including duplicates of the required 1100 + vacuum tubes, and other ancillary equipment, has been informally estimated at a staggering 80 tons. The official government designation for this station was the RC-220-T1 Terminal.

The use of pseudorandom keys in combination with speech bandwidth compression and re-expansion into a pulse-code modulation (PCM) format to utilize the full bandwidth of a telephone channel could be viewed as a SR spectrum-spreading system. The intended application, admirably achieved, was unbreakable enciphered telephony; the system "contained no anti-jamming provisions, and for the most part, did not require any" [55]. Surprisingly, synchronization of the turntables was not as difficult as expected, because of the use of excellent frequency standards.

Both Claude Shannon of Bell Laboratories and Alan Turing of the Government Code and Cypher School at Bletchley Park, England, independently verified the security of the X-System [3], [225]. Robert C. Mathes [57] and Ralph K. Potter [58] applied for patents on X-System apparatus in 1941, the disclosures being immediately placed under secrecy order. Thirty more patents, with X-System applications, followed [225, p. 297].

BTL continued its work on key-stream generation and in the mid-1940s filed for patents on all-electronic key generators which combined several short keys of relatively prime lengths to produce key streams possessing long periods [59], [60]. Such schemes also had been studied by Shannon [30] at BTL, but his comments on these were deleted before republication of his declassified report on secrecy systems in the *Bell System Technical Journal*. All of these BTL patent filings remained under secrecy order until the 1970s when the orders were rescinded and the patents issued.

Alan Turing, who had seen the enormous amount of equipment required by the X-System and heard Shannon explain the sampling theorem, considered the vocoder a non-essential part of its electronics. This led to his conception of the Delilah, a secure voice communication system named for the biblical "deceiver of men" [231]. The idea in Turing's design was to perform modular addition of a random key directly to time samples of speech, at a rate high enough to allow eventual speech reconstruction at the receiver, as indicated by the Whittaker-Shannon sampling theorem. The Delilah signal processors were built and laboratory tested by the end of 1944, with random noise from an antennaless radio wired directly to the transmitter and receiver key inputs during the tests. Turing then devised a sophisticated key-stream generator for the system (rather than accept the key distribution problems associated with one-time pads), and tackled the key synchronization problem. The effort was not completed in time for use during the war.

Gustav Guanella, the Swiss inventor who held pre-World War II patents on radars and direction finders employing noise-like signals [13], [14], also applied his innovative talents in the related field of speech scrambling systems [61]. A 1946 NDRC report [56] in fact discusses his designs and the level of security which they provide. No direct mention of the ultra-secret X-System was included in this report, which does include descriptions of noise-masking systems and multiplicative signal-modifying systems, as well as methods for protecting the reference signals in TR systems. (An example of good TR-technique is the subject of the next vignette.) It was further noted that if a short-period signal was used as a multiplicative signal in an effort to disguise a speech waveform, then the transmission was quite vulnerable to "cryptanalysis." Several long-period code generators were discussed.

The historically useful reports [27], [56], [62], [63] of the NDRC were undoubtedly precursors of the postwar demobilization of that civilian organization. Military officials, recognizing the major contributions of the academic world during the war just ended, and realizing the need for a continued strong research base, initiated a program of support for advanced research in the nation's universities [64, p. 98]. First installed at M.I.T., Harvard, Columbia, and Stanford, this joint Services Electronics Program now sponsors research at many major U.S. institutions.

One can view the advanced Telefunken system as an avatar of a TR system since specialized signals are transmitted to solve the disk synchronization problem. Another novel variation of TR voice communication was conceived in the U.S. during the war years by W. W. Hansen. This Sperry/M.I.T. Radiation Laboratory scientist is noted for his invention of the microwave cavity resonator and for his joint effort with the Varian brothers in originating the Klystron. In a 1943 patent application [65], Hansen describes a two-channel system (see Figure 2.4) with the reference channel used solely for the transmission of noise, and the intelligence channel bearing the following signal (in complex notation):

$$\exp\left\{ j \int^{t} [\omega_1 + An(t')] \, dt' \right\} \cdot \exp\left\{ j \int^{t} [\omega_2 + Bv(t')] \, dt' \right\}$$

where $n(t)$ is a filtered version of the noise communicated via the reference channel, $v(t)$ is the voice signal, and assuming $n(t)$ and $v(t)$ are at comparable levels, $A \gg B$. The intelligence signal is the result of combining a wide-swing noise-modulated FM waveform with a narrow-swing voice-modulated FM waveform in a device "similar in principle of operation to the mixers used in superheterodyne receivers."

At the receiver, the reference channel signal is used to reconstruct the first of the above factors, and that in turn is mixed with the received intelligence signal to recover the voice-modulated waveform represented by the second factor. This receiver mixer appears to be similar in many respects to Marchand's bandpass correlator.

To overcome some of the fundamental weaknesses of TR systems (see Chapter 1, Section 3), Hansen threw in an additional twist: The filtering of

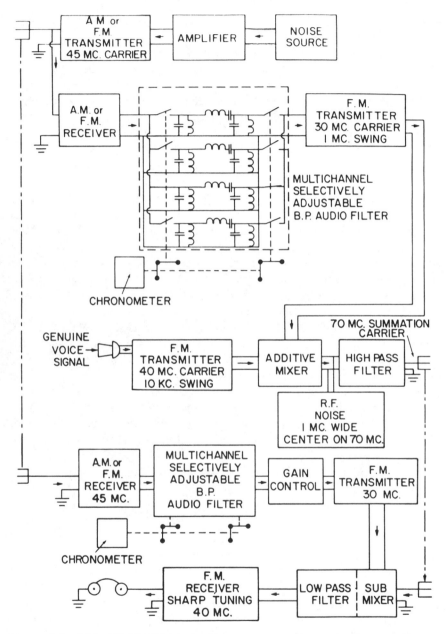

Figure 2.4. This TR-FM-SS block diagram redrawn from Hansen's patent [65] illustrates one method for denying an unauthorized listener direct access to the reference signal used by the receiver to detect the information bearing transmission.

the reference channel signal, used to generate $n(t)$, was made time dependent, with transmitter and receiver filters required to change structure in virtual synchronism under the control of a chronometer. This structural change could not be detected in any way by observing the reference channel.

When presenting his design along the TR-FM-SS lines, Hansen notes that the intelligence signal cannot be heard by unauthorized narrow-band receivers because "such wide-swing modulations in effect tune the transmitted wave outside the frequency band of the unauthorized listener's receiver for the greater portion of the time and thus make such a receiver inoperative." Concerned about the fact that a wide-band FM receiver might conceivably recover the signal $An(t) + Bv(t)$, he also concludes that "if therefore the noise $[n(t)]$ has important components throughout the range of signal frequencies and if the swing due to the noise is large compared to the swing due to the signal $[v(t)]$, deciphering is impossible."

Curiously enough, as a result of the use of an exponential form of modulation, Hansen's design is constructed as a TR-FM-SS communication system at radio frequency (RF), but equivalently at demodulated baseband, it is simply a "typical noise masking" add/subtract TR system. (This latter appraisal of [65] is from the case file—open to the public as for any issued patent—in Crystal City on an SR-SS invention [49] of major importance to a later period in this history.) Moreover, except for its TR vulnerabilities, Hansen's system is good AJ design, and as he points out, a large amount of additional noise can be injected at the RF output of the transmitter's intelligence channel for further masking without seriously degrading system performance.

Surprisingly, without the spectral spreading and chronometer-controlled reference signal filters, Hansen's system would bear a strong resemblance to a TR-FM system described in 1922 by Chaffee and Purington [66]. Hence, the concept of transmitting a reference signal to aid in the demodulation of a disguised information transmission is over sixty years old!

E. M. Deloraine's history [67] of the early years at ITT contains segments recounting the efforts and risks undergone in France during World War II to protect valuable information, and to eventually transport it to the United States. In a daring move, Louis Chereau, then Director of Patents and Information under Deloraine at ITT's Paris Laboratories, indicates [2] that he visited the French Patent Office in Vichy just before the Germans crossed the demarcation line, for the purpose of removing Henri Busignies' MTI patent application on "Elimination of fixed echoes," filed in Lyon on May 27, 1942, and also Ferdinand Bac's LORAN application. The MTI application was based on a Busignies memorandum, dated October 24, 1940, and the memorandum in turn was motivated by a "brainstorming session" initiated by Chereau who had put forth the idea of echo cancellation. The memorandum was written shortly before Deloraine and Busignies, along with Emile Labin and Georges Chevigny and their families, escaped via

Marseilles, Algiers, Casablanca, Tangier, and Lisbon to New York, arriving on New Year's Eve, the last day of 1940. Busignies filed his application in the United States on March 5, 1941, and was granted U.S. Patent 2 570 203.

Busignies, a remarkably prolific inventor who over his lifetime was granted about 140 patents, soon collaborated with Edmond Deloraine and Louis deRosa in applying for a patent [68] on a facsimile communication system with intriguing anti-jam possibilities here set forth:

> [The system uses a transmitter which sends each character] a plurality of times in succession, [and a receiver in which the character signals are visually reproduced,] one on top another... to provide a cumulative effect. [If] the interference signals are not transmitted to provide such a cumulative effect, the interference will form only a bright background but will not prevent the signals from being read.

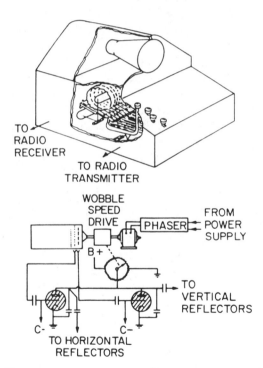

Figure 2.5. The core of Deloraine, Busignies, and deRosa's time-wobbling system consisted of a drum containing a slot track for each transmittable character. When the drum is turned at a constant speed, the signal produced by reading a track with a light beam was that required to create a character on a CRT screen. For security, the drum's rotation rate was "wobbled" and compensating variations in the CRT beam deflection voltages were inserted, so that interference, being wobbled on receive only, could not easily produce a bright screen signal. (Redrawn from [68], [69].)

From a jamming viewpoint, the real novelty in the the disclosure is in the fact that the mechanisms which read the characters at the transmitter and write the characters at the receiver synchronously vary in rate of operation (see Figure 2.5). Thus, attempts to jam the system with periodic signals which might achieve the "cumulative effect" at the receiver output will be unsuccessful.

In a sequel patent filed six weeks later [69], it is specified that the facsimile pulse modulation should have a low average duty cycle, be characterized by steep wavefronts, and have high peak-to-average power, in order to attain superior protection. This time-wobbling system is obviously an early relative of modern TH-SS systems. Concurrently with these efforts, deRosa covered similar applications in the field of radar by filing what may be the first patent on random jittering of pulse repetition frequencies [70].

Another study of protected communications was launched when ITT submitted Proposal 158A to the NDRC for consideration. Although the original proposal only suggested the use of redundancy in time or frequency as a possible AJ measure, a 1944 report [71] stated with regard to jamming that

> The enemy can be forced to maintain a wide bandwidth if we use a coded frequency shifting of our narrower printer bandwidth so that it might at any time occupy any portion of a wider band.

This clear suggestion of FH-SS signaling was not explored further in the last year of the contract. Several different tone signalling arrangements were considered for communication to a printer at rates on the order of one character per second. Synchronization of these digital signalling formats was accomplished in open-loop fashion using precision tuning forks as reference clocks.

> These forks are temperature compensated over a wide range and are mounted in a partial vacuum, so that their rate is not affected by the low barometric pressures encountered at high altitudes. Their accuracy is of the order of one part in a million, so that once the receiving distributor was phased with the transmitted signal, it remained within operable limits for two hours or more. A differential gear mechanism, operated by a crank handle on the front panel, was provided for rephasing the receiving distributor whenever this became necessary.

The receiving distributor controlled the reinitializing of L-C tank circuits tuned to detect transmitted tones. Because of their high Q, these circuits performed an integrate-and-dump operation during each distributor cycle. This detector was a significant improvement over the prior art, a fact indeed recognized intuitively by ITT, rather than derived from correlation principles.

ITT's printer communication system was tested at Rye Lake Airport on February 21, 1945. The printer performed well in the presence of jamming

11 dB stronger than the desired signal, and under conditions where voice on the same channel was not intelligible [72]. The interference in this test consisted of an AM radio station. Test results of the printer system are mentioned briefly in a 1946 NDRC Division 15 report [62] which also points out in a radar context that

> There is factual evidence that tunability is foremost as an AJ measure. Frequency spread of radars, which serves the same function, is a corollary and equally important. [With regard to communications,] RF carrier frequency scrambling and time modulation of pulses with time scrambling [are possible communication anti-jam measures].

The report's final recommendations state that "any peacetime program to achieve protection against jamming should not be concerned with the type of equipment already in service, but should be permitted an unrestricted field of development." This was sensible advice to follow, when practical, in the post-war years.

An explicit description of an FH system was put forward in early January 1943, when U.S. Army Signal Corps officer Henry P. Hutchinson applied for a patent on FH signaling for "maintaining secrecy of telephone conversations" or for "privately transmitting information" [73]. His scheme employed on-line cryptographic machines to produce a pseudorandom hopping pattern on demand. Although the subject of the patent is secrecy and privacy, Hutchinson has stated [3] that he was aware of the advantage his concept could have for avoiding interference. The patent application, which received official review [74], was held under secrecy order by the U.S. Patent Office until 1950.

As will soon be seen, Hutchinson's landmark patent was preceded in time, and eclipsed in human interest, by another true SR-FH-SS patent application filed by a rising Hollywood star!

2.1.5 Remote Control and Missile Guidance

Interest in radio guidance dates back to the turn of the century. Notable pioneering efforts took place at the Hammond Radio Research Laboratory, a privately organized research group. Its founder, John Hays Hammond, Jr., an inventor whose "...development of radio remote control served as the basis for modern missile guidance systems," also "...developed techniques for preventing enemy jamming of remote control and invented a radio-controlled torpedo..." [75]. By 1914, he was actively exploring means to "...greatly minimize the possibility of an enemy determining the wave lengths used in the control of the craft and thereupon interfering with the control thereof" [76]. In the radio guidance of remote vehicles, Hammond was following in the footsteps of Adm. Bradley A. Fiske and of Nikola Tesla [77, p. iii]. (Note: See [44, ch. XXIX] for more background on Hammond, Fiske, and Tesla.) The efforts against jamming were, initially, pursued by Benjamin F. Miessner, Hammond's sole coworker during 1912, who had invented a primitive form of SS signaling [77, p. 32] which was quite likely the earliest to be thus "... transmitted in a peculiar way" [77, p. 65].

The Miessner-Hammond carrier broadening was incorporated into a military transmitter, which with its associated receiver (that additionally introduced amplitude-limiting action) was delivered to the U.S. Army in France just before World War I ended. There, Maj. Edwin H. Armstrong verified the Hammond system's ability to communicate in the face of powerful enemy interference, a challenge officially recognized as "... one of the most important matters connected with the war" [78, footnote 48].

From the First World War on into World War II, the furtherance of wideband techniques continued at the Hammond Laboratory, with concentration on frequency-wobbling methods to send secret signals which sounded like "... some new kind of man-made static..." [78, p. 1203]. The anti-jamming attribute of spectrum spreading became obvious as the frequency deviation of the wobbling increased so that its "... wider swing reduced the amount of time that a given narrow-band disturbance could affect the intermediate-frequency circuit of the receiver..." [78, p. 1202]. Contributions of this kind, originating in that era from the Hammond group (additionally to [66]), are documented in patents on TR and SR wobbling which were filed, respectively, by Emory L. Chaffee [79] in 1922 and Ellison S. Purington [39] in 1930.

The Chaffee patent claims that his invention, involving "rapid and erratic" wobbling, is useful for "secret radio telephony" and for "... telegraphic signals or radio-dynamic control." With respect to the Purington patent [39] (which seems somehow to have been overlooked in the chronicles [78] and [80]), that SR invention for obtaining "secrecy in which a rapid variation rate is necessary..." for the wobbling was a sequel to Hammond/Navy experiments of 1921–1922 which "... established points of interest regarding information theory when the interference greatly exceeds the signal" [78, p. 1202]. Also, circa 1920, the Hammond company had contracts from the U.S. Army to show that "noninterferable characteristics" could be secured for the radio control of aircraft [80, p. 1262].

During World War II, the NDRC entered the realm of guided missiles with a variety of projects [81] including the radio control in azimuth only (AZON) of conventionally dropped bombs (VB's) which trailed flares for visibility, radar-controlled glide bombs (GB's) such as the Pelican and the Bat, and the remotely controlled ROC VB-10 using a television link. Now documented mostly through oral history and innocuous circuit patents, one of several secure radio guidance efforts took place at Colonial Radio, predecessor of the Sylvania division at Buffalo, NY. This project was under the direction of Madison Nicholson, with the help of Robert Carlson, Alden Packard, Maxwell Scott, and Ernest Burlingame. The secret communications system concept was stimulated, so Carlson thinks, by talks with Navy people who wanted a system like the "Flash" system which the Germans used for U-boat transmissions. However, it wasn't until the Army Air Force at Wright Field posed the following problem that the Colonial Radio effort began seriously.

The airfoil surfaces of the glide bombs were radio controlled by a mother plane some distance away, sometimes with television display (by RCA) relayed back to the plane so that closed-loop guidance could be performed. It was feared that soon the Germans would become adept at jamming the control. To solve this problem Colonial Radio developed a secure guidance system based on a pulsed waveform which hopped over two diverse frequency bands. This dual band operation led to the system's nickname, Janus, after the Roman god possessing two faces looking in opposite directions. Low duty cycle transmission was used, and although the radio link was designed to be covert, the system could withstand jamming in one of its two frequency bands of operation and still maintain command control.

The Colonial Radio design's transmitter for the mother aircraft was designated the AN/ARW-4, and the corresponding glide bomb receiver was the AN/CRW-8. Testing of the radio guidance system took place at Wright Field in 1943, under the direction of Lt. Leonhard Katz, Capts. Walter Brown and Theodore Manley, and Project Engineer Jack Bacon. The contract, including procurement of two transmitters and seven receivers, was completed by June 1944 [82].

ITT also participated in these World War II guidance programs, notably with a system called Rex [63]. One patent, evidently resulting from this work and filed in 1943 by Emile Labin and Donald Grieg [83], is interesting because it suggests CDMA operation in pulse code modulation (PCM) systems by slight changes in the pulse repetition frequency. In addition, the patent notes the jammer's inherent problem of trying to deliver its interference to the victim receiver in synchronism with the transmitted pulse train. However, the notion of multiplicity factor or spectrum spreading is not mentioned.

A third guidance system for the control of VB's and GB's was proposed by the Hammond Laboratory. The Hammond system used a complicated modulation format which included a carrier wobbled over 20 kHz to protect against tone interference, and FM control signals amplitude modulated onto this frequency-modulated carrier [63]. More notable in this history than the system itself is the fact that Ellison Purington in 1948 came close to describing a TH and FH carrier for a radio control system in a patent application [84] (see Figure 2.6). The actual details describe a TH-SS system with control signals coded into the transmission using frequency patterns. Magnetic or optical recording "on a rotating member driven by a constant speed motor" was one suggestion for the storage of different time hopping patterns, while another possibility mentioned involves delay line generation of pulse train patterns. Control keys are hidden in the way that the patterns

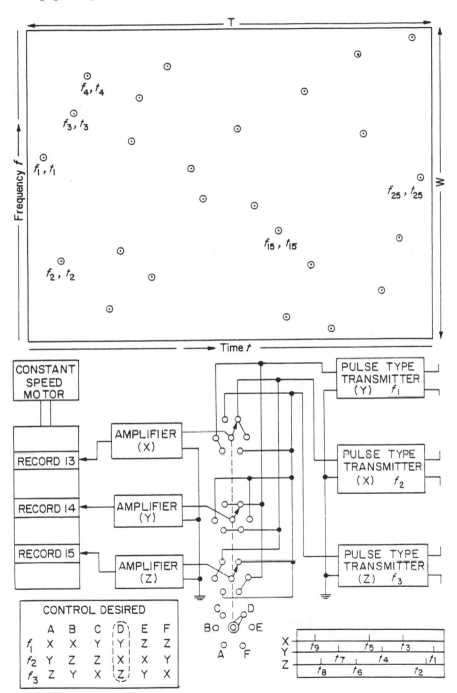

Figure 2.6. Purington's time-frequency chart definitely suggests a hybrid TH and FH system. However, the illustrated transmitter uses the assignment of TH patterns to frequency tracks as a method for signal encoding. (Redrawn from [84]).

are mapped onto different frequencies to create "radiations...randomly distributed in time and in frequency."

Other salient patents, based on World War II AJ and command/control efforts, include those of Hoeppner [85] and Krause and Cleeton [86].

In mid-1941 an application for the FH patent [87] was filed by Hedy K. Markey (see Figure 2.7) and George Antheil. Neither inventor was an engineer, conversant with the prior art, e.g., the broad FH method [88] invented a dozen years earlier. Being at that time a recent ex-spouse of Hollywood scriptwriter Gene Markey, Hedy Markey had been baptized Hedwig Eva Maria Kiesler. Growing up in Austria, this only child of a prominent Vienna banker had shown, at age 16, a flair for innovation by letting herself be filmed in total nudity while starring in the Czech-produced classic, *Ecstasy* (the fifth of her many motion pictures [89]). Several years later she was one of the small minority among non-Jewish Austrians who saw great danger to the world in Germany's early-1938 *Anschluss* of her homeland. That year, permanently leaving her country and her munitions-magnate husband, Friedrich A. "Fritz" Mandl, as washouts to Hitler, the actress came to the United States on a seven-year contract from Metro-Goldwyn-Mayer. There, she legalized her stage-renaming (by Louis B. Mayer) to Hedy Lamarr. The now-escaped wife brought with her memories

Figure 2.7. Hedy Lamarr, inventress of the first frequency-hopping spread-spectrum technique explicitly conceived for anti-jamming communications. (Photo courtesy of Kenneth Galente, The Silver Screen, New York.)

of company films, which she had witnessed, of difficulties Mandl and his factory managers were encountering in getting their "aimlessly" unguided torpedoes to hit evasive targets.

Once settled in southern California but still greatly concerned by the war then impending for the United States, Lamarr sought out the versatile and volatile symphony composer Antheil [90, ch. 32]. Quickly stimulating a new application of his creative talents, she led him to their joint conception of a radio-control scheme in which the transmitted carrier frequency would jump about via a prearranged, randomized, and nonrepeating FH code. A torpedo carrying a properly synchronized receiver could thereby be secretly guided from its launch site all the way to its target. Hedy Lamarr and George Antheil thought such a stealthy "dirigible craft" capability, for missiles as well as torpedoes, would soon be needed by Germany's opponents.

Drawing large diagrams while stretched out on Lamarr's carpeted floor, she and Antheil concentrated on them for weeks until they arrived at a secure and feasible FH-SS concept. The system design took special advantage of the composer's know-how, in their plan to synchronize the radio transmission and reception frequencies by means of twin, identically crypto-code slotted, paper music rolls like those used in player piano, audio-frequency (!) mechanisms (see Figure 2.8). Indeed, Antheil had already achieved such synchronization precisely in his multi-player-piano opus of the 1920s, *Ballet Mécanique* [90, p. 185]. Their invention disclosure points out that an FH repertoire of eighty-eight radio frequencies could readily be accommodated.

The Lamarr-Antheil invention promised to be "sturdy and foolproof," and well within the manufacturing capabilities of the 1940s, and its FH secrecy features were enhanced by the inventors' advocacy of short-pulse transmission to provide low detectability. But what seems, for that day, to

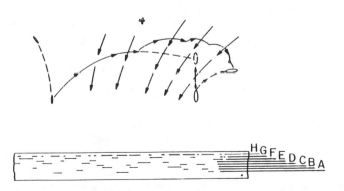

Figure 2.8. Redrawn from the Markey-Antheil patent [87]. Upper figure shows midcourse corrections to a torpedo course, made by FH-SR-SS communication from an observing aircraft. A "piano-roll" recording of the frequency hopping sequence is illustrated in the lower drawing.

be most perceptive in the initial installment [91] of the invention disclosure is presented quite boldly:

> ...it is veritably impossible for an enemy vessel to 'jam' or in any way interfere with the radio-direction of such a previously synchronized torpedo because, simply, no ship may have enough sending stations aboard or nearby to 'jam' every air-wavelength possible or to otherwise succeed except by barest accident to deflect the course of the oncoming radio controlled torpedo —unless, of course, it happened to have the exact synchronization pattern between sender-ship and torpedo.

(Minor, mostly typographical improvements have been made here within this quote while dropping its originally all-capitals lettering; in the patent itself [87], "block control" is said rather than "jam.")

Lamarr next brought her and Antheil's secret system concept to the attention of the then newly government-established National Inventors Council, which soon (quoting from [92]) "... classed Miss Lamarr's invention as in the 'red hot' category. The only inkling of what it might be was the announcement that it was related to remote control of apparatus employed in warfare." That is how it was guardedly publicized by the U.S. Department of Commerce after being scrutinized by Charles F. Kettering, the noted General Motors inventor who was also a pioneer (with Elmer A. Sperry) in remotely piloted vehicles.

Despite this first reaction of enthusiasm mixed with caution, the Lamarr (Markey)-Antheil patent appears to have been routinely issued and published, curiously without imposition of a Secrecy Order. It may be that such potential restraint from the U.S. Patent Office was precluded by the fact that Lamarr's continuing (until 1953) Austrian citizenship rendered her an "enemy alien" during most of World War II. Although this personal circumstance is indeed reflected in the patent case file [93], it seems to have been a mere technicality, which did not impair her screen-actress career nor her image to the American public.

At one of the many war-bond rallies through which she expressed her loyalty, Hedy Lamarr told a crowd of 10,000 in Elizabeth, NJ, that "... she knew what Nazism would mean to this country because she knew what it did to her native country, Austria. 'I'm giving all I can because I have found a home here and want to keep it'" [94]. She had certainly tried to contribute to the war effort, too, by her inventiveness.

Lamarr and Antheil seem, however, to have been more than a score of years ahead of their time, considering that FH-SS evidently was not used operationally against intentional jamming until the 1963 exercise, by the U.S. Navy, of the Sylvania BLADES system. It appears that no coded-paper-hole implementation ever resulted from their FH invention, and that in the decades since the issuance of its patent, whatever scant notice it has received has been confined to the popular press [95].

As far as technology is concerned, all of the above communication systems share a common propensity for the use of electromechanical devices, especially where signal storage and synchronization are required. Undoubtedly in the 1940s the barriers to be overcome in the development of SS communications were as much technological as they were conceptual. The emergence of missile guidance as a potential weapon delivery technique requiring light-weight, rugged construction, did much to drive communication technology toward all-electronic and eventually all-solid state systems. Furthermore the use of guided weapons placed communication jamming and AJ in a more serious light [52]:

> radio jamming, in World War II generally did not require, and did not receive, much attention. Everyone was far more concerned [for intelligence analysis] with listening in. However, when... inhuman radio-guided missiles put in their terrifying appearance, the electromagnetic frequencies employed by the new military engines became suddenly too dangerous to neglect.

2.2 EARLY SPREAD-SPECTRUM SYSTEMS

The following accounts of early SS developments are given to some extent as system genealogies. As we shall see, however, the blood lines of these system families are not pure, there being a great deal of information exchange at the conceptual level despite the secrecy under which these systems were developed. Approximate SS system time lines for several of the research groups tracked here are shown in Figure 2.9. Since the SS concept was developed gradually during the same period that Shannon's work on information theory became appreciated, J.R. Pierce's commentary [96] on the times should be borne in mind:

> It is hard to picture the world before Shannon as it seemed to those who lived in it. In the face of publications now known and what we now read into them, it is difficult to recover innocence, ignorance, and lack of understanding. It is easy to read into earlier work a generality that came only later.

2.2.1 WHYN

Many of the roots of SS system work in the U.S.A. can be traced back to the pioneering of FM radar by Major Edwin Armstrong during the early phases of World War II. The Armstrong technique involved transmitting a sinusoidally modulated wide-band FM signal, and then heterodyne-mixing the return from the target with the transmitted signal at a frequency offset. When the frequency of the modulation was properly adjusted so that the round-trip propagation delay corresponded to one modulation period, the output of the mixer was very narrow-band and the one-period difference between the transmitted modulation and that of the replica then gave a measure of the two-way propagation delay to the target. Certainly, this created a bandwidth expansion and compression methodology, primitive though it was, since the FM wobbulation was simply a sine wave. Sylvania's

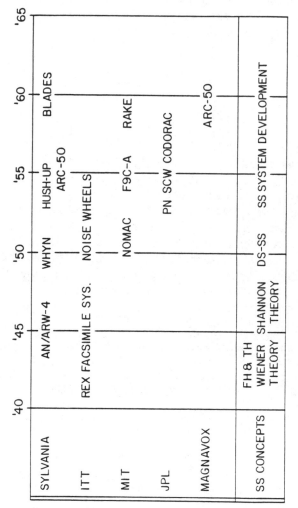

Figure 2.9. Approximate time lines for the systems and concepts feature in this history.

Bayside Laboratories on Long Island received the contract in World War II to continue development of the Armstrong radar.

In 1946, Sylvania received a subcontract from Republic Aviation under Army Air Force Project MX-773 to develop a guidance system for a 500–1500 mile surface-to-surface missile. Although celestial and inertial navigation were possibilities, it was decided that a radio-controlled system using FM ranging would be the most easily realized.

Accurate high-frequency (HF) ranging requires that the receiver extract the ground wave propagation and ignore the potentially strong skywave multipath, as well as ambient noise and jamming. The MX-773 subcontract specifications called for satisfactory discrimination against intereferences of the following types:

1. Skywave, identical in modulation to the ground wave guidance signal but forty times greater in amplitude and delayed 100–250 μs.
2. Other guidance signals identical in modulation, but fifteen times greater in amplitude and differing in arrival time by 50–2000 μs.
3. Unmodulated, pulse, or noise-modulated interference up to twenty times the guidance signal in amplitude.

These time resolution and interference rejection requirements eventually motivated the use of wide-band modulation techniques.

Norman Harvey, the leader in this development effort, states:

> The MX-773 project started out in a strictly study phase, with no special linkage to the FM radar project. It was the specification of 0.5 mile accuracy at 1500 miles that initially gave us the greatest concern. At the time, DECCA (England) was operating a pure cw, cycle-matching navigation system that attracted our interest because it promised that kind of accuracy under ideal conditions. Bob Bowie, John Wilmarth of Republic Aviation, and I went to England in the spring of 1946 to attend the Provisional International Conference on Air Organization (PICAO), to look particularly at DECCA, but also to see if there were other systems that might be adapted to meet our requirements.
>
> Not long after that, the idea of combining a DECCA-type cycle matching system (for accuracy) with FM modulation techniques (for resolution), occurred to me, and the WHYN concept was born.

At least two classes of navigation systems were studied, the first being a circular-navigation, two-ground station system in which the range to each station was determined separately. The second technique measured the relative delay between two identically modulated ground station transmissions (see Figure 2.10). Measured accurately, this information located the receiver on a hyperbolic curve. The acronym WHYN, coined by Harvey for this latter system, stands for Wobbulated HYperbolic Navigation. From the receiver's viewpoint, the WHYN system might be considered a primitive form of TR-FM-SS communication, with information contained in the relative signal delays.

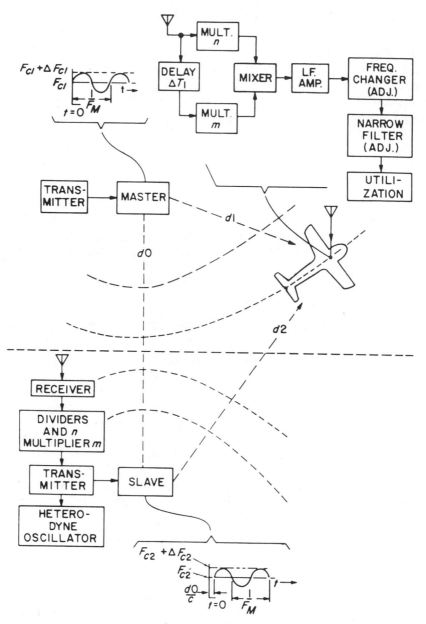

Figure 2.10. Harvey's WHYN system concept is shown here with a plane flying along a hyperbolic curve. Knowing the appropriate ΔT_1 setting for its destination, the aircraft could employ this curve-following tactic to navigate, using only two ground-station transmissions. A third station would be required to provide unambiguous location information. Note that the mixer-i.f. amplifier circuitry acts as the correlating mechanism in the aircraft's receiver. (Diagram abstracted from [107]).

The Bayside engineering team, headed by Harvey, Walter Serniuk, and Meyer Leifer, and joined in 1947 by Nathan Marchand, felt that an FM signal with a more complex modulation than Armstrong's would satisfy requirements. The concept was bench tested via analog simulation with perfect guidance signal synchronization being wired in. Using multiple tone modulation under a maximum frequency deviation constraint of 10 kHz, no simple multitone FM modulation satisfying the contractual constraints could be found. However, low-frequency noise modulation was shown on the bench test to give "an excellent discrimination function with no secondary peaks."

The Sylvania team recognized that noise modulation was "very appealing from the antijamming and security aspects," but its utility in WHYN was questionable since the recording and reproduction requirement in the actual system would be severe. Accordingly, electronic generation of a reproducible multitone modulation function remained the preferred approach. Although the above arc quoted from [97], these revealing results were in classified print by October 1948 [98], simultaneously with Shannon's open publication of pseudorandom signalling.

When Republic Aviation's missile development was discontinued, Sylvania work proceeded on WHYN under the auspices of the Air Force's Watson Laboratories [later to become the Rome Air Development Center (RADC)] with this support spanning the 1948–1952 time frame. Noise modulation never made it into the WHYN system, but correlation detection certainly did. In fact, it was noted [99] in 1950 that "Had the full significance of cross-correlation been realized [at the beginning], it is probable that the name [WHYN] would be different." Advocacy (see Figure 2.11) of correlation detection reached an artistic peak when the following classified Sylvania jingle was heard at a 1950 autumn meeting in Washington.

> *Correlation is the best,*
> *It outdoes all the rest,*
> *Use it in your guided missile*
> *And all they'll hear will be a whistle.*
> *Whistle, whistle, whistle...*

Sung to the tune of a popular Pepsi-Cola commercial, this bit of creativity may have been inspired by the arrival, at Sylvania's helm, of Pepsi's chief executive.

The earliest public disclosure of the concept which had evolved in the first WHYN study appears circumspectly in the last paragraph of an October 1950, article by Leifer and Marchand in the *Sylvania Technologist* [101]:

... The factors determining signal bandwidth and receiver noise bandwidth are entirely different; in the former it is resolution and in the latter, rate of flow of information. A signal that provides good resolution and, hence, has fairly large bandwidth, should be made more complex in nature within this bandwidth for anti-jamming characteristics. Finally, it is important to note

Figure 2.11. Nathan Marchand, organizer and first chairman of the IRE Information Theory Group, early practitioner of correlation techniques, and codesigner of the WHYN system. He was the lyricist of the classified jingle presented at a Washington meeting where radar correlation-detection methodology was discussed [100]. (Photo courtesy of N. Marchand.)

that nowhere has the type of modulation of the signals been specified; the conclusions apply equally to pulse-, frequency-, and phase-modulated signals.

Similar views are expressed by Harvey in a companion paper [102]. Ideas and analyses which were prompted by the Sylvania Bayside work appeared in the literature [103]–[106], in two Harvey patents, the first on WHYN [107] and the second on a collision warning radar [108] which could employ noise modulation, and in another patent [109] on spectrum shaping for improving TOA measurement accuracy in correlation detectors. With continued study, the need for bandwidth expansion to improve system performance became even more apparent, and it was declared that [110]

Jamming signals which are noise modulated or non-synchronous cw or modulated signals are rejected to the same extent that general noise is rejected, the improvement in signal over interference in terms of power being equivalent to the ratio of the transmission bandwidth to the receiver bandwidth.

This improvement property of SS systems is usually referred to as processing gain, which nominally equals the multiplicity factor of the system. By suitably setting these bandwidth parameters, acceptable receiver operation from 40 to 60 dB interference-to-signal ratio was reported in laboratory tests, and navigation receivers operating at −25 dB SNR were predicted [111].

2.2.2 A Note on CYTAC

WHYN was one of the competitors in the development of LORAN (LOng RAnge Navigation), a competition which was eventually won by Sperry Gyroscope Company's CYTAC [112]. Developed in the early 1950s, the CYTAC system and its CYCLAN predecessor had many of the attributes of WHYN, but signal-wise, CYTAC was different in two regards. First, pulse modulation was used so that earliest arriving skywaves could be rejected by gating, and second, phase coding of the pulses was innovated to reject multihop skywaves. These same properties, designed into the system and later patented by Robert Frank and Solomon Zadoff [113], were also used to discriminate between signals from different LORAN stations. The polyphase codes originally designed for CYTAC's pulse modulation were patented separately by Frank [114], but were eventually replaced in LORAN-C by biphase codes to reduce complexity [115]. A certain degree of receiver mismatching also was employed for enhancing time resolution, a similar strategem having been used for the WHYN system [109].

Since narrow-band interference was a potential problem in LORAN, the anti-interference capabilities of this pulse-compression type of signaling were appreciated and reported in 1951 [116]. To further improve performance against in-band CW interference, manually tuned notch filters were added to CYTAC in 1955, and automatic anti-CW notch filters [117], [118] were added to LORAN-C in 1964. To indicate progress, Frank notes that LORAN receivers with four automatically tunable notch filters are now on the market, some for under $1600.

2.2.3 Hush-Up

In the summer of 1951, Madison Nicholson (see Figure 2.12) of Sylvania Buffalo headed a proposal effort for the study of a communication system which he called "Hush-Up." Undoubtedly, the SS ideas therein were distilled versions of those brought to Colonial Radio from the WHYN project by Norman Harvey shortly before that subsidiary lost its identity and was absorbed by Sylvania in February 1950. Nicholson coaxed his old colleague, Robert M. Brown, who had worked at Bayside on the Armstrong radar in World War II while Nicholson had led the AN/ARW-4 team at Colonial, back to Sylvania to work with him and Allen Norris for the duration of the proposal effort. Harvey, by then chiefly responsible for commercial television work, left the realm of military communications research and development. In due course Wright Air Development Center (WADC) gave Sylvania a contract beginning in May 1952, and Nicholson's team went "behind closed doors" to begin work.

Having boned up on Sylvania Bayside's WHYN reports, the engineers at Buffalo set out to verify that a noise-like signal could be used as a carrier, and received coherently, without causing insoluble technical problems.

Figure 2.12. One of the last snapshots of Madison "Mad" Nicholson, at age 51, on a cold Easter Sunday in 1958. As a tribute to this dedicated scientist who died suddenly in Mid-January, 1959, the library at Sylvania's Amherst Laboratory was named the Madison G. Nicholson, Jr., Memorial Library. (Photo courtesy of Dana Cole.)

Independently adopting a pattern of experimentation which was being pursued secretly by other researchers at the time, detector operation was initially examined in the laboratory using a broad-band carrier whose source was thermal noise generated in a 1500 Ω resistor. This wide-band carrier signal was wired directly to the receiver as one input of the correlation detector, thereby temporarily bypassing the remaining major technical problem, the generation of a noise-like carrier at the transmitter and the internal production of an identical, synchronous copy of the same noise-like carrier at the receiver.

In 1953, as the follow-on contract for Hush-Up commenced, James H. Green was hired specifically to develop digital techniques for producing noise-like carriers. John Raney, a Wright Field Project Engineer who had worked on WHYN, also joined Nicholson as System Engineer in early 1953. Nicholson and Raney almost certainly deserve the credit for coining the now universally recognized descriptor "spread spectrum," which Sylvania termed their Hush-Up system as early as 1954, (see [3], footnote 2).

During the second contractual period, which lasted into 1957, Green and Nicholson settled on the form of noise-like carrier which Hush-Up would employ in place of WHYN's FM, namely, a pseudorandomly generated binary sequence PSK-modulated (0 or 180 degrees) onto an RF sinusoid. Such binary sequences with two-level periodic correlation were called "perfect words" by Nicholson. In the end, a variety of perfect word known as an m-sequence was advocated for implementation (more on m-sequences later). Synchronization of the DS-SS signal was accomplished by a sinusoidally dithered tau tracker (τ = delay). Nicholson and Green's tau tracker invention has been, until recently, under patent secrecy order [119] (see Figure 2.13).

As development progressed, the system evolving from the Hush-Up effort was officially designated the ARC-50. Sylvania engineer Everard Book fabricated the ARC-50/XA-2 "flying breadboards." In 1956, flight testing began at Wright-Patterson Air Force Base (WPAFB) with WADC Project Engineers Lloyd Higginbotham and Charles Arnold at the ground end of the ARC-50/XA-2 test link and Capt. Harold K. Christian in the air. The assigned carrier frequency for the tests was the WPAFB tower frequency; the ground terminal of the ARC-50 was about 100 yards from the tower antenna, and communication with the airborne terminal was acceptable at ranges of up to 100 miles. Vincent Oxley recalls that tower personnel, and the aircraft with whom they were conducting normal business, were never aware of ARC-50 transmissions. While the tests were successful, it must have been disheartening to Buffalo engineers when Sylvania failed to win the development/production contract for the ARC-50.

2.2.4 BLADES[2]

In the mid-1950s, Madison Nicholson spent part of his considerable creative energies in the development of methods for generating signals having selectable frequency deviation from a reference frequency. Nicholson achieved this goal with notable accuracy by creating an artificial Doppler effect using a tapped delay line. Even though patent searches uncovered similar frequency-synthesis claims by the Hammond Organ Company, the resulting inventions [120], [121] were a breakthrough for Sylvania engineers working on SS systems.

In addition to being used to slew the time base in the Hush-Up receiver, Nicholson's "linear modulator" (or "cycle adder") was an essential part of another system which Jim Green named the Buffalo Laboratories Application of Digitally Exact Spectra, or BLADES for short. Initiated with company funds in 1955, and headed by Green and Nicholson, the BLADES effort was originally intended to fill Admiral Raeburn's Polaris submarine communications requirements.

Perhaps concern for the serious distortions that multipath could produce in long-range HF communication caused the ARC-50 DS configuration to

---

[2]It is convenient to recount this Sylvania system next, even though chronologically it would belong toward the end of this chapter.

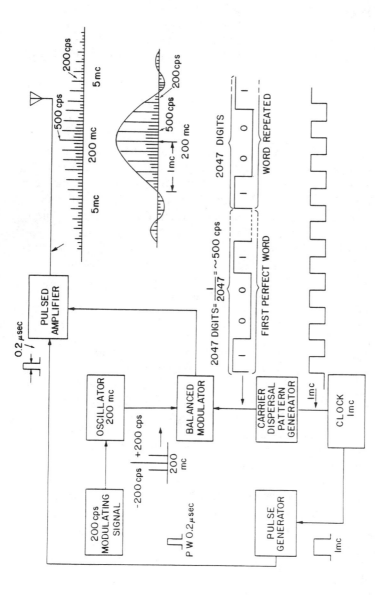

Figure 2.13 (a). Green and Nicholson's SS transmitter contains an *m*-sequence generator, shown here as a "carrier-dispersal-pattern" generator. Data is keyed onto this SS carier by means of a 200 Hz oscillator. The lower of the two displayed output amplitude spectra corresponds to PSK suppressed carrier modulation by the 1 MHz-clocked *m*-sequence. The upper spectrum shows the effect of further spreading by modulating the same carrier dispersal pattern onto .2 µsec pulses. This latter technique for band-spreading probably was employed, since 1950s technology could not reliably support higher-speed *m*-sequence generators.

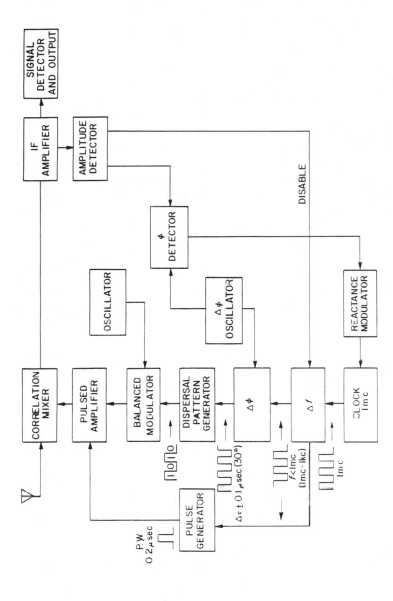

Figure 2.13 (b). The receiver's τ-dither tracker is built around a correlator which compares a locally generated replica of the transmitted carrier to the received signal. Superimposed on the 1 MHz clock driving the carrier dispersal pattern generator is a sinusoidal phase variation controlled by the Δφ oscillator, which in turn produces a corresponding oscillation of the correlator's output amplitude. The phase of this amplitude oscillation, relative to that of the Δφ oscillator, provides an error signal for clock correction. (Both diagrams are abstracted from [119].)

be abandoned in favor of an FH-SS system. In 1957, a demonstration of the breadboard system, operating between Buffalo, NY, and Mountain View, CA, was given to a multiservice group of communications users. Vincent Oxley was system engineer on this development, as well as for the follow-on effort in 1958 to produce a packaged prototype.

The original breadboard contained only an FH-SS/FSK anti-jam mode. The system achieved its protection ratio (Sylvania's then current name for processing gain) by using the code generator to select two new frequencies for each baud, the final choice of frequency being dictated by the data bit to be transmitted. To be effective, a jammer would have to place its power on the other (unused) frequency, or as an alternative, to place its power uniformly over all potentially usable frequencies. Because of the possibility that a jammer might put significant power at the unused frequency, or that the selected channel frequency might be in a fade, a $(15, 5)$ error-correcting code was developed and implemented for the prototype, and was available as an optional mode with a penalty of reducing the information transmission rate to one-third.

While apparently no unclassified descriptions of BLADES are available, glimpses of the system can be seen in several "sanitized" papers and patents produced by Sylvania engineers. Using the results of Pierce [122], Jim Green, David Leichtman, Leon Lewandowski, and Robert Malm [123] analyzed the performance improvements attainable through the diversity achieved by FH combined with coding for error correction. Sylvania's expertise in coding at that time is exemplified by Green and San Soucie's [124] and Fryer's [125] descriptions of a triple-error-correcting $(15,5)$ code (see Figure 2.14), Nicholson and Smith's patent on a binary error-correction system [126], and Green and Gordon's patent [127] on a selective calling system. All are based on properties of the particular perfect words called m-sequences, which were investigated in Sylvania Buffalo's Hush-Up studies. Also involved in BLADES development were R. T. Barnes, David Blair, Ronald Hileman, Stephen Hmelar, James Lindholm, and Jack Wittman at Sylvania, and Project Engineers Richard Newman and Charles Steck at the Navy's Bureau of Ships.

The prototype design effort was aimed at equipment optimization. Extremely stable, single quartz crystal, integrate-and-dump filters were developed. Based on their success, a bank of thirty-two "channel" filters was implemented for an M-ary FSK optional mode to transmit a full character (five bits) per baud. Loss of a single baud in this case meant loss of a full character because the $(15,5)$ decoder could only correct three bit errors per codeword. A "noodle slicer" was implemented to avoid this problem by interleaving five different codewords, so that each baud carried one bit from each word. This interleaving technique was the subject of a patent filed in 1962 by Sylvania engineers Vincent Oxley and William De Lisle [128]. Noodle slicing was never employed in the FH binary FSK mode.

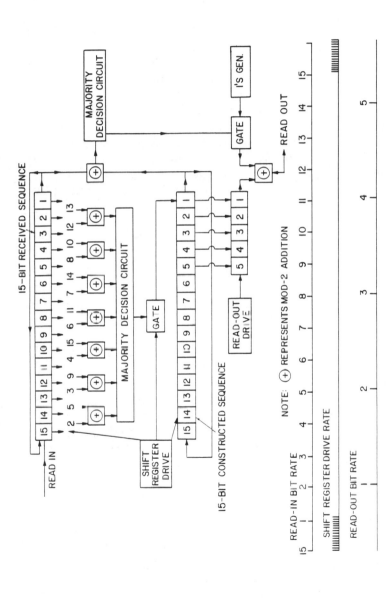

Figure 2.14. The (15, 5) error-correcting code used in BLADES consisted of a (15, 4) code and its complement. The (15, 4) code in turn was composed of the all-zeros 15-tuple and the fifteen cyclic shifts of a length-15 m-sequence. Fryer's description of the (15, 5) decoder (shown above) indicated that the received word was majority-logic decoded as a member of the (15, 4) code, using the two 15-stage registers, and the result compared to the received code word. The complement of the decoded word was assumed transmitted if and only if more than seven bit changes were made in the (15, 4) decoding process. (Diagram from [125].)

BLADES occupied nearly 13 kHz of bandwidth in its highest protection mode. In addition to being a practical AJ system, as Vincent Oxley recalls, during initial breadboard on-air tests, the system also served very well as an unintentional jammer, efficiently clearing all other users from the assigned frequency band.

After considerable in-house and on-air testing between the Amherst Laboratories at Williamsville, NY, and San Juan, PR, the packaged prototype was finally delivered for shipboard testing in 1962. Such a system was evidently carried into the blockade associated with the Cuban missile crisis, but a radio silence order prevented its use. In 1963, BLADES was installed on the command flagship Mt. McKinley for operational development tests. Successful full-duplex field trials over intercontinental distances were observed by Sylvania engineer Gerry Meiler, who disembarked at Rota, Spain, leaving the system in the hands of Navy personnel. Further into the Mediterranean, intentional jamming was encountered, and BLADES provided the only useful communication link for the McKinley. Thus, BLADES was quite likely the earliest FH-SS communication system to reach an operational state.

2.2.5 Noise Wheels

At the end of World War II, ITT reorganized and constructed a new facility at Nutley, NJ, incorporated as Federal Telecommunication Laboratories (FTL), with Henri Busignies as Technical Director. There, in 1946, a group of engineers in Paul Adam's R-16 Laboratory began working on long-range navigation and communication techniques to meet the requirements of the expanding intercontinental air traffic industry. In the available frequency bands, it was expected that multipath generated by signal ducting between the ionosphere and the earth would cause significant distortion, while the prime source of independent interference at the receiver would consist of atmospheric noise generated for the most part by lightning storms in the tropics. A major effort was initiated to study the statistical properties of the interference and to learn how to design high performance detectors for signals competing with this interference.

This was the situation in 1948 when Shannon's communication philosophy, embracing the idea that noise-like signals could be used as bearers of information, made a distinct impression on FTL engineers. Mortimer Rogoff, one of the engineers in R-16 at the time, was an avid photographic hobbyist. He conceived of a novel experimental program using photographic techniques for storing a noise-like signal and for building a nearly ideal cross correlator. Supported by ITT funds and doing some work in a makeshift home lab, Rogoff prepared a 4 in. × 5 in. sheet of film whose transmissivity varied linearly in both directions, thus creating a mask whose transmission characteristic at every point (X, Y) was proportional to the product XY. Two signals then were correlated by using them as the X and Y inputs to the oscilloscope, reading the light emitted from the masked oscilloscope face with a photomultiplier, and low-pass filtering the resultant output.

Rogoff's noise-like carrier came straight from the Manhattan telephone directory. Selecting at random 1440 numbers not ending in 00, he radially plotted the middle two of the last four digits so that the radius every fourth of a degree represented a new random number (see Figure 2.15). This drawing was transferred to film which, in turn, when rotated past a slit of light, intensity-modulated a light beam, providing a stored noise-like signal to be sensed by a photocell.

In initial experiments, Rogoff mounted two identical noise wheels on a single axle driven by a Diehl 900 rpm synchronous motor (see Figure 2.16). Designed and assembled by Rogoff and his colleague, Robert Whittle, separate photocell pickups were placed on each wheel, one stationary and one on an alidade, so that the relative phase between the two signals could be varied for test purposes. Using time-shift keying (an extra pickup required) to generate MARK or SPACE, one noise wheel's signal was modulated and then combined with interference to provide one correlator input, while the other input came directly from the second noise wheel. These baseband experiments, with data rates on the order of a bit per second and, hence, a multiplicity factor of well over 40 dB, indicated that a noise-like signal hidden in ambient thermal noise could still accurately convey information.

In another part of FTL, highly compartmentalized for security purposes, Louis deRosa headed the R-14 Electronic Warfare Group. DeRosa, who earlier had collaborated with Busignies and Deloraine, and who had exchanged many friendly arguments with Nathan Marchand concerning the

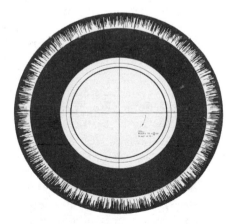

Figure 2.15. Only two copies of Rogoff's secret noise wheel, shown here, were made to support ITT's early research on spread-spectrum systems. The noise wheel concept was revived briefly in 1963 when two more wheels were produced and tested in a system at ITT. (Photo from [130], courtesy of ITT.)

Figure 2.16. ITT's equipment constructed for bench-testing a communication system based on noise-like carriers stored on wheels. (Photo from [130], courtesy of ITT.)

merits of IF correlation (à la Marchand [38]) versus baseband correlation via homodyning (deRosa's favorite), held an umbrella contract through Dr. George Rappaport, Chief of the Electronic Warfare Branch at WADC, to pursue a variety of electronic countermeasures and counter-countermeasures. The contract, codenamed Project Della Rosa, spanned the 1947–1951 time frame and, hence, was concurrent with Rogoff's work.

The first written indication of deRosa's visualization of an SR-SS technique occurs in one of this prolific inventor's patents, filed in January 1950, with L. G. Fischer and M. J. DiToro [129], and kept under secrecy order for some time. The fine print of this patent calls out the possibility of using an arbitrarily coded waveform generated at the transmitter and an identical, synchronous, locally generated waveform at the receiver to provide a reference for a correlation detector, to reliably recover signals well below the noise level.

On August 1, 1950, de Rosa gave a laboratory demonstration of Rogoff's noise wheels to visiting U.S. Air Force personnel, with the system extracting signals 35 dB below the interfering noise. Later the same month, deRosa and Rogoff produced a secret proposal [130] outlining Rogoff's work and proposing several refinements including PSK data modulation, wider bandwidth carrier generation (either by scaling Rogoff's original system or by introducing flying spot scanners reading a pseudorandom image), and quicker-response drives for the receiver's noise wheel synchronizing servo. This proposal also contains a performance analysis resulting in a processing gain representation as a ratio of transmission bandwidth to data bandwidth.

Whittle recalls that in mid-1951 the wheels were separated by about 200 yards in the first test of his synchronization system for the noise wheel drives. During these tests, Bing Crosby's crooning on radio station WOR provided the jamming as Morse code was successfully transmitted at -30 dB SNR. Tapes of the test were made and taken on unsuccessful Washington, DC, marketing trips, where there was considerable interest but evidently the government could not grasp the full significance of the results.

In 1952, an FTL Vice President, retired General Peter C. Sandretto, established relations between deRosa and Eugene Price, then Vice President of Mackay Marine, whereby Mackay facilities in Palo Alto, CA, were made available for transcontinental tests of the FTL equipment. Testing began in late November and ended before Christmas 1952, with Whittle and Frank Lundburg operating an ARC-3 Collins transmitter at the Mackay installation, and deRosa and Frank Bucher manning the receiver at Telegraph Hill, NJ.

Coordination of these field trials was done by telephone using a codeword jargon, with

"crank it" = bring up transmitter power,

"take your foot off" = reduce transmitter power,

"ring it up" = advance the sync search phase,

"the tide is running" = severe fading is being encountered,

"go north" = increase transmission speed.

Initial synchronization adjustments typically took 3 5 min. Matched tuning forks, ringing at a multiple of 60 Hz, provided stable frequency sources for the drives with the receiver synchronizer employing a war-surplus Bendix size-10 selsyn resolver for phase-shifting purposes. Rogoff's original noise wheels were retained for the transcontinental tests, as was his photo-optical multiplier, although the multiplier was improved to handle both positive and negative inputs. Using ionospheric prediction charts, transmission was near the maximum usable frequency (where multipath is least), in the 12–20 MHz range, without FCC license. The system bandwidth was fixed at 8 kHz, the data rate varied down to a few bits per second, and the transmitter power was adjustable between 12 and 25 W.

Although documentation of the test results has not yet been made available, Whittle recalls that during a magnetic storm that happened to occur, a 50 kW Mackay transmitter could not communicate with the East Coast using its conventional modulation, while FTL's test system operated successfully on 25 W. Often the noise-wheel system communicated reliably, even while interference in the same frequency band was provided by the high-power Mackay transmitter. Air Force observer Thomas Lawrence, Project Engineer on Della Rosa and Chief of the Deceptive Countermeasures Section at WADC (another WADC team member was Frank Catanzarite), also recalls witnessing these capabilities.

However, some problems were encountered. In addition to the FTL system once being detected out-of-band (probably in the vicinity of the transmitter), propagation effects apparently caused trouble at times. The signals received at Telegraph Hill were preserved by a speed-lock tape recorder which had been built from scratch to have adequate stability. In the months following the transcontinental tests, John Groce performed correlation recovery experiments on the taped signals, experiencing considerable difficulty with multipath.

Government-decreed project isolation prevented Rogoff from being told about the above tests of his noise wheels. In fact, Rogoff could not follow developments after 1950, except to participate in a patent application with deRosa, for which [130] served as the disclosure. With the help of patent attorney Percy Lantzy, the application, which described a full-fledged SR-SS single-sideband communication system based on Rogoff's noise wheels, was filed in March 1953 (see Figure 2.17). The original patent claims placed few restrictions on the DS modulation technique to be employed, but subsequently these were struck out in favor of single-sideband specification.

In June 1953, the Bureau of Ships placed a secrecy order against the application, which stood until July 1966, when the Navy recommended recision of the order and issuance of the patent. Technically, this was accomplished in November 1966, but before the printing presses in the U.S. Patent Office had begun to roll, a civil servant at the National Security Agency (NSA) noted the invention and was able to get secrecy reimposed. This order stood until 1978, when NSA permitted wholesale recision on scores of patents including at least a dozen on SS techniques. The deRosa-Rogoff patent [49] was finally awarded in November 1979, nearly thirty years after the invention's conception.

The emphasis in both invention and early experimental work at FTL was on covert communication and on suppressing atmospheric noise. It is impossible to determine exactly when FTL engineers appreciated the fully robust AJ capabilities of their system. In 1950, they suspected that broadband noise jamming would be the best attack against the receiver's signal processor, while the receiver itself might be disabled by any strong signal if it did not possess sufficient dynamic range. The deRosa-Rogoff patent, although using the phrase "secrecy and security" several times, never specifically claims AJ capabilities. However, during the course of their work, FTL engineers coined the term "chip" to denote an elementary pulse which is modulated by a single random or pseudorandom variable, and they realized that high performance against atmospheric noise, or when hiding beneath a strong signal like radio station WOR, required many chips per data bit of transmission.

For unknown reasons, FTL was unable to capitalize significantly on this early entrance into the SS field. When in June 1970, as an Assistant

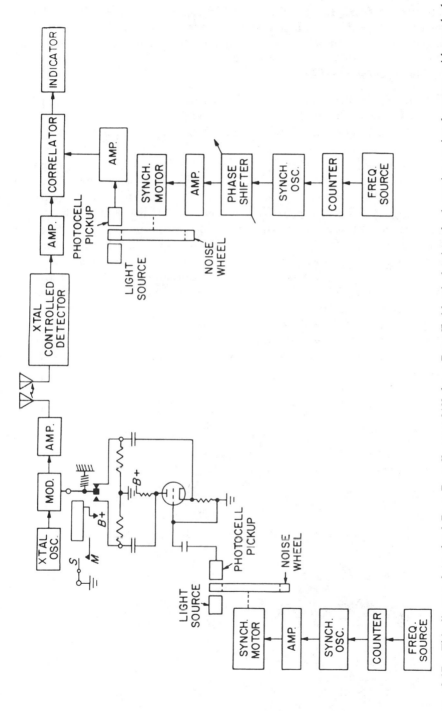

Figure 2.17. This diagram from the deRosa-Rogoff patent [49] shows Rogoff's identical noise wheels serving as signal storage, with matched frequency sources used to control the speed of their synchronous motor drives.

Secretary of Defense, Louis deRosa (see Figure 2.18) was asked about later developments involving the FTL system, he mentioned only Project Dog, a U.S. Navy covert communications operation in the North Korean theater.

2.2.6 The Hartwell Connection

In January 1950, the Committee on Undersea Warfare of the National Research Council addressed a letter to Admiral C. B. Monsen, Assistant Chief of Naval Operations, in which the committee urged the determination of a long-range program against submarines [131]. This was the beginning of a sequence of events which led to the formation of a classified study program known as Project Hartwell, held at M.I.T. in June through August 1950. Under the direction of Prof. Jerrold Zacharias, the study brought together highly qualified experts from the military, industry, and universities, to find new ways to protect overseas transportation.

A subsequent history [132] of the Research Laboratory of Electronics (RLE) at M.I.T. indicates that Hartwell was possibly the most successful of M.I.T.'s summer study projects, motivating the development of "the Mariner class of merchant vessels; the SOSUS submarine detection system; the atomic depth charge; a whole new look at radar, sonar and magnetic detection; and a good deal of research on oceanography." This 1966 history omitted (perhaps because of classification) the fact that transfer of an important concept in modern military communications took place at Hartwell.

Figure 2.18. Louis deRosa remained with ITT until 1966 when he joined the Philco-Ford Corporation as Director of Engineering and Research. In 1970, he left a Corporate Vice President position at Philco-Ford to be sworn in (above) by Melvin Laird as Assistant Secretary of Defense for Telecommunications, the first holder of that office. He died unexpectedly in 1971 after a long workout on the tennis court. (Photo courtesy of Mrs. Louis deRosa, standing next to Secretary Laird.)

One of the many ideas considered was the possibility of hiding fleet communication transmissions so that enemy submarines could not utilize them for direction finding. Appendix G of the secret final report on Project Hartwell suggested that a transmitter modulated by a wide band of noise be employed, reducing the energy density of the transmitted signal "to an arbitrarily small value." If at the same time the actual intelligence bandwidth was kept small, covert communications should be possible in certain situations.

Three systems for accomplishing covert communications were described in the report. One, acknowledged to be the suggestion of FTL's Adams and deRosa (Adams alone was an attendee), was an SR-SS system. A second system, attributed to J. R. Pierce of BTL, used very narrow pulses to achieve frequency spreading, pulse pair spacing to carry intelligence, and coincidence detection at the receiver. It was noted that if synchronized (random) pulse sources were available at transmitter and receiver, then cryptographic-like effects were possible, presumably by transmitting only the second of each pulse pair.

A third system, with no proponent cited, is the only one described by a block diagram in the final report (see Figure 2.19). To avoid the synchronization problems inherent in stored reference systems, it was proposed that the noise-like carrier alone be transmitted on one channel, and that an information-bearing delay-modulated replica of the carrier also be trans-

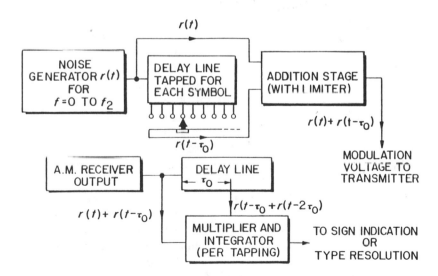

Figure 2.19. Block diagram of a basic transmitted reference system using pure noise as a carrier and time-shift keying for data modulation. No effort has been made to separate the data and reference channels in this system proposed by the East German Professor F. H. Lange. His configuration is nearly identical to that secretly suggested in Project Hartwell a decade earlier. (Redrawn from [12].)

mitted at either the same frequency or at an offset frequency. A cross-correlation receiver still would be employed in this TR-SS system, but the carrier storage and synchronization problems of an SR-SS system would be traded for the headaches of a second channel.

The Hartwell report noted that the SR system was cryptographically more secure than the TR system, which transmitted a copy of the wide-band carrier in the clear. Furthermore, it would be improper to transmit the intelligence-free wide-band carrier on the same channel as the intelligence-modulated carrier with a fixed delay τ between them, since this delay-line addition would impose a characteristic $\cos(\pi f \tau)$ periodic ripple on the power spectral density of the transmitted signal. This ripple might be detectable on a panoramic receiver, compromising the covertness of the transmission. Although not mentioned in the report, it was realized at about the same time that multipath could produce a similar delay-line effect with similar results on any wide-band signal, including SR-SS transmissions.

To close this revealing discussion of noise modulation, the Hartwell report suggested that several of these kinds of systems, using different wide-band carriers, could operate simultaneously in the same band with little effect on each other. This concept, which, it is noted, would eliminate the cooperative synchronization required in time-division multiple-access (TDMA) systems, is one of the earliest references to CDMA operation.

While the authorship of Appendix G is not indicated explicitly in the Hartwell report, it is clear that the main contributors to this portion of the study effort were Jerome B. Wiesner, Edward E. David, Harald T. Friis, and Ralph K. Potter. Wiesner and David went on from M.I.T. to become science advisers to Presidents Kennedy and Nixon, respectively. Friis and Potter were already well-known for their accomplishments at BTL, the latter being a prime innovator of the X-System [58]. There is considerable evidence that Wiesner, then professor of electrical engineering and Associate Director of RLE at M.I.T., was the author of Appendix G, and most likely the source of the TR-SS and SS-CDMA concepts set forth in this report [3].

Concerning Wiesner's place in the development of modern communications, it was later said by an M.I.T. professor [133], "Perhaps one might put it that Wiener preached the gospel and Wiesner organized the church. Jerry's real strength . . . lies in his ability to spot the potential importance of an idea long before others do."

Certainly Wiesner appreciated the possibilities of the wide-band communication systems discussed at Hartwell. Shortly after Hartwell, Wiesner met Robert Fano in a hallway near the Building 20 bridge entrance to the RLE secret research area and told Fano of a "Navy study idea" for using a noise-modulated carrier to provide secure military communications. Even though Fano was familiar with Shannon's precepts and had been an early contributor to the new field of information theory, this made a profound impression on him. He in turn discussed the concept with Wilbur Davenport, a then recent recipient of the Sc.D. degree from M.I.T. They decided to split the research possibilities, with Fano studying radar applications and

Davenport developing the communication applications. This was a fortunate juxtaposition with radar work alongside communications since covertness could not be maintained in radar applications and jamming was always a possibility. The AJ potential of SS systems was appreciated immediately and reported in a series of RLE secret Quarterly Progress Reports.

The year 1951 saw another secret summer study, known as Project Charles, in action at M.I.T. Under the direction of F. W. Loomis of the University of Illinois, Project Charles investigated air defense problems, including electronic warfare. Appendix IV-1 of the Charles Report [134], written by Harry Nyquist of BTL, suggests that carrier frequencies be changed in accordance with a predetermined random sequence, and that by using this FH pattern over a wide band, the effects of jamming could be minimized. In the next section of Appendix IV, the Charles Report proposes that a ground wave radar use a noise-modulated CW carrier to achieve security against countermeasures, and indicates that M.I.T. is investigating this technique (this is over a decade after Gustav Guanella's original conception).

2.2.7 NOMAC

Correlation methodology is so basic to modern communications that it may be difficult to imagine a time when the technique was not widely accepted. Fano, commenting on that era, has said, "There was a heck of a skepticism at the time about crosscorrelation... it was so bad that in my own talks I stopped using the word crosscorrelation. Instead I would say, 'You detect by multiplying the signals together and integrating.'" Nevertheless, by the outset of the 1950s, M.I.T. researchers had become leading proponents of correlation techniques, and were finding more and more problems which correlation might help solve [40]–[42], [151], [229], [230].

It was into this climate that Wiesner brought the noise-like wide-band carrier concept from Project Hartwell to M.I.T. researchers. Within a year of this event, Lincoln Laboratory received its organizational charter and commenced operation, its main purpose being the development of the SAGE (Semi-Automatic Ground Environment) air defense system defined by Project Charles. Soon thereafter, the classified work at RLE was transferred to Lincoln Laboratory and became Division 3 under the direction of William Radford. There, fundamental SS research was performed, to a significant extent by M.I.T. graduate students, guided by Group Leaders Fano and Davenport. The acronym NOMAC, classified confidential at the time and standing for "NOise Modulation And Correlation," was coined by one of these students, Bennett Basore, to describe the SS techniques under study. The term "spread spectrum" was never heard at M.I.T. in those days.

Basore's secret Sc.D. thesis [135], the first on NOMAC systems, was completed under Fano, Davenport, and Wiesner in 1952. Basore recalls that his research was motivated by a desire to "examine how the probability of error for correlation-detection compared with the probability computed by Rice [228], which Rice showed approached zero under conditions consistent with Shannon theory." His thesis, which documented the results of this effort, consisted of a comparison of the performances of transmitted- and stored-reference systems operating in the presence of broad-band Gaussian noise. An RF simulation of a NOMAC system with multiplicity factors up to 45 dB was used to back up theoretical analyses. As in Nicholson's and Rogoff's initial experiments, the synchronization problem of the SR system was bypassed in the experimental setup. The carrier was obtained by amplifying thermal and tube noise, while the interfering noise was produced by some old radar RF strips made originally for M.I.T.'s Radiation Laboratory. Data were on-off keyed. A bandpass correlator was employed in which two inputs at offset frequencies were inserted into an appropriate nonlinearity, the output signal then observed at the difference frequency through a narrow bandpass integrating filter, and the result envelope-detected to recover correlation magnitude. Basore's conclusion was that the effect of noise in the reference channel was to reduce the receiver's output SNR by the ratio of the signal power level to the signal-plus-noise power level in the reference channel.

While the advantages of TR systems have since dwindled as a result of the development of synchronization techniques for the SR system, the disadvantages of TR systems are to a great extent fundamental. Considerable experimental work on TR-NOMAC systems was performed at M.I.T. in the 1950–1952 time frame. Davenport's Group 34 at Lincoln Laboratory developed several TR-SS systems, including one called the P9D. An HF version of the P9D was tested between Lincoln Laboratory and a Signal Corps site in New Jersey and, according to Davenport, worked "reasonably well." This led to the development of a VHF version intended for an ionospheric scatter channel to a Distant Early Warning (DEW) radar complex near Point Barrow, AK. Since the need for LPI and AJ was marginal, SS modulation was not considered necessary and the DEW-Line link was eventually served by more conventional equipment.

A TR system study also was carried out by U.S. Army Signal Corps Capt. Bernard Pankowski in a secret Master's degree thesis [136], under the direction of Davenport. Published at the same time as Basore's thesis, Pankowski's work details several ideas concerning jamming, multiplexing, and CDMA operation of TR-NOMAC systems. In particular, it noted that jamming a TR system is accomplished simply by supplying the receiver with acceptable alternative reference and data signals, e.g., a pair of sine waves in the receiver's pass-bands at the appropriate frequency separation.

Bernie Pankowski offered three possible solutions to the jamming problem, namely, going to the MF or SR systems which others were studying at the time, or developing a hybrid pure noise-TR, FH-SR system with one of the two channels frequency hopped to deny offset frequency knowledge to the jammer. Similarly, CDMA operation was achieved by assigning each transmitter-receiver pair a different frequency offset between their data and reference channels. Laboratory experiments on various single-link TR system configurations with two multiplexed circuits sharing the same reference channel were carried out for a channel bandwidth of 3000 Hz and a data bandwidth of 50 Hz.

There were several exchanges of ideas with other research groups during the period following Basore's and Pankowski's theses. For example, at Lincoln Laboratory on October 2, 1952, Sylvania, Lincoln, and Air Force personnel participated in discussions led by Meyer Leifer and Wilbur Davenport on the subject of secure communications [137]. In February 1953, Sylvania, Lincoln and Jet Propulsion Laboratory researchers attended the (Classified) RDB Symposium on the Information Theory Applications to Guided Missile Problems at the California Institute of Technology [111], [138]. Detailed records of these kinds of exchanges appear to be virtually nonexistent. (RDB: the Pentagon's Research and Development Board.)

As Group 34 studied the TR approach, it became apparent that the SR approach had advantages that could not be overlooked. The task of solving the key generation and synchronization problems for an SR system was given to another of Davenport's Sc.D. candidates, Paul Green. Green's secret thesis [139] is a clearly written comparison of several NOMAC system configurations, the aim of which is to determine a feasible SR design. Comparisons are based on the relationship between input and output signal-to-noise (or jamming) ratios for the receiver's signal processor, and the degradations in this relationship as a result of synchronization error and multipath. Green deduced that correlation at baseband would require a phase-locked carrier for good correlator performance, while correlation at IF à la Basore, with the correlator output being the envelope of the bandpass-filtered IF signal, would require SS carrier sync error to be bounded by the reciprocal of the SS carrier bandwidth.

Green then designed and built (see Figure 2.20) a digitally controlled SS carrier generator in which five stagger-tuned resonant circuits were shock-excited by pseudorandom impulse sequences which in turn were generated from fifteen stored binary sequences of lengths 127, 128, and 129 (see Figure 2.21). The resultant signal had a long period and noise-like qualities in both the time and frequency domains, yet was storable and reproducible at an electronically controlled rate at both ends of a communication link. The proposed SS carrier synchronization procedure at the receiver was quite similar to then contemporary tracking-radar practice, progressing through search, acquisition, and track modes with no change in signal structure. Tracking error was sensed by differencing correlator outputs for slightly different values of clock oscillator phase. Based on Green's results which indicated that an SR system was feasible, and on jamming tests which confirmed TR system vulnerability [140], Group 34 resources were turned toward prototyping an SR system. This marked the end of TR system research at Lincoln Laboratory.

Figure 2.20. These two racks of equipment constitute the transmitter and receiver used to carry out the experimental portion of Paul Green's secret Sc.D. dissertation. The SS carrier generators occupy the upper half of each rack, with the plug boards allowing the operator to change the structure of the 15 stored binary sequences. Later in the F9C system, these plug boards were replaced by punched card readers. (Photo from [139], courtesy of M.I.T. Lincoln Laboratory.)

2.2.8 F9C-A / Rake

The prototype SR-NOMAC system developed for the Army Signal Corps by Lincoln Laboratory was called the F9C. Its evolution to a final deployed configuration, which spanned the 1953–1959 time frame, was carried out in cooperation with the Coles Signal Laboratory at Ft. Monmouth, in particular with the aid of Harold F. Meyer, Chief of the Long Range Radio Branch, and Bernard Goldberg, Chief of the Advanced Development Section, and also Lloyd Manamon and Capt. H. A. ("Judd") Schulke, all of that laboratory. This effect had the whole-hearted support of Lt. Genl. James D. O'Connell, then Chief Signal Officer of the U.S. Army.

Paul Green remained at Lincoln Lab after completing his thesis, and was placed in charge of building and testing F9C equipment. Included in the group of engineers contributing to the development of the F9C were Bob Berg, Bill Bergman, John Craig, Ben Eisenstadt, Phil Fleck, Bill McLaughlin, Bob Price, Bill Smith, George Turin, and Charles Wagner (originator of the Wagner "code," the simplest version of the Viterbi algorithm).

The F9C system [141] occupied 10 kHz of bandwidth and originally employed frequency-shift data modulation at a rate of approximately 22

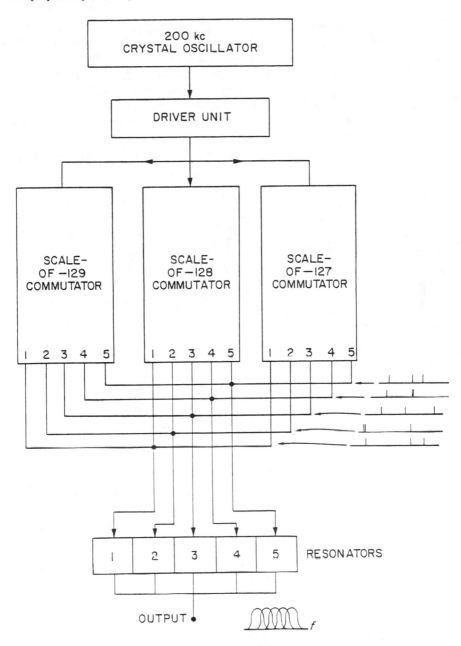

Figure 2.21. The boxes in the above diagram of Paul Green's SS signal generator are located so that they correspond to the physical layout in the equipment racks of Figure 2.20. An SS signal generator similar to the one shown here, combining waveforms of relatively prime periods, was chosen for the F9C system. (Redrawn from [139].)

ms/bit. This resulted in a multiplicity factor greater than 200. The F9C radioteletype system was intended for long-range fixed-plant usage over HF channels. Initially, SS signal generation was accomplished by combining the 28 outputs of four 7-stage counters (fixed to have periods 117, 121, 125, and 128) using an array of AND gates, and driving a bandpass filter with the resultant pulse train. For security against jamming, the gate array connections were controlled by changeable punched cards and this served the role of a key for the system. At the time, there were discussions concerning the possibility of making the SS signal provide cryptographic security as well, but this idea was eventually dropped in favor of conventional data encryption before modulation.

Both the SS signal generator and data modulation technique were later modified to improve spectral and correlation characteristics and change the SS signal period, thereby increasing AJ and privacy capabilities [142]. (For a discussion of the effects encountered in combining sequences of different periods, see [143]–[145].) Also, in a 1955 secret report [226], Price proposed improving DS-SS by resorting to error correction coding in combination with soft or hard decisions against CW or pulse jammers. This first conceptualization of error correction in an AJ strategy was not implemented in the F9C system.

At the suggestion of Signal Corps Capt. John Wozencraft, the bandpass filter in Basore's bandpass correlator was replaced by an active filter [146] employing a diode-quenched high-Q L-C tank circuit, thereby attaining true IF integrate-and-dump correlation operation. A different circuit achieving the same matched-filter-type improvement on sinusoids was developed independently by M. L. Doelz for the Collins Radio Company [147].

Synchronization of the SS signal was accomplished initially by sending a tone burst at a preagreed frequency to start the four 7-stage counters in near synchronism. A fine search then began to bring the receiver's SS modulation clock into precise alignment with the received modulation. When synchronism was achieved, a tracking loop was closed to maintain sync. The fine search was conducted at a rate of 1000 s for each second of relative delay being swept. The frequency standards used in the system were stable enough that even with propagation variations, a disablement of the tracking loop for a day would cause a desynchronization of at most 10 ms. Eventually it was demonstrated [148] that the tone burst was not necessary and the four 7-stage clocks were approximately aligned by time of day at 5 min intervals in initial search situations.

Transcontinental field trials of the F9C system commenced in August, 1954 [142]. The transmitter was located in Davis, CA, and the receiver in Deal, NJ, to provide an eastbound HF link for F9C tests. A conventional teletype link was supplied for westbound communication (see Figure 2.22). Initial tests verified what many suspected, namely, that multipath could severely reduce the effectiveness of SS systems. While at low data rates an ordinary FSK receiver would operate based on the energy received over all propagation paths, the high time resolution inherent in an SS receiver would force the receiver to select a single path for communication, resulting in a considerable loss in signal level. Based on these early trials, several of the previously mentioned modifications were made, and in addition, it was decided to add diversity to the system to combat multipath. Two receivers with antennas displaced by 550 feet were used for space diversity tests, and two correlators were employed to select signals propagated by different paths, in tau-diversity (time delay) tests.

A second set of field trials began in February 1955 to determine the effects of these changes on performance of the transcontinental link. Results showed that an ordinary FSK system with space diversity and integrate-and-dump reception still significantly outperformed the F9C, with tau-diversity showing some hope of improving F9C performance. Both local and remote jamming tests were conducted in this second series, the interfering signal being an in-band FSK signal with MARK and SPACE frequency spacing identical to that of the F9C data modulation. The remote jammers were located at Army Communication Station ABA in Honolulu, HI, and at the Collins Radio Company in Cedar Rapids, IA. With tau-diversity, the F9C achieved a rough average of 17 dB improvement over FSK against jamming in the presence of multipath, justifying transition to an F9C-A production phase.

```
RGR
PHIL MOVED YESTERDAY MOST OF HIS JUNK THAT IS THOUGH HE IS STILL AT
ROBINSON RD
THERE IS NOT TOO MUCH DOING AT THE LAB AT THE MOMENT
WE ARE TRYING TO FIND EQUIPMENT FOR THE ATLATIC PULSE TESTS
THE IPSWICH SITE IS COMING ALONG SLOWLY BECAUSE OF THE LONG DELAY IN
STARTING
MARIE WILL BE FLYING OUT TO SFO ON MONDAY
SHE SAYS R  THAT SHE WILL GIVE YOU A CALL WHEN SHE GETS TR  THERE
I DONT KNOW IF SHE HAS A JOB YET

RGR I WAS WONDERING WHEN SHE WOULD ARRIVE OUT HERE AND ALSO IF I COULD
BE OF ANY ASSISTANCE TO HERE  I WILL WAIT FOR HER CALL AT EDL ON MONDAY
IN REGARD TO MY OWN PLANS I PLAN TO STAY OUT HERE UNTIL THE FIRST EQUIP
AT EDL IS SHOWN TO BE OK AND THE IF TIME PERMITS WILL GO BACK TO THE

LAB FOR A SHORT STAY  IT WILL BE SHORT SINCE I AM GOING UP TO TACOME
TO BE MARRIED ON THE 25 OF AUGUST AND THEN I PLAN TO GO TO SWEDEN ON
THE 28 OF SEPTEMBER FOR AT LEST A MONTH
SO I WILL BE ON VACATION FOR ABOUT THREE MONTHS  GA

OH REALLY ??
YES
WE ARE ALL BREAKING OUR NECKS TRYING TO GT  GET AT THIS

MACHINE TO SAY CONGRATULATIONS!

AND HER I EXPECTLD PITY  MMMMMM

WELL THIS IS RATHER SUDDEN AS THE GIRLS SAY

YEP

HELLO BOB THIS IS P GREEN     I KNEW IT WOULD COME TO THIS SOME DAY

CHEER UP THE SORST IS YET TO COME     CONGRATULZYIONS

CI
    YES I UNDERST,$ 5-5
IBATISREEDSETTWOBESHOULD HOPE FOR THE BEST AND BE SATE SATISFIR SATISB
H
AND BE SATISFIED WITH THE WORST  GA
```

Figure 2.22. This duplex teletype output, made during coast-to-coast tests of the F9C system, includes undoubtedly the first wedding announcement afforded the security of spread-spectrum communications. (Copy courtesy of the announcer at the West Coast station, Robert Berg.)

While initially the F9C MARK-SPACE modulation was FSK, this was eventually changed to another, equally phase-insensitive form of orthogonal signaling, which might be called the "mod-clock" approach. The mod-clock format, conceived by Neal Zierler and Bill Davenport, consisted of either transmitting the SS code in its original form (MARK), or transmitting it with every other pulse from the SS code generator inverted (SPACE) (see Figure 2.23).

Perhaps it was a case of serendipity that several years earlier Fano had suggested communication-through-multipath as an Sc.D. thesis topic to Bob Price. In any event, after a particularly frustrating day of field tests in which they encountered highly variable F9C performance, Price and Green got together in their Asbury Park boarding house to discuss multipath problems. Price already knew the optimal answers to some questions that were to come up that evening. Since receiving his doctorate and having been rehired by Davenport after trying his hand at radio astronomy in Australia, he had been polishing his dissertation with "lapidary zeal" (Green's witticism). Price had, in fact, statistically synthesized a signal processing technique for minimum-error-probability reception of signals sent over a channel disturbed by time-varying multipath as well as noise [150].

Green separately had been trying to determine how to weight the outputs of a time-staggered bank of correlators in order to improve F9C performance, and, acting on Jack Wozencraft's suggestion, had decided to choose weights to maximize the resultant transversal filter's output signal-to-noise ratio. Of course, the TOA resolution capability of the F9C was sufficient to guarantee that the outputs of different correlators in the bank represented signals arriving via different paths. Thus, the problem was one of efficiently recombining these signals. It took little time for Price and Green to realize that the results of their two approaches were nearly identical, and from that evening onward, the "Rake" (coined by Green) estimator-correlator became part of their plans for the F9C. Price took charge of building the Rake prototype, with the assistance of John Craig and Robert Lerner.

Related to Wiesner and Lee's work on system function measurements using cross correlation [151], Brennan's work on signal-combining techniques [152], and Turin's multipath studies [153], the Rake receiver could in turn be viewed as a predecessor of adaptive equalizers [154]. The Rake processor [155]–[157] (patented at Davenport's prompting) is adaptive in the sense that the weight on each MARK-SPACE tap pair is determined by the outputs of that MARK-SPACE tap pair, averaged over a multipath stability time constant. (See Figure 2.24). In its ultimate form, the magnetostrictive tapped delay line (patented by Nicholson [158]), around which the processor was built, contained fifty taps spanning 4.9 ms, the spacing being the reciprocal of the NOMAC signal bandwidth.

In addition to solving the multipath dilemma and thereby securing the full 23 dB of potential processing gain, Rake also allowed the sync search rate to be increased so that only 25 s were necessary to view one second of

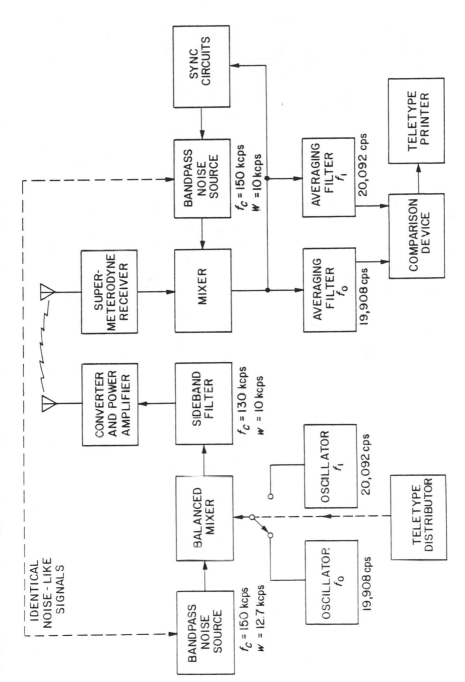

Figure 2.23 (a). Original F9C block diagram with FSK data modulation. (Redrawn from [141], [142].)

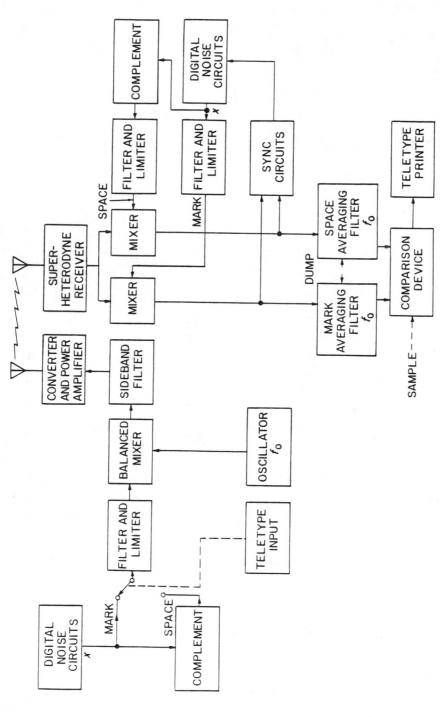

Figure 2.23 (b). Revised F9C block diagram with alternate chip inversion for the creation of orthogonal MARK and SPACE signals. (Redrawn from [142].)

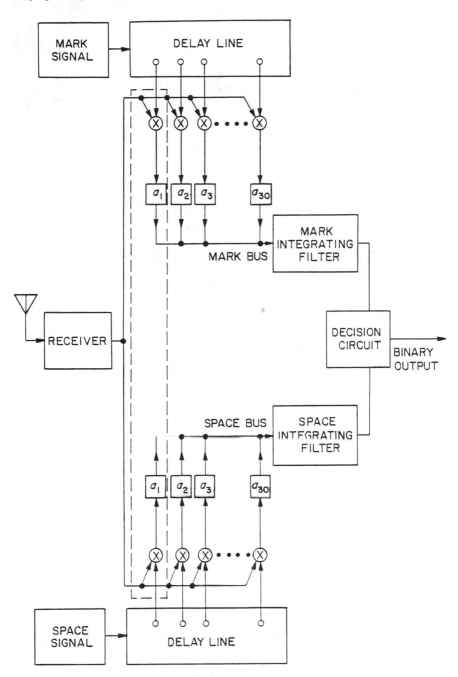

Figure 2.24 (a). This two-delay line version of Rake shows how signals arriving via different path delays are recombined for MARK and SPACE correlation detection. In practice, a single delay line configuration was adopted.

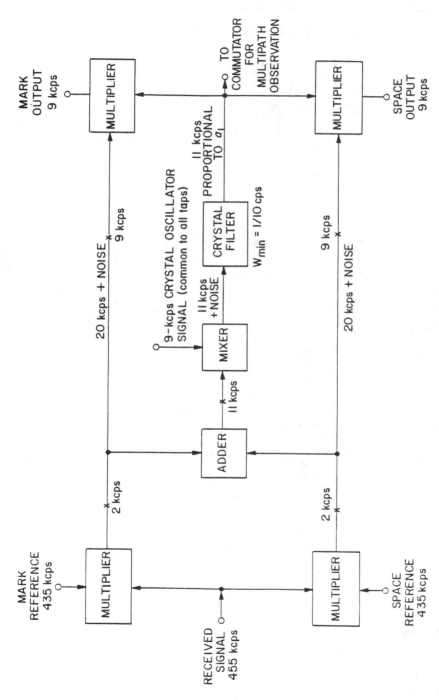

Figure 2.24 (b). The tap unit diagrammed here includes a long-time-constant crystal filter whose output signal envelope is proportional to the combining weight (a_i). This processing corresponds to that shown in the dashed box in (**a**). Rejection traps to eliminate undesirable cross products are shown by $\times$'s.

Figure 2.24 (c).   This Rake rack contains thirty tap units, two helical magnetostrictive delay lines, and a commutator chassis. (Diagrams redrawn from [155], photo courtesy of M.I.T. Lincoln Laboratory.)

delay uncertainty [148]. (Readers of this early literature should note that to prevent disclosure of the actual F9C SS signal structure, all unclassified discussions of Rake, e.g., [156], invoked m-sequences for signal spreading. In addition, mod-clock MARK-SPACE modulation was never mentioned in this open literature.)

The F9C-A production contract was let to Sylvania Electronic Defense Laboratory (EDL) at Mountain View, CA, in 1955, with Judd Schulke acting as Project Engineer for the Signal Corps, and Bob Berg as Lincoln Lab's representative, resident at EDL. By December 1956, the first training manuals had been published [159]. Originally 16 F9C-A transmitter-receiver pairs were scheduled to be made, but funds ran out after production of only six pairs. The first installation was made for Washington, DC, near Woodbridge, VA/La Plata, MD. Worldwide strategic deployment commenced with the installation in Hawaii in January 1958, and was followed by installations in Germany (Pirmasens/Kaiserslautern, February 1958), Japan, and the Philippines. With the threat of a blockade of Berlin, the equipment assigned to Clark Field in the Philippines was moved in crates of Philippine mahogany to Berlin in the spring of 1959.

Rake appliques for the F9C-A receivers were fabricated later by the National Company of Malden, MA. These were produced with an improved, yet simplified, circuit configuration, invented at General Atronics [160], which employed tap units having a full 10 kHz of internal bandwidth instead of being structured as in Figure 2.24 (b). Additionally, the F9C-

A/Rake appliques introduced a novel method of ionospheric multipath display, in which the multipath-matched tap-combining weights were successively sensed by a short pulse traveling along the magnetostrictive delay line, the pulse duty cycle being low enough to have negligible effect on the Rake signal processing. Bernie Goldberg was the Project Director for this effort and Robert L. Heyd served as the Project Engineer. Together they also developed Goldberg's innovative "stored ionosphere" concept [161] in which the F9C-A/Rake's multipath measurement function was used to record ionospheric channel fluctuations for their later re-creation in testing short-wave apparatus. This measurement capability was also employed to assess multipath effects, between Hawaii and Tokyo, of a high altitude nuclear detonation in the Pacific in July 1962.

The F9C-A/Rake is no longer on-site, operational or supported by the Army.

2.2.9 A Note on PPM

As the Hartwell report indicated, J. R. Pierce of BTL had suggested that covertness be achieved by using extremely narrow pulses for communication, thereby spreading the transmission spectrum. This idea was undoubtedly based on BTL's postwar work on pulse position modulation (PPM) [96]. After discussing the CDMA idea generally in a 1952 paper, Pierce and Hopper make the following observations:

> There are a number of ways in which this sort of performance could be achieved. One way has been mentioned: the use of random or noise waveforms as carriers. This necessitates the transmission to or reproduction at the receiver of the carrier required for demodulation. Besides this, the signal-to-noise ratio in such a system is relatively poor even in the absence of interference unless the bandwidth used is many times the channel width... In the system discussed here, the signal to be sent is sampled at somewhat irregular intervals, the irregularity being introduced by means of a statistical or 'random' source. The amplitude of each of the samples is conveyed by a group of pulses, which also carries information as to which transmitter sent the group of pulses. A receiver can be adjusted to respond to pulse groups from one transmitter and to reject pulse groups from other transmitters.

This early unclassified reference [162] not only mentions an unpublished, noise-carrier-CDMA notion proposed by Shannon in 1949, but also indicates a PPM technique for achieving the CDMA property of an SS system. PPM systems evidently remained of interest to BTL engineers for some time (e.g., see [163]), and also formed the basis for some Martin Company designs [164], [165].

2.2.10 CODORAC

In 1952, the Jet Propulsion Laboratory (JPL) of the California Institute of Technology was attempting to construct a radio command link for the purpose of demonstrating remote control of the Corporal rocket. The two groups most closely connected with the formation of a system for accomplishing this task were the Telemetry and Control Section under Frank Lehan and the Guidance and Control Section under Robert Parks, both reporting to William Pickering.

One novel concept was formulated by Eberhardt Rechtin, a recent Caltech Ph.D. under Parks, who decided that the current radio design approach, calling for the IF bandwidth to match the Doppler spread of the signal, could be improved dramatically. Rechtin's solution was to adjust the receiver's local oscillator automatically to eliminate Doppler variations, thereby significantly reducing the receiver's noise bandwidth. This automated system used a correlator as its error detector, with the correlator inputs consisting of the received signal and the derivative of the estimate of the received signal. The resultant device, called a phase-locked loop (PLL), with its characteristics optimized for both transient and steady-state performance [166], was a key ingredient of all later JPL guidance and communication systems. Surprisingly, when attempts were made to patent an advanced form of PLL, the prior claim which precluded the award did not come from television, which also had synchronization problems, but came instead from a 1905 patent on feedback control. In retrospect, Eb Rechtin feels that perhaps his greatest contribution in this area consisted of "translating Wiener's 'Yellow Peril' into English," and converting these elegant results into practice.

In struggling with blind-range problems occurring in the integration of a tracking-range radar into the Corporal guidance system, Frank Lehan realized that his problems were due to the shape of the radar signal's autocorrelation function. The thought that the autocorrelation function of broad-band noise would be ideal led Lehan to formulate the concept of an elementary TR-SS communication system using a pure noise carrier. In May 1952, Lehan briefly documented his partially developed ideas and their potential for LPI and AJ in a memo to Bill Pickering. Lincoln Laboratory's NOMAC work was quickly discovered, since both JPL's and Lincoln's were sponsored by the Army, and the wealth of information contained in Lincoln's detailed reports was made available to JPL researchers.

By the spring of 1953, JPL had decided upon a DS-SS configuration for the Corporal guidance link, and Rechtin, noting applications for his tracking loop theory in SS code synchronization, transferred to Lehan's section to head a group studying this problem. Seeing the value of the M.I.T. documentation, JPL began a series of bimonthly progress reports in February 1953, these later being combined and annotated for historical purposes in 1958 [167] (see Figure 2.25).

The term "pseudonoise" with its abbreviation "PN" was used consistently from 1953 onward in JPL reports to denote the matched SS signals used in a DS system. Two PN generators initially were under consideration (see Figure 2.26), the first being a product of twelve digitally generated (± 1) square waves having relative periods corresponding to the first eleven primes. This type-I generator was eventually dropped because of its exces-

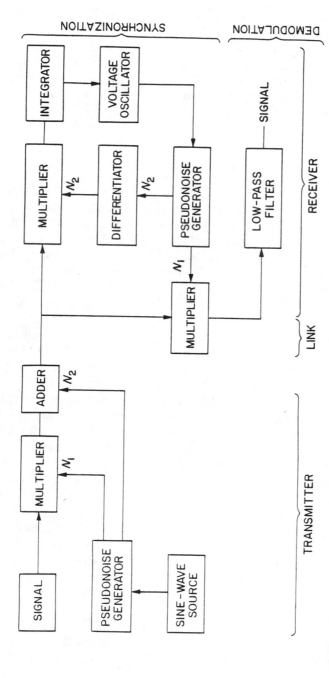

Figure 2.25. JPL's first attempt at a stored reference spread-spectrum design is shown here. This particular system uses one unmodulated noise signal N_2 for synchronizing the receiver's pseudonoise generator and another N_1 for carrying data. Most SR-SS systems do not use a separate signal for SS signal synchronization. (Redrawn from [167].)

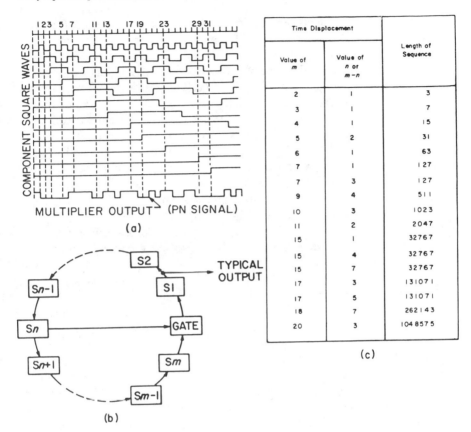

	Time Displacement		
	Value of m	Value of n or $m-n$	Length of Sequence
	2	1	3
	3	1	7
	4	1	15
	5	2	31
	6	1	63
	7	1	127
	7	3	127
	9	4	511
	10	3	1023
	11	2	2047
	15	1	32767
	15	4	32767
	15	7	32767
	17	3	131071
	17	5	131071
	18	7	262143
	20	3	1048575

(c)

Figure 2.26(a). The type-I PN generator uses a multiplier to combine the outputs of binary ($+1$ or -1) signal shapers which in turn are driven by the outputs of relatively prime frequency dividers operating on the same sinusoid. The component square waves and the resultant PN product signal are shown here. **(b)** JPL's type-II generator was an m-stage linear-feedback shift register which produced binary (0 or 1) sequences of maximum period. The output of the mth and nth stages are added modulo 2 to produce the input of the first stage S1. **(c)** This first list of connections for the type-II generator was produced at JPL by hand and computer search. (Diagrams and table redrawn from [167].)

sive size and weight. The type-II PN generator was

...based on the equation

$$x(t+m) = x(t)x(t+n)$$

where t represents time, m and n are integers (m represents a time displacement greater than n), and the functions $x(t+m)$, $x(t)$, and $x(t+n)$ may equal ± 1 only.... If the correct values of m and n are chosen, the period before repeat is $2^m - 1$.... The correlation function of all type-II PN generators consists of a triangle of height unity and of a width equal to twice the shift time standing on a block of height $(2^m - 1)^{-1}$.

This origination of an almost perfect spike-like autocorrelation function, accompanied by descriptions of shift register hardware, positive results of baseband synchronization experiments at -20 dB SNR's and a table of suitable values of m and n for values of m up to 20, was reported as progress through August 1953 [167], [168]. In later works by other researchers, these PN sequences were called shift-register sequences or linear-recurring sequences as a reminder of their particularly convenient method of generation, and were also termed m-sequences since their period is maximal.

On January 18, 1954, a JPL PN radio system was operated over a 100-yard link and two independent commands were communicated. Initial synchronization was achieved with the aid of a land line which was disconnected after sync acquisition. The system was able to withstand jammer-to-signal power ratios of 15–20 dB before losing lock, against a wide variety of jamming threats. This test was the assurance that JPL engineers needed regarding the practicality of SR-SS communications.

At this point, work on the command system was temporarily dropped and a major effort was begun to optimize a pure ranging system, called the Scrambled Continuous Wave (SCW) system, which consisted of a "very narrow-band CW system scrambled externally by a PN sequence." On July 27, 1954, Corporal round 1276-83 carrying an SCW transponder was launched at White Sands Proving Ground. The transponder operated successfully from takeoff to near impact seventy miles away, providing range and range rate without loss of lock in the synchronization circuitry. Rechtin, engineer Walter Victor, and Lehan (who left JPL in 1954) later filed an invention disclosure based on the SCW system results from this test, and called the system a COded DOppler RAdar Command (CODORAC) system. This acronym was used to describe the radio guidance systems developed for the Sergeant and later the Jupiter missiles in the 1954–58 time frame.

Throughout this period one of the major problems in establishing one-way communication to a missile was to make the PN generator tough enough to withstand high temperatures and vibrations as well as small and light enough to fit into the missile design. A variety of devices (e.g., subminiature hearing aid tubes) and potting compounds were tested. In 1954, Signal Corps liaison official G. D. Bagley was able to obtain approximately 100 of the Western Electric type 1760 transistors (the first available outside BTL) for use by JPL engineer Bill Sampson in the construction of a PN generator. The resulting circuitry was an interesting combination of distributed-constant delay lines and transistor amplifiers and logic, chosen because it minimized the number of active elements required [169]. This general method of construction remained the norm at JPL through 1958.

Late in 1954, a separate group under Sampson was formed for the purpose of investigating possible countermeasures against the SCW system equipment designed by a group headed by Walt Victor. Created to make this phase of the program as objective as possible, this organization brought forth a thoroughly designed system with high countermeasures immunity. Here are three issues on which significant progress was made.

1. It was hoped that repeater jamming would be ineffective as a result of the high TOA resolution capability of SS and the excess propagation delay incurred by the repeater. The period of the PN sequence was made longer than the missile flight time so that it would be impossible for a repeater to store a PN coded signal for nearly a full period and deliver it to the victim receiver in synchronism with the directly transmitted PN sequence one period later. A weakness in this regard still existed in a simple m-sequence generator based on a linear recursion. Specifically, these sequences possessed a "cycle-and-add" property (see Chapter 5, Section 5.4.2) by which the modulo 2 sum of a sequence and a delayed version of that sequence results in the production of the same sequence at still another delay. The equivalent "shift-and-multiply" property for the ± 1 version of these m-sequences, satisfying the equation quoted earlier in this subsection, conceivably could be used by a jammer to produce an advance copy of the sequence without waiting a full period. In an effort to completely rule out this possibility, Caltech graduate student Lloyd Welch was hired in 1955 to study the generation of sequences which avoid the cycle-and-add property by resorting to nonlinear recursions [171]. Although laboratory system work continued to use linearly generated PN sequence for test purposes, final designs were to be based on nonlinear generators.

2. Initial jamming tests revealed weaknesses in the SCW system when confronted by certain narrow-band jammers. Most of these were due to problems in the mechanization of the multiplications required in the PN scrambler and correlator descrambler. For example, if the descrambler effectively mechanizes a multiplication of the jamming signal by a constant plus the receiver's PN sequence replica (the constant representing a bias or imbalance in the multiplicaton/modulation process), then the multiplier output will contain an unmodulated replica of the jamming signal which has a free ride into the narrowband circuitry following the descrambler. The sure cure for this problem is to construct better balanced multipliers/modulators since the processing gain achievable in an SS system is limited by the "feedthrough" (or bias) in its SS multipliers. In the mid-1950s, JPL was able to build balanced modulators which would support systems with processing gains near 40 dB.

3. Another major concern of system designers was the decibel range and rates of variation of signal strength, as a result of missile motion and pulsed or intermittent jamming. At the circuit level, the two approaches to controlling signal levels in critical subassemblies were automatic gain control (AGC) and limiting. The AGC approach suffers from the possibility that its dynamics may make it susceptible to pulse jamming, while limiters, although instantaneous in nature, generate harmonics which

might be exploited by a jammer. The eventual design rule-of-thumb was that limiters could be used when necessary on narrow-band signals (e.g., prior to PLL phase detectors), and that AGC techniques should be used in the wide-band portions of the system. Analytical support for this work came from JPL's own studies of AGC circuits [172], [173], and from Davenport's classic paper on limiters [174]. It was not realized until much later that in some instances the limiter theory was not appropriate for coherent signal processing analyses [175]. For further discussion of AGC capture by jamming, see [176].

Many of these kinds of problems remain with the designer today, the differences being in the technology available to solve them.

Both the Sergeant and Jupiter guidance programs were terminated when decisions were made to choose all-inertial jam-proof guidance designs as the baseline for those missile systems. However, CODORAC technology survived in the JPL integrated telemetry, command, tracking, and ranging system for the Deep Space Program, and in the later projects of subcontractors who had worked for JPL in the Jupiter program. A modified version of CODORAC became the Space Ground Link Subsystem (SGLS) now used routinely in U.S. Department of Defense missile and space instrumentation.

2.2.11 *M*-Sequence Genesis

The multiplicative PN recursion given in [167] and its linear recursion counterpart in modulo 2 arithmetic, namely

$$y(t + m) = y(t) \oplus y(t + n)$$

were among those under study by 1954 at several widely separated locations within the United States. Lehan recalls that the idea of generating a binary sequence recursively came out of a discussion which he had with Morgan Ward, Professor of Mathematics at Caltech, who had suggested a similar decimal arithmetic recursion for random number generation. It is hard to determine if this idea was mentioned at the (Classified) RDB Symposium held at Caltech in February 1953. Lincoln Laboratory's Bill Davenport remembers that the first time that he had seen a PN generator based on the above recursion was in Lehan's office on one of his trips west. This generator, built to Rechtin's specifications, was used to extend Rechtin's hand-calculated table of *m*-sequence generators from a shift register length of at most 7, to lengths up to 20 (see Figure 2.26).

Sol Golomb, then a summer hire at the Glenn L. Martin Company in Baltimore, MD, first heard of shift register-generated sequences from his supervisor, Tom Wedge, who in turn had run across them at a 1953 M.I.T. summer session course on the mathematical problems of communication theory. (This meeting included an elite group of the founding fathers of information theory and statistical communications. See Figure 2.27.)

On the other hand, Neal Zierler, who joined Lincoln Laboratory in 1953, recalls discovering shift register generation of PN sequences while looking for ways to simplify the SS signal generators used for the F9C system. Golomb's [170], [177], [178] and Zierler's [179]–[181] work established them as leading theorists in the area of pseudonoise generation. However, Zierler's shift register-generated sequences were never used in the F9C-A system since they possessed cryptanalytic weaknesses. Golomb's work gained further recognition after he joined JPL in August 1956.

Madison Nicholson's early attempts at PN sequence design date back to 1952 [137]. Nicholson's first exposure to the pseudorandomness properties of linearly recurring sequences probably came from Allen Norris, who remembers relating to Nicholson ideas developed from lectures by the noted mathematician, A. A. Albert, of the University of Chicago. Coworkers recollect that Nicholson used paper-and-pencil methods for finding a perfect word of length 43. Jim Green, in due course, joined in this exploration, and

Figure 2.27. Cut from a 1953 photograph of a summer session class on mathematical problems of communication theory, this picture salutes (from left to right) Yuk Wing Lee, Norbert Wiener, Claude Shannon, and Robert Fano. It is ironic that Wiener could not prevent the transfer of his theories, through meetings like the one at which this picture was taken, to the military research which he refused to support after World War II. (Photo courtesy of M.I.T.)

built demonstration hardware. Bob Hunting was assigned to investigate the generation of long *m*-sequences and spent a considerable amount of time exercising Sylvania's then-new UNIVAC 1 in the Corporate Computer Center at Camillus, NY, and eventually produced an extensive list of "perfect word" generators. R. L. San Soucie and R. E. Malm developed nonlinear sequence-combining techniques for the BLADES prototype, the result being an SS carrier with a period of about 8000 centuries. Oliver Selfridge of Lincoln Laboratory's Group 34 became the government representative whose approval was required on Sylvania's SS code designs for Air Force contracts, but was not involved with the Navy's BLADES effort.

Early work by others on linear-feedback shift registers includes that of Gilbert [182], Huffman [183], and Birdsall and Ristenbatt [184]. Additional insights were available from the prewar mathematical literature, especially from Ward [185], Hall [186], and Singer [187], [188]. Of course, in the top secret world of cryptography, key-stream generation by linear recursions may very well have been known earlier, particularly since Prof. Albert and others of similar stature were consultants to NSA. But it is doubtful that any of these had a direct impact on the pioneering applications to SS communication in 1953–54.

2.2.12 AN/ARC-50 DEVELOPMENT AT MAGNAVOX

In 1953, a group of scientists interested in the design of computers left the University of California at Los Angeles and formed a research laboratory under an agreement with the Magnavox Corporation. Their first contact with SS systems came when JPL approached them with the problem of building DS-SS code generators for the Jupiter missile's proposed radio navigation link. This exposure to JPL's work on PN sequences and their application to radio guidance paid dividends when Lloyd Higginbotham at WADC became interested in getting high-speed, long-period generators for the ARC-50 system which was emerging from the Hush-Up study at Sylvania Buffalo. At Sylvania, Hush-Up had started out under the premise of radio silence, and was aimed for an application to the then-new air-to-air refueling capability developed by the Strategic Air Command (SAC). After a demonstration of the wired system at Sylvania, a SAC representative made the "obvious" statement, "When you are in radar range, who needs radio silence?" From that time onward, the design was based on AJ considerations.

The AJ push resulted in NSA being brought into the program for their coding expertise. However, because of their nature, NSA passed technical judgment rather than providing any concrete guidance. The NSA view was that the SS codes had to be cryptographically secure to guarantee AJ capability, and Lincoln Laboratory had established that the proposed ARC-50 SS PN code was easily breakable. On this point, Lloyd Higginbotham says, "At that time we felt we were being treated unfairly because the system was still better than anything else then in existence."

By March 1958 Magnavox had parlayed their knowledge of high-speed PN generators into a development contract for the ARC-50 system, won in competition with Sylvania. Magnavox Research Laboratories operated out of a garage on Pico Boulevard in Santa Monica in those early days, with Jack Slattery as general manager and Ragnar Thorensen as technical director. From their few dozen employees, a team was organized to design the code generators and modem, while RF equipment was built at Magnavox's Fort Wayne facility. Shortly thereafter, Magnavox Research Laboratories moved to Torrance, CA, into a new facility sometimes referred to as "the house the ARC-50 built." Harry Posthumus came from Fort Wayne as program manager and teamed with system designers Tom Levesque, Bob Grady, and Gene Hoyt; system integrator Bob Dixon; and Bill Judge, Bragi Freymodsson, and Bob Gold.

Although retaining the spirit of the DS-SS system developed at Sylvania, technologically the design evolved through several more phases at Magnavox. Nowhere was this more obvious than in the design of the SS code generators, the heart of the system. The earliest Magnavox code generators were built using a pair of lumped constant delay lines, run in syncopated fashion to achieve a rate of 5 Mchips/s. This technology was expensive with a code generator costing about $5000, and was not technically satisfactory. The first improvement in this design came when the delay lines were transistorized, and a viable solution was finally achieved when 100 of the first batch of high-β, gold-doped, fast rise-time 2N753 transistors made by Texas Instruments were received and used to build a single-register code generator operating at 5 Mchips/s.

Originally to facilitate SS code synchronization, the system employed a synchronization preamble of 1023 chips followed by an m-sequence produced by a 31 stage shift register. Register length 31 was chosen because the period of the resultant m-sequence, namely 2,147,483,653, is prime, and it seemed unlikely that there would exist some periodic substructure useful to a jammer. Lacking knowledge of the proper connections for the shift register, a special machine was built which carried out a continuing search for long m-sequences. Problems were encountered involving false locks on correlation sidelobe peaks in the sync preamble (sometimes it seemed that a certain level of noise was necessary to make the system work properly), and concerning interference between different ARC-50 links as a result of poor cross-correlation properties between SS codes.

The ARC-50 was configured as a fully coherent system in which the SS code was first acquired, and the sinusoidal carrier was then synchronized using PLL techniques. Because of apprehension that jamming techniques might take advantage of coupling between the RF oscillator and the code chip clock, these two signals were generated independently in the transmitter. The receiver's PLL bandwidth was constrained by the fact that no frequency search was scheduled in the synchronization procedure; the assumption being that the pull-in range of the PLL was adequate to overcome both oscillator drifts and Doppler effects. Being a push-to-talk voice system which could operate either as a conventional AM radio or in an SS mode, a 5 s sync delay was encountered each time the SS modem was activated. Ranging up to 300 miles was possible with the measurement time taking about 40 s. To retain LPI capability in this AJ system, transmitter power was adjustable from minute fractions of a watt up to 100 W.

Testing of the Magnavox ARC-50 began in 1959. Bob Dixon, joined by John G. Smith and Larry Murphy of Fort Wayne, put the ARC-50 through preliminary trials at WPAFB, and later moved on to the Verona site at RADC. One radio was installed in a C131 aircraft and the other end of the link resided in a ground station along with a 10 kW, CW jammer (the FRT-49). Testing consisted of flying the aircraft in the beam of the jammer's 18 dB antenna while operating the ARC-50. Limited results in this partially controlled environment indicated that the receiver could synchronize at jammer-to-noise ratios near those predicted by theory.

Shortly after these flight tests, an upgraded version of the ARC-50 was developed with significantly improved characteristics. To alleviate SS-code correlation problems, a new design was adopted, including an m-sequence combining procedure developed by Bob Gold [189], [190] which guaranteed low SS-code cross correlations for CDMA operation. The SS sync delay was

Figure 2.28 (a).   Examples of early- and mid-1960s technology. **(a)** SS code generator portion of a TH system developed by Brown, Boveri, and Company for surface-to-air missile guidance. (Photo courtesy of I. Wigdorovits of Brown, Boveri, and Co.)

Figure 2.28 (b). 1965 picture of the MX-118 applique, a member of the ARC-50 radio family and the first to use Gold codes. (Photo courtesy of Robert Dixon.)

reduced to one second and an improved ranging system yielded measurements in two seconds.

Even though the ARC-50 possessed obvious advantages over existing radios such as the ARC-27 or ARC-34, including a hundredfold improvement in mean time between failures, there was Air Force opposition to installing ARC-50's in the smaller fighter aircraft. The problem revolved around the fact that pilots were accustomed to having two radios, one being a backup for the other, and size-wise a single ARC-50 would displace both of the prior sets.

Certainly, the ARC-50 was a success, and Magnavox became an acknowledged leader in SS technology. Among the descendants of the ARC-50 are the GRC-116 troposcatter system which was designed free of a sync preamble, and the URC-55 and URC-61 ground-station modems for satellite channels. An applique, the MX-118, for the Army's VRC-12 family of VHF-FM radios was developed, but never was procured, partly as a result of inadequate bandwidth in the original radios (see Figure 2.28).

2.3 BRANCHES ON THE SS TREE

Many designs of the 1940s and 1950s have not yet been mentioned, but those described thus far seem in retrospect to have been exemplary pioneering efforts. It is time now to take notice of several SS systems left out of the previous accounting, some of which were never even prototyped.

2.3.1 Spread-Spectrum Radar

With the exception of the 1940s state-of-art descriptions of technology, we have made a distinction between the use of SS designs for communication and their use for detection and ranging on noncooperative targets, and have omitted a discussion of the latter. The signal strength advantage which the target holds over the radar receiver in looking for the radar's transmission versus its echo means that LPI is very difficult to achieve. Moreover, the fact that an adversary target knows *a priori* its relative location means that even with pure noise transmission the radar is vulnerable to false echo creation by a delaying repeater on the target.

Nonetheless, SS signaling has some advantages over conventional low time-bandwidth-product radar signaling: in better range (TOA) resolution for a peak-power-limited transmitter (via pulse compression techniques), in range ambiguity removal, in greater resistance to some nonrepeater jammers [7], and in a CDMA-like capability for sharing the transmission band with similar systems. Modern uses of SS radars include fusing (for a patent under wraps for twenty-four years, see [191]) and pulse compression, the latter's applications extending to high-resolution synthetic-aperture ground mapping.

2.3.2 Other Early Spread-Spectrum Communication Systems

Despite the security which once surrounded all of the advances described in previous sections, the SS system concept could not be limited indefinitely to a few companies and research institutions. The following notes describe several other early SS design efforts.

Phantom: An MF-SS system developed by General Electric (GE) for the Air Force, this system was built around tapped delay line filters. As shown in Costas and Widmann's patent [192], the tap weights were designed to be varied pseudorandomly for the purpose of defeating repeater jammers (see Figure 2.29). Constructed in the late 1950s, the Phantom spread its signal over 100 kHz. As with the F9C-A, this system was eventually used also to measure long-haul HF channel properties. For a description of other SS-related work performed at GE in the 1951–54 time frame, under the direction of Richard Shuey, see [193].

WOOF: This Sylvania Buffalo system hid an SS signal by placing within its transmission bandwidth a high-power, friendly, and overt transmitter. Thereby, the SS transmission would be masked by the friendly transmitter, either completely escaping notice or at least compounding the difficulties encountered by a reconnaissance receiver trying to detect it.

RACEP: Standing for Random Access and Correlation for Extended Performance, RACEP was the name chosen by the Martin Company to describe their asynchronous discrete address system that provided voice service for up to 700 mobile users [164]. In this system, the voice signal was first converted to pulse position modulation, and then each pulse in the resultant signal was in turn converted to a distinctive pattern of three narrow pulses and transmitted at one of a possible set of carrier frequencies. With the patterns serving also as addresses, this low duty cycle format possessed some of the advantages of SS systems.

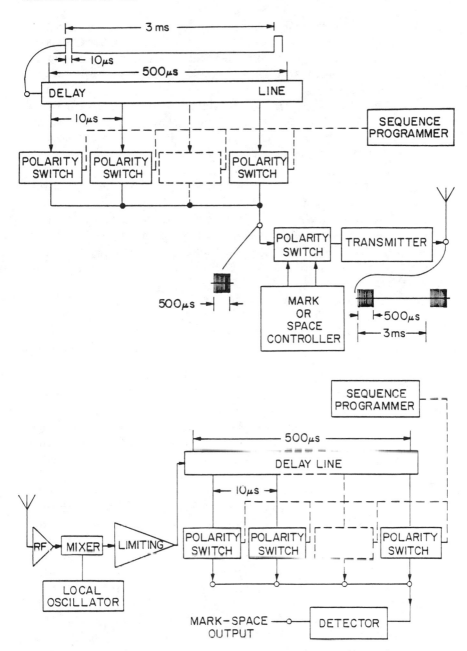

Figure 2.29. Costas and Widmann's Phantom system employs a pulsed delay line with pseudorandomly controlled taps summed to provide an SS signal for modulation. An identically structured system with a synchronous replica of the tap controller is used to construct a matched pseudorandom filter for data detection at the receiver. (Diagrams modified from [192].)

Cherokee: Also by the Martin Company, this was a PN system with a transmission bandwidth of nearly a megahertz and a processing gain of about 16 dB [164]. Both RACEP and Cherokee were on display at the 15th Annual Convention of the Armed Forces Communications and Electronics Association in June 1961.

MUTNS: Motorola's Multiple User Tactical Navigation System was a low frequency, hyperbolic navigation system employing PN signalling. Navigation was based on stable ground wave propagation with the SS modulation used to discriminate against the skywave, as it was in Sylvania's WHYN. Motorola, a subcontractor to JPL on the Jupiter CODORAC link, began Army-supported work on MUTNS in 1958. The first complete system flight test occurred on January 23, 1961 [194], [195].

RADA: RADA(S) is a general acronym for Random Access Discrete Address (System). Wide-band RADA systems developed prior to 1964 include Motorola's RADEM (Random Access DElta Modulation) and Bendix's CAPRI (Coded Address Private Radio Intercom) system, in addition to RACEP [196].

WICS: Jack Wozencraft, while on duty at the Signal Corps Engineering Laboratory, conceived WICS, Wozencraft's Iterated Coding System (see Figure 2.30). This teletype system was an SR-FH-SS system employing 155 different tones in a 10 kHz band to communicate at fifty words/min. Each bit was represented by two successively transmitted tones generated by either the MARK or the SPACE pseudorandomly driven frequency programmer. Bit decisions were made on detecting at least one of the two transmitted frequencies in receiver correlators, and parity checking provided further error correction capability. The subsequent WICS development effort by Melpar in the mid-1950s contemplated its tactical usage as an applique to radios then in inventory [148]. However, just as in ITT's early system concepts, the intended generation of pseudorandom signals via recording [197] did not result in a feasible production design.

Melpar Matched Filter System: A more successful mid-1950s development, this MF-SS design was largely conceived by Arthur Kohlenberg, Steve Sussman, David Van Meter, and Tom Cheatham. To transmit a MARK in this teletype system, an impulse is applied to a filter composed of a pseudorandomly selected, cascaded subset of the several hundred sections of an all-pass lumped-constant linear-phase delay line. The receiver's MARK matched filter is synchronously composed of the remaining sections of the delay line. The same technique was used to transmit SPACE [148] (see Figure 2.31). Patents [198], [199] filed on the system and its clever filter design, the latter invented by Prof. Ernst Guillemin who was a Melpar consultant, were held under secrecy along with the WICS patent until the mid-1970s. An unclassified discussion of an MF-SS system for use against multipath is given in [200].

Kathryn: Named after the daughter of the inventor, William Ehrich, and developed by General Atronics, Kathryn's novel signal processing effected

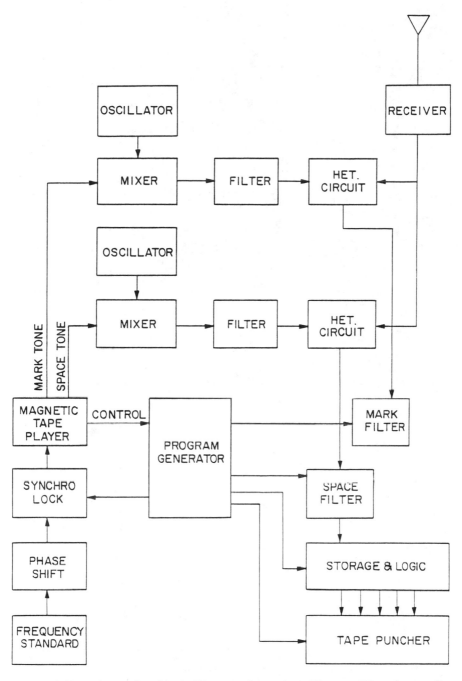

Figure 2.30. The receiver block diagram redrawn from Wozencraft's patent application [197] is shown here with magnetic tape used for storage of independent pseudo-random FH signals for MARK and SPACE reception.

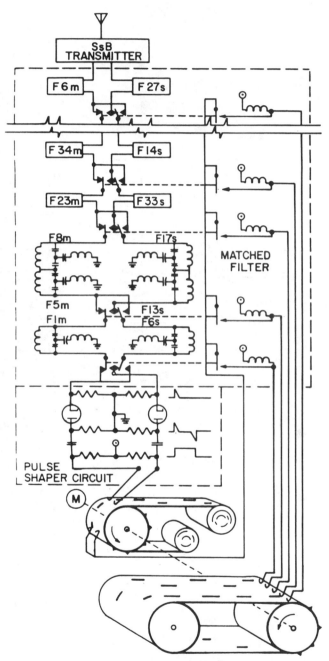

Figure 2.31. Guillemin's patented filter system design [199] is shown here imbedded in the transmitter's modulation generator described in the patent [198] of Kohlenberg, Sussman, and Van Meter. Filter sections were switched in and out according to a schedule recorded on an endless punched tape (and as shown here, punched tape was also used as a binary data source). The receiver contained a corresponding matched filter, synchronously controlled by an identical tape.

the transmission of the Fourier transform of a time-multiplexed set of channel outputs combined with a PN signal. Upon reception, the inverse transform yielded the original PN $\times$ multiplexed-signal product, now multiplied by the propagation medium's system function, thereby providing good or bad channels in accordance with that function. When jamming is present, the data rate is reduced by entering the same data bit into several or all data channels. In this case, a Rake-like combiner is used to remerge these channels at the output of the receiver's inverse Fourier transformer [148], [201]. The modern SS enhancement technique of adaptive spectral nulling against nonwhite jamming was at least implicitly available in this system.

Lockheed Transmitted Reference System: Of the several TR-SS systems patented, this one designed by Jim Spilker (see Figure 2.32) made it into production in time to meet a crisis in Berlin, despite the inherent weaknesses of TR systems [202]. The interesting question here is, "What circumstances would cause someone to use a TR system?" Evidently, extremely high chip rates are part of the answer. For an earlier TR patent that spent almost a quarter-century under secrecy order, see [203].

NOMAC Encrypted-Voice Communications: In 1952, Bob Fano and Bennett Basore, with the help of Bob Berg and Bob Price, constructed and briefly tested an IF model of a NOMAC-TR-FM voice system. At first surprised by the clarity of communication and lack of the self-noise which typifies NOMAC-AM systems, Basore soon realized that SS-carrier phase noise was eliminated in the heterodyne correlation process and that SS-carrier amplitude noise was removed by the limiting frequency-discriminator. Little more was done until years later when, in 1959, John Craig of Lincoln Laboratory designed an experimental SR-SS system based on low-deviation phase modulation of a voice signal onto an F9C-like noise carrier. The system provided fair quality voice with negligible distortion and an output SNR of about 15 dB, the ever-present noise deriving from system flaws. Simulated multipath caused problems in this low-processing-gain system, and it was postulated that Rake technology might alleviate the problem [204], [205], but the work was abandoned.

NOMAC Matched Filter System: In approximately October 1951, Robert Fano performed a remarkable acoustic pulse experiment involving high time-bandwidth-product matched filters (see Figure 2.33). At that time he disclosed a multiple matched-filter communication system to his colleagues [206]. Based on Fano's research, an MF-SS teletype communication system was suggested in 1952 [207]. Research at Lincoln Laboratory on this SS communication system type was confined to exploring a viable filter realization. This communication approach apparently was dropped when full-scale work began on the F9C system. Fano later patented [208] the wide-band matched filter system concept, claiming improved performance in the presence of multipath.

While Fano's invention, which originally suggested a reverse-driven magnetic-drum recording for signal generation, basically employed analog sig-

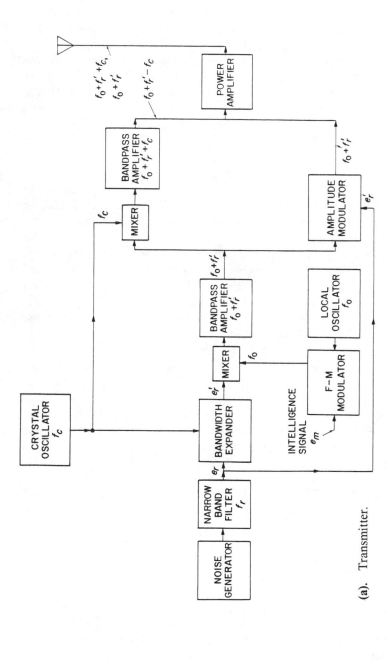

(a).   Transmitter.

Figure 2.32.   Spilker's patent on a TR-SS system uses a novel bandwidth expansion mechanism to generate a wideband reference signal e_r' from a random narrowband key signal e_r. Both transmitted signals (frequency offset by f_c Hz), contain the intelligence signal as FM modulation, and the signal at $f_0 + f_r'$ Hz additionally carries the key as AM modulation.

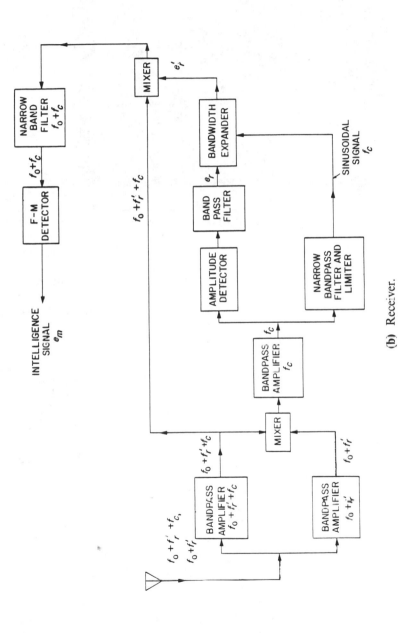

(b) Receiver.

Figure 2.32 (continued). To demodulate at the receiver, the two received signals are cross-correlated to recover the AM modulated key, which is then bandwidth-expanded to produce the wideband reference signal needed to demodulate the intelligence. (Diagrams from [202].)

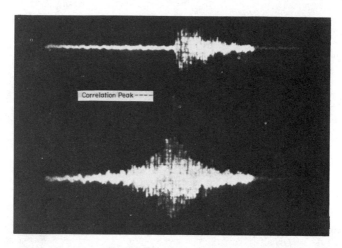

Figure 2.33. Fano's elegant matched filter experiment consisted of transmitting an acoustic pulse into a chamber containing many reflectors. The upper signal shown here represents the sound sensed by a microphone in the room, and tape recorded. The tape was then reversed (not rewound) and replayed into the chamber, the microphone this time sensing the lower of the above two signals, specifically intended to be the autocorrelation of the upper signal. Fano recalls being startled by his inability to see at first the extremely narrow peak of the autocorrelation function on the oscilloscope screen. The peak was soon discovered when the light were turned off. (Photo courtesy of Robert Fano.)

nals, another then contemporary matched-filter invention by Ronald Barker [227] definitely used digital signals. Barker's design employs the binary patterns, which now bear his name, as frame sync markers in digital data streams. While this application is not inherently bandwidth expanding, the waveform correlation design objectives in frame sync applications are quite similar to those for SS-MF communication applications, as well as to those for radar pulse-compression.

Spread Eagle: Philco Corporation developed this secure interceptor-control data link for WADC in the late 1950s. Eight hundred chips of a complex binary waveform, transmitted in 200 μsec, were used to represent each data bit, providing 29 db of improvement against jamming. The two possible waveforms, one for MARK and one for SPACE, were detected by a non-coherent MF receiver with in-phase and quadrature channels both containing synchronous matched filters.

The delay lines in the MF implementation were limited to 100 μsec, evidently for economy of size and weight. Hence, the 200 μsec waveforms actually consisted of repeated 100 μsec waveforms, the MF output being sampled twice in the detection process [176].

SECRAL: This ITT missile guidance system development of the late 1950s was a DS-SS design.

Longarm and Quicksilver: These are both early FH anti-multipath systems built by Hughes Aircraft Company, under the leadership of Samuel Lutz and Burton Miller, and sponsored by Edwin McCoppin of WPAFB.

2.3.3 Spread-Spectrum Developments Outside the United States

This historical review has concentrated on SS development in the United States for several reasons.

1. The theories of Wiener, and especially Shannon, which propounded the properties of and motivated the use of random and pseudorandom signals, were available in the U.S. before such basics were appreciated elsewhere (with the exception of Guanella). This gave U.S. researchers a significant lead time, an important factor near the outset of the Cold War when the Voice of America was being jammed intensively. Additional impetus for SS development came in urgent response to the threats posed by the onslaught of the Korean War and the tense confrontation over Berlin.
2. SS development occurred just after the Second World War, at a point in time when many of the world's technological leaders had suffered tremendous losses in both manpower and facilities, and additionally in Germany's case, political self-control. Research and industry in the U.S., on the other hand, were unscathed and the U.S. became the home for many leading European scientists, e.g., Henri Busignies and Wernher von Braun, to name two among many.
3. The available unclassified literature (virtually all the references in this history are now unclassified) points to the earliest SS developments having arisen in the United States.

We will now look at evidence of some SS beginnings outside the U.S.A.

Bill Davenport remembers a secret interchange with a visiting British delegation in which pre-Rake NOMAC concepts were discussed. Later, he was informed that the British had not pursued that approach to secure, long-range communication because they envisioned major problems from multipath [209]. Frank Lehan recalls a discussion with a British scientist who told him that the British had studied PN sequences several years before JPL developed the idea. Bob Dixon dates Canada's experimental Rampage system to the early 1950s, with no further details yet available [210]. So it seems that the closest friends of the U.S. were at least cognizant of the SS concept, knew something of PN generation, and to some extent had experimented with the idea. Further information on these early efforts has not been uncovered.

In neutral Switzerland, Brown, Boveri, and Company developed, starting in the late 1950s, an SS guidance system (see Figure 2.28). This was no doubt stimulated by Gustav Guanella, the pioneering inventor of noise-modulated radar [13] and of encryption schemes which the NDRC had sought to decipher during World War II. He quickly appreciated, and may well have seen, the true significance of the Rake concept upon its publication. Now, an intriguing question is, "When did the Soviet bloc become privy to the SS concept and realize its potential?"

In the mid-1950s some members of a high-level U.S. task force were convinced that the Russians knew about SS techniques and in fact might also be using them. For example, Eugene Fubini personally searched the U.S. Patent Office open files to see what a foreign country might be able to learn there of this new art; nomenclature was a problem and he had to look under "pulse communications" as well as many other patent classifications. (This difficulty was eased recently when the Patent Office created a special subclass 375-1 entitled *Spread Spectrum*.) Also curious about this issue, Paul Green determined to try to find out for himself the status of Russian knowledge about NOMAC techniques. After studying the language, he examined the Russian technical literature, surveying their work in information theory and attempting to uncover clues that might lie there to noise modulation concepts. Green came to believe that there was no plausible reason to suspect that the Soviets were then developing spread-spectrum systems, partly because of lack of technology and possibly because there was no perceived need for AJ communications capability.

Later Paul Green visited the Soviet Union and gave a talk in Russian on the use of Rake to measure properties of the ionosphere, which seemingly was accepted at face value. Because of this contact and his literature scrutiny, in the mid-1960s Green decided to postpone his plans to write an unclassified account of Lincoln Laboratory's NOMAC work, toward which full military clearances had already been granted.

The earliest Soviet reference (as cited in, e.g., [211]) proposing noise-like, intelligence-bearing signals is a 1957 publication by Kharkevich [212] on amplitude or frequency modulation of pure noise. Like Goldsmith's [46], Kharkevich's work is missing a key ingredient, namely, the attainment of synchronous detection via correlation with a stored or transmitted reference. Within a few months of the approved 1958 publication of the Rake concept for using wide-band signals ostensibly to counter multipath, that paper was translated into Russian, and hardly a year later an exposition of Rake appeared in Lange's first book *Korrelationselektronik* [213]. Thus began the revelation of the SS concept in the U.S. literature from scientific journals and conference proceedings to magazines such as *Electronics*, *Electronic Design*, and *Aviation Week*. Here is a small sample of U.S. open papers referenced in the Soviet literature:

1. March 1958. Rake remedy for multipath, using wideband signals [156].
2. December 1959. Use of wideband noise-like signals, CDMA, and jamming [214], [215].
3. Fall 1960. PN-controlled TH-SS command link for missile guidance [165].

4. January 1961. Analysis of a pure noise (TR) communications system [216].
5. March 1961. Discussion of RADA systems [217].
6. 1963. 200 Mcps PN generator construction [218].
7. December 1963. Wideband communication systems including Rake, RACEP, and RADEM [196].

It is clear from these citations and other evidence that the Russians were studying SS systems no later than 1963 [219], and by 1965 had carefully searched and reported [220] on the U.S. open literature discussing Rake, Phantom, and the various RADA systems. Between 1965 and 1971, the Soviets published several books [211], [221]–[224] concerned with SS principles and their applications to secure communication, command, and control.

2.4 A VIEWPOINT

One can paint the following picture of the development of spread-spectrum communications. During World War II the Allies and the Axis powers were in a desperate technological race on many fronts, one being secure communications. Jamming of communication and navigation systems was attempted by both sides and the need for reliable communication and accurate navigation in the face of this threat was real. One major AJ tactic of the war was to change carrier frequency often and force the jammer to keep looking for the right narrow band to jam. While this was possible to automate in the case of radar, communication frequency hopping was carried out by radio operators, in view of the major technological problem of providing an accurate synchronous frequency reference at the receiver to match the transmitter. Thus, at least frequency hopping and, to a similar extent, time hopping were recognized AJ concepts during the early 1940s.

Many of the early "secure" or "secret" non-SS communication systems seem to have been attempts to build analog equivalents of cryptographic machines and lacked the notion of bandwidth expansion (e.g., the Green Hornet, the Telefunken dual wheels system). The initial motivation for direct sequence systems appears, on the other hand, to have come from the need for accurate and unambiguous time-of-arrival measurements in navigation systems (e.g., WHYN and CODORAC), and from the desire to test or extend Shannon's random-signaling concept and, thus, communicate covertly (e.g., Rogoff's noise wheels experiment). The DS concept followed the FH and TH concepts by several years partly because the necessary correlation detection schemes were just emerging in the late 1940s.

Who first took these diverse system ideas and recognized the unifying essential requirements of a spread-spectrum system (e.g., high carrier-to-data bandwidth ratio, an unpredictable carrier, and some form of correlation detection)? From the available evidence, it appears that Shannon certainly had the insight to do it but never put it in print, and that two close friends, Nathan Marchand and Louis deRosa, both key figures in the formation of the IRE's Group on Information Theory, led Sylvania Bayside and FTL, respectively, toward a unified SS viewpoint. It seems that Sylvania Bayside had all the ingredients of the direct sequence concept as early as 1948, but did not have the technology to solve some of the signal processing problems. It remained for Mortimer Rogoff to provide a method for storing pseudo-noise (a technique reminiscent of Telefunken's wheels), giving ITT the complete system assembled and tested under the Della Rosa contract and documented to a government agency.

Meanwhile, the idea either was propagated to, or was independently conceived by, several research and design groups, notably at M.I.T. in 1950 and at JPL in 1952. Group 34 at M.I.T. Lincoln Laboratory, sparked by Bill Davenport, Paul Green, and Bob Price, is generally credited with building the first successful SS communication system for several reasons.

1. The Rake system was the first wide-band pseudorandom-reference system to send messages reliably over the long-range HF multipath channel.
2. The F9C-A system, soon followed by the Rake applique, was probably the first deployed (nonexperimental) broad-band communication system which differed in its essentials from wide-deviation FM, PPM, or PCM.
3. The Rake system was the first such SR communication system to be discussed in the open literature, other than information theoretic designs.

JPL's radio control work, in competition with inertial guidance systems, did not reach a deployment stage until suitable applications appeared in the Space Program. In addition to opening new vistas in the development of PN generation techniques, JPL's contribution to SS technology has been the innovation of tracking loop designs which allow high-performance SS systems to be placed on high-speed vehicles with results comparable to those of stationary systems. Both the M.I.T. and JPL programs have left a legacy of excellent documentation on spread-spectrum signal processing, spectral analysis, and synchronization, and have provided some of the finest modern textbooks on communications.

A very successful long-term SS system investigation began at Sylvania Buffalo under Madison Nicholson and later under Jim Green, and ended up merging with some JPL-based experience at Magnavox in the development and production of the ARC-50 family of systems. The ARC-50 was the first deployed SS system with any of the following characteristics:

1. avionics packaging,
2. fully coherent reception (including carrier tracking),
3. a several megahertz chip rate, and
4. voice capability.

Figure 2.34. VIP's at the IEEE NAECON '81 included Robert Larson, Wilbur Davenport, Paul Green, B. Richard Climie, Mrs. Mortimer Rogoff, Mortimer Rogoff, Mrs. Louis deRosa, and Robert Price. Featured at this meeting was the presentation of the Pioneer Award to deRosa (posthumously), Rogoff, Green, and Davenport for their ground-breaking work in the development of spread-spectrum communications. (Photo courtesy of W. Donald Dodd.)

Although losing the ARC-50 final design and production contract to Magnavox, Sylvania continued on to develop BLADES, the earliest FH-SS communication system used operationally. Moreover, BLADES represented, by publication (e.g., [124]) and actual hardware, the start of real-world application of shift-register sequences to error correction coding, an algebraic specialty that would flourish in coming years.

Since the 1950s when the SS concept began to mature, the major advances in SS have been for the most part technological, with improvements in hardware and expansion in scope of application continuing to the present day. Now with the veil of secrecy being lifted, the contributions of some of the earliest pioneers of SS communications are being recognized (see Figure 2.34). We hope that this historical review has also served that purpose by highlighting the work of the many engineers who have figured prominently in the early conceptual development and implementation of spread-spectrum systems.

2.5 REFERENCES

[1] R. A. Scholtz, "The origins of spread-spectrum communications," *IEEE Trans. Commun.*, COM-30, pp. 822–854, May 1982 (Part I).

[2] R. A. Scholtz, "Notes on spread-spectrum history," *IEEE Trans. Commun.*, COM-31, pp. 82–84, Jan. 1983.

[3] R. Price, "Further notes and anecdotes on spread-spectrum origins," *IEEE Trans. Commun.*, COM-31, pp. 85–97, Jan. 1983.

[4] E. V. Appleton and M. A. F. Barnett, "On some direct evidence for downward atmospheric reflection of electric rays," *Proc. Roy. Soc. Ser. A.* vol. 109, pp. 621–641, Dec. 1, 1925.

[5] D. G. C. Luck, *Frequency Modulated Radar*, New York: McGraw-Hill, 1949.

[6] S. L. Johnston, "Radar ECCM history," *Proc. NAECON*, May 1980, pp. 1210–1214.

[7] G. R. Johnson, "Jamming low power spread spectrum radar," *Electron. Warfare*, pp. 103–112, Sept.–Oct. 1977.

[8] D. O. North, "An analysis of the factors which determine signal/noise discrimination in pulsed carrier systems," RCA Lab., Princeton, NJ. Rep. PTR-6C, June 25, 1943; see also *Proc. IEEE*, vol. 51, pp. 1015–1028, July 1963.

[9] J. H. Van Vleck and D. Middleton, "A theoretical comparison of the visual, aural, and meter reception of pulsed signals in the presence of noise," *J. Appl. Phys.*, vol. 17, pp. 940–971, Nov. 1946.

[10] P. M. Woodward, *Probability and Information Theory, with Applications to Radar*, New York: Pergamon, 1953.

[11] F. E. Terman, "Administrative history of the Radio Research Laboratory," Radio Res. Lab., Harvard Univ., Cambridge, MA. Rep. 411-299, Mar. 21, 1946.

[12] F. H. Lange, *Correlation Techniques*, Princeton, NJ: Van Nostrand, 1966.

[13] G. Guanella, "Distance determining system," U.S. Patent 2 253 975, Aug. 26, 1941 (filed in U.S. May 27, 1939; in Switzerland Sept. 26, 1938).

[14] _____, "Direction finding system," U.S. Patent 2 166 991, July 25, 1939 (filed in U.S. Nov. 24, 1937; in Switzerland Dec. 1, 1936).

[15] J. M. Whittaker, *Interpolatory Function Theory* (Cambridge Tracts in Mathematics and Mathematical Physics, no. 33), New York: Cambridge Univ. Press, 1935.

[16] E. T. Whittaker, "On the functions which are represented by the expansions of the interpolation theory," *Proc. Roy. Soc. Edinburgh*, vol. 35, pp. 191–194, 1915.

[17] J. McNamee, F. Stenger, and E. L. Whitney, "Whittaker's cardinal function in retrospect," *Math. Comput.*, vol. 25, pp. 141–154, Jan. 1971.

[18] A. J. Jerri, "The Shannon sampling theorem—Its various extensions and applications: A tutorial review," *Proc. IEEE*, vol. 65, pp. 1565–1596, Nov. 1977.

[19] R. V. L. Hartley, "The transmission of information," *Bell Syst. Tech. J.*, vol. 7, pp. 535–560, 1928.

[20] H. Nyquist, "Certain topics in telegraph transmission theory," *AIEE Trans.*, vol. 47, pp. 617–644, Apr. 1928.

[21] V. A. Kotelnikov, "Carrying capacity of 'ether' and wire in electrical communications" (in Russian), *Papers on Radio Communications, 1st All-Union Conv. Questions of Technical Reconstruction of Communications*, All-Union Energetics Committee, USSR, 1933, pp. 1–19.

[22] C. E. Shannon, "Communication in the presence of noise," *Proc. IRE*, vol. 37, pp. 10–21, Jan. 1949.

[23] D. Slepian, "On bandwidth," *Proc. IEEE*, vol. 64, pp. 292–300, Mar. 1976.

[24] A. H. Nuttall and F. Amoroso, "Minimum Gabor bandwidth of M orthogonal signals," *IEEE Trans. Inform. Theory*, vol. IT-11, pp. 440–444, July 1965.

[25] D. Gabor, "Theory of communication," *J. Inst. Elec. Eng.* (*London*), vol. 93, part 3, pp. 429–457, Nov. 1946.

[26] N. Wiener, "Generalized harmonic analysis," *Acta Math.*, vol. 55, pp. 117–258, May 9, 1930; reprinted in *Generalized Harmonic Analysis and Tauberian Theory* (*Norbert Wiener: Collected Work*, Vol. 2), P. Masani, ed. Cambridge, MA: M.I.T. Press, 1979.

[27] "Gunfire control," Nat. Defense Res. Committee, Office Sci. Res. Develop., Washington, DC, Summary Tech. Rep., Div. 7, 1946 (AD 200795).

[28] N. Wiener, *Extrapolation, Interpolation, and Smoothing of Stationary Time Series with Engineering Applications*, Cambridge, MA: M.I.T. Press, 1949.

[29] N. Levinson, "The Wiener rms (root mean square) error criterion in filter design and prediction," *J. Math. Phys.*, vol. 25, pp. 261–278, Jan. 1947.

[30] C. E. Shannon, "A mathematical theory of cryptography," Bell Tel. Lab., memo., Sept. 1, 1945; later published in expurgated form as "Communication theory of secrecy systems," *Bell. Syst. Tech. J.*, vol. 28, pp. 656–715, Oct. 1949.

[31] _____, "A mathematical theory of communication," *Bell Syst. Tech. J.*, vol. 27, pp. 379–423, July, and 623–656, Oct. 1948.

[32] W. L. Root, "Some notes on jamming, I," M.I.T. Lincoln Lab., Tech. Rep. 103, Jan. 3, 1956 (AD 090352; based on Lincoln Lab. Group Rep. 34–47, Dec. 15, 1955).

[33] _____, "Communications through unspecified additive noise," *Inform. Contr.*, vol. 4, pp. 15–29, 1961.

[34] N. M. Blachman, "Communication as a game," in *IRE Wescon Rec.*, San Francisco, CA, Aug. 20–23, 1957, part 2, pp. 61–66 (Note here a coincidental juxtaposition with a survey of U.S.S.R. literature, pp. 67–83—carried out *sub rosa* re SS by P. E. Green, Jr., as mentioned in Section 2.3.3). See also: N. M. Blachman, "On the effect of interference: prevarication vs. redundancy," Electronic Defense Laboratory, Sylvania Electric Products, Mountain View, CA, Tech. Memo EDL-M104, May 1, 1957 (AD 136152).

[35] _____, "On the capacity of a band-limited channel perturbed by statistically dependent interference," *IRE Trans Inform. Theory*, IT-8, pp. 48–55, Jan. 1962.

[36] L. R. Welch, "A game-theoretic model of communications jamming," Jet Propulsion Lab., Pasadena, CA, Memorandum No. 20-155, April 4, 1958.

[37] *Report of Proceedings, Symp. Inform. Theory*, London, England, Sept. 26–29, 1950; reprinted in *Trans. IRE Professional Group Inform. Theory*, vol. PGIT-1, Feb. 1953.

[38] N. Marchand, "Radio receiver," U.S. Patent 2 416 336, Feb. 25, 1947 (filed May 21, 1942).

[39] F. S. Purington, "Single side band transmission and reception," U.S. Patent 1 992 441, Feb. 26, 1935 (filed Sept 6, 1930).

[40] Y. W. Lee, J. B. Wiesner, and T. P. Cheatham, Jr., "Apparatus for computing correlation functions," U.S. Patent 2 643 819, June 30, 1953 (filed Aug. 11, 1949).

[41] Y. W. Lee, T. P. Cheatham, Jr., and J. B. Wiesner, "The application of correlation functions in the detection of small signals in noise," M.I.T. Res.

Lab. Electron., Tech. Rep. 141, Oct. 13, 1949 (ATI 066538, PB 102361).

[42] H. E. Singleton, "A digital electronic correlator," M.I.T. Res. Lab. Electron., Tech. Rep. 152, Feb. 21, 1950.

[43] S. L. Johnston, *Radar Electronic Counter-Countermeasures*, Dedham, MA: Artech House, 1979.

[44] L. S. Howeth, *History of Communications-Electronics in the United States Navy*. Washington, DC: U.S. Gov. Print. Off., 1963.

[45] J. D. O'Connell, A. L. Pachynski, and L. S. Howeth, "A summary of military communication in the United States—1860 to 1962," *Proc. IRE*, vol. 50, pp. 1241–1251, May 1962; see also [44].

[46] A. N. Goldsmith, "Radio signalling system," U.S. Patent 1 761 118, June 3, 1930 (filed Nov. 6, 1924).

[47] C. B. H. Feldman, "Wobbled radio carrier communication system," U.S. Patent 2 422 664, June 24, 1947 (filed July 12, 1944).

[48] P. Kotowski and K. Dannehl, "Method of transmitting secret messages," U.S. Patent 2 211 132, Aug. 13, 1940 (filed in U.S. May 6, 1936; in Germany May 9, 1935).

[49] L. A. deRosa and M. Rogoff, "Secure single sideband communication system using modulated noise subcarrier," U.S. Patent 4 176 316, Nov. 27, 1979 (filed Mar. 20, 1953); reissue appl. filed Sept. 4, 1981, Re. Ser. No. 299 469.

[50] D. Kahn, *The Codebreakers*, New York: Macmillan, 1967.

[51] L. Cranberg, "German Optiphone equipment 'Li-Spr-80.'" Memo., "Captured enemy equipment report no. 26." Signal Corps Ground Signal Agency, Bradley Beach, NJ, May 5, 1944 (PB 001531).

[52] G. R. Thompson and D. R. Harris, *The Signal Corps: The Outcome (Mid. 1943 Through 1945) (United States Army in World War II*, Vol. 6, Part 5: *The Technical Services*, vol. 3). Washington, DC, Off. Chief of Military History, U.S. Army, 1966.

[53] M. L. Marshall, ed., *The Story of the U.S. Army Signal Corps*, New York: Franklin Watts, 1965.

[54] S. V. Jones, "After 35 years, secrecy lifted on encoded calls," *New York Times*, p. 27, July 3, 1976.

[55] W. R. Bennett, "Secret Telephony as a historical example of spread-spectrum communication," *IEEE Trans. Commun.*, COM-31, pp. 98–104, January, 1983.

[56] "Speech and facsimile scrambling and decoding," Nat. Defense Res. Committee. Off. Sci. Res. Develop., Washington, DC. Summary Tech. Rep. Div. 13, vol. 3, 1946 (AD 221609); reprinted by Aegean Park Press, Laguna Hills, CA.

[57] R. C. Mathes, "Secret telephony," U.S. Patent 3 967 066, June 29, 1976 (filed Sept. 24, 1941).

[58] R. K. Potter, "Secret telephony," U.S. Patent 3 967 067, June 29, 1976 (filed Sept. 24, 1941).

[59] A. J. Busch, "Signalling circuit," U.S. Patent 3 968 454, July 6, 1976 (filed Sept. 27, 1944).

[60] A. E. Joel, Jr., "Pulse producing system for secrecy transmission," U.S. Patent 4 156 108, May 22, 1979 (filed Jan. 21, 1947).

[61] G. Guanella, "Methods for the automatic scrambling of speech," *Brown Boveri Review*, pp. 397–408, Dec. 1941.

[62] "Radio countermeasures," Nat. Defense Res. Committee, Office Sci. Res.

Develop., Washington, DC, Summary Tech. Rep. Div. 15, vol. I, 1946 (AD 221601).

[63] "Guided missiles and techniques," Nat. Defense Res. Committee, Office Sci. Res. Develop., Washington, DC, Summary Tech. Rep. Div. 5, vol. I, 1946 (AD 200781).

[64] H. A. Zahl, *Electrons Away, or Tales of a Government Scientist*, New York: Vantage, 1968.

[65] W. W. Hansen, "Secret communication," U.S. Patent 2 418 119 Apr. 1, 1947 (filed Apr. 10, 1943).

[66] E. Chaffee and E. Purington, "Method and means for secret radiosignalling," U.S. Patent 1 690 719, Nov. 6, 1928 (filed Mar. 31, 1922).

[67] M. Deloraine, *When Telecom and ITT Were Young*, New York: Lehigh, 1976; first published in French as *Des Ondes et des Hommes Jeunesse des Telecommunications et de l'ITT*, Paris, France: Flammarion, 1974.

[68] E. M. Deloraine, H. G. Busignies, and L. A. deRosa, "Facsimile system," U.S. Patent 2 406 811, Sept. 3, 1946 (filed Dec. 15, 1942).

[69] _____, "Facsimile system and method," U.S. Patent 2 406 812, Sept. 3, 1946 (filed Jan. 30, 1943).

[70] L. A. deRosa, "Random impulse system," U.S. Patent 2 671 896, Mar. 9, 1954 (filed Dec. 13, 1942).

[71] E. M. Deloraine, "Protected communication system," Fed. Radio Tel. Lab., New York, NY, Rep. 937-2, Apr. 28, 1944 (from the National Archives, Record Group 227; this report was written to Division 15 of the National Defense Research Committee, Office of Scientific Research and Development on Project RP-124) (ATI 014050).

[72] H. Busignies, S. H. Dodington, J. A. Herbst, and G. R. Clark, "Radio communication system protected against interference," Fed. Tel. Radio Corp., New York, NY, Final Rep. 937-3, July 12, 1945 (same source as [71]).

[73] H. P. Hutchinson, "Speech privacy apparatus," U.S. Patent 2 495 727, Jan. 31, 1950 (filed Jan. 7, 1943).

[74] D. O. Slater, "Speech privacy and synchronizing system devised by Captain Henry P. Hutchinson," Bell Tel. Lab., Rep. 20 under Project C-43 of NDRC (OSRD 4573B), July 31, 1943 (as cited in [56, p. 126]; available from U.S. Nat. Archives, Washington, DC).

[75] "John Hays Hammond, Jr." *Micropaedia*, vol. IV, *The New Encyclopaedia Britannica*, Chicago, IL: Encyclopaedia Britannica, 1975, pp. 877–878.

[76] J. H. Hammond, Jr., "System of aeroplane control," U.S. Patent 1 568 972, Jan. 12, 1926 (filed Mar. 7, 1914); see also Hammond's U.S. Patent 1 420 257 (filed 1910), and "Security of radio control," [78, pp. 1193–1197].

[77] B. F. Miessner, *On the Early History of Radio Guidance*. San Francisco, CA: San Francisco Press, 1964.

[78] J. H. Hammond, Jr. and E. S. Purington, "A history of some foundations of modern radio-electronic technology," *Proc. IRE*, vol. 45, pp. 1191–1208, Sept. 1957.

[79] E. L. Chaffee, "System of radio communication," U.S. Patent 1 642 663, Sept. 13, 1927 (filed Aug. 11, 1922).

[80] L. Espenschied *et al.*, "Discussion of 'A history of some foundations of modern radio-electronic technology,'" *Proc. IRE*, vol. 47, pp. 1253–1268, July 1959.

[81] B. Gunston, *Rockets and Missiles*, New York: Crescent, 1979.

[82] Air Force Corresp. File on Contr. W535-sc-707 with Colonial Radio Corp., May 1943–June 1944.

[83] E. Labin and D. D. Grieg, "Method and means of communication," U.S. Patent 2 410 350, Oct. 29, 1946 (filed Feb. 6, 1943).

[84] E. S. Purington, "Radio selective control system," U.S. Patent 2 635 228, Apr. 14, 1953 (filed June 2, 1948).

[85] C. H. Hoeppner, "Pulse communication system," U.S. Patent 2 999 128, Sept. 5, 1961 (filed Nov. 4, 1945).

[86] E. H. Krause and C. E. Cleeton, "Pulse signalling system," U.S. Patent 4 005 818, Feb. 1, 1977 (filed May 11, 1945).

[87] H. K. Markey and G. Antheil, "Secret communication system," U.S. Patent 2 292 387, Aug. 11, 1942 (filed June 10, 1941).

[88] W. Broertjes, "Method of maintaining secrecy in the transmission of wireless telegraphic messages," U.S. Patent 1 869 659, Aug. 2, 1932 (filed Nov. 14, 1929; in Germany, Oct. 11, 1929).

[89] C. Young, *The Films of Hedy Lamarr*, Secaucus, NJ: Citadel, 1978, pp. 92–97.

[90] G. Antheil, *Bad Boy of Music*, Garden City, NY: Doubleday, 1945.

[91] "Designs and ideas for a radio controlled torpedo," Dec. 23, 1940, contained in [93].

[92] *New York Times*, p. 24, Oct. 1, 1941.

[93] Case file for U.S. Patent Serial No. 397 412, supplied courtesy of J. D. McConaghy, Esq., Lyon and Lyon, Los Angeles, CA (per authorization of H. Lamarr–not publicly available).

[94] *New York Times*, p. 32, Sept. 6, 1942.

[95] *Time*, vol. 103, p. 52, Feb. 18, 1974; *Parade* (e.g., in *Boston Sunday Globe*), p. 16, Aug. 1, 1982. (Both items indicate only that the patent is in military or secret communications; no mention is made of frequency hopping or anti-jamming.)

[96] J. R. Pierce, "The early days of information theory," *IEEE Trans. Inform. Theory*, IT-19, pp. 3–8, Jan. 1973.

[97] "The WHYN guidance system," Phys. Lab., Sylvania Elec. Products, Bayside, NY, Final Eng. Rep., Modulation Wave Form Study & F-M Exciter Develop., Contr. W28-099ac465, June 1949 (AD895816).

[98] "The WHYN guidance system," Phys. Lab., Sylvania Elec. Products, Flushing, NY, Interim Eng. Rep. 5, Contr. W28-099ac465, Oct. 1948 (ATI 44524).

[99] "The WHYN guidance system," Phys. Lab., Sylvania Elec. Products, Bayside, NY, Final Eng. Rep., Equipment Develop. & East Coast Field Test, Contr. W28-099ac465, June 1950 (AD 895815).

[100] "Cross-correlation radar," Phys. Lab., Sylvania Elec. Products, Bayside, NY. Rep. YD-51-5, Feb. 1951.

[101] M. Leifer and N. Marchand, "The design of periodic radio systems," *Sylvania Technologist*, vol. 3, pp. 18–21, Oct. 1950.

[102] N. L. Harvey, "A new basis for the analysis of radio navigation and detection systems," *Sylvania Technologist*, vol. 3, pp. 15–18, Oct. 1950.

[103] N. Marchand and H. R. Holloway, "Multiplexing by orthogonal functions," presented at the IRE Conf. Airborne Electron., Dayton, OH, May 23–25, 1951.

[104] N. Marchand and M. Leifer, "Cross-correlation in periodic radio systems,"

presented at the IRE Conf. Airborne Electron., Dayton, OH, May 23–25, 1951.

[105] W. E. Budd, "Analysis of correlation distortion," M.E.E. thesis, Polytech. Inst. Brooklyn, Brooklyn, NY, May 1955.

[106] N. L. Harvey, M. Leifer, and N. Marchand, "The component theory of calculating radio spectra with special reference to frequency modulation," *Proc. IRE*, vol. 39, pp. 648–652, June 1951.

[107] N. L. Harvey, "Radio navigation system," U.S. Patent 2 690 558, Sept. 28, 1954 (filed Feb. 4, 1950).

[108] _____, "Collision warning radar," U.S. Patent 2 842 764, July 8, 1958 (filed Feb. 21, 1951).

[109] H. C. Harris, Jr., M. Leifer, and D. W. Cawood, "Modified cross-correlation radio system and method," U.S. Patent 2 941 202, June 14, 1960 (filed Aug. 4, 1951).

[110] "The WHYN guidance system," Phys. Lab., Sylvania Elec. Products, Bayside, NY, Final Eng. Rep., Equipment, Syst. Lab. Tests & Anal., Contr. W28-099ac 465, June 1953 (AD 024044).

[111] M. Leifer and W. Serniuk, "Long range high accuracy guidance system," presented at the RDB Symp. Inform. Theory Appl. Guided Missile Problems, California Inst. Technol., Pasadena, Feb. 2–3, 1953.

[112] W. P. Frantz, W. N. Dean, and R. L. Frank, "A precision multi-purpose radio navigation system," in *IRE Nat. Conv. Rec.*, New York, NY, Mar. 18–21, 1957, part 8, pp. 79–98.

[113] R. L. Frank and S. Zadoff, "Phase-coded hyperbolic navigation system," U.S. Patent 3 099 835, July 30, 1963 (filed May 31, 1956).

[114] R. L. Frank, "Phase-coded communication system," U.S. Patent 3 099 795, July 30, 1963 (filed Apr. 3, 1957).

[115] W. Palmer and R. L. Frank, in "1971 Pioneer Award," *IEEE Trans. Aerosp. Electron. Syst.*, vol. AES-7, pp. 1015–1021, Sept. 1971.

[116] [R. L. Frank and S. A. Zadoff], "Study and field tests of improved methods for pulse signal detection," Sperry Gyroscope, Great Neck, NY, Final Eng. Rep., Contr. AF28(099)-333, Sperry Eng. Rep. 5223-1245, June 1951 (ATI 150834).

[117] E. J. Baghdady, "New developments in FM reception and their application to the realization of a system of 'power-division' multiplexing," *IRE Trans. Commun. Syst.*, vol. CS-7, pp. 147–161, Sept. 1959.

[118] R. L. Frank, "Tunable narrowband rejection filter employing coherent demodulation," U.S. Patent 3 403 345, Sept. 24, 1968 (filed July 19, 1965).

[119] J. H. Green, Jr., and M. G. Nicholson, Jr., "Synchronizing System," U.S. Patent 4 361 890, Nov. 30, 1982 (filed June 17, 1958).

[120] M. G. Nicholson, Jr., "Apparatus for generating signals having selectable frequency deviation from a reference frequency," U.S. Patent 2 972 109, Feb. 14, 1961 (filed June 11, 1956).

[121] _____, "Generator of frequency increments," U.S. Patent 2 923 891, Feb. 2, 1960 (filed July 6, 1956).

[122] J. N. Pierce, "Theoretical diversity improvement in frequency shift keying," *Proc. IRE*, vol. 46, pp. 903–910, May 1958.

[123] J. H. Green, Jr., D. K. Leichtman, L. M. Lewandowski, and R. E. Malm, "Improvement in performance of radio-teletype systems through use of ele-

ment-to-element frequency diversity and redundant coding," presented at the 4th Annu. Symp. Global Commun., Washington, DC, Aug. 1–3, 1960.

[124] J. H. Green, Jr., and R. L. San Soucie, "An error correcting encoder and decoder of high efficiency," *Proc. IRE*, vol. 46, pp. 1741–1744, Oct. 1958.

[125] R. G. Fryer, "Analytical development and implementation of an optimum error-correcting code," *Sylvania Technologist*, vol. 13, pp. 101–110, July 1960.

[126] M. G. Nicholson, Jr., and R. A. Smith, "Data transmission systems," U.S. Patent 3 093 707, June 11, 1963 (filed Sept. 24, 1959).

[127] J. H. Green, Jr., and J. Gordon, "Selective calling system," U.S. Patent 3 069 657, Dec. 18, 1962 (filed June 11, 1958).

[128] V. C. Oxley and W. E. De Lisle, "Communications and data processing equipment," U.S. Patent 3 235 661, Feb. 15, 1966 (filed July 11, 1962).

[129] L. A. deRosa, M. J. DiToro, and L. G. Fischer, "Signal correlation radio receiver," U.S. Patent 2 718 638, Sept. 20, 1955 (filed Jan. 20, 1950).

[130] L. A. deRosa and M. Rogoff, Sect. I (Communications) of "Application of statistical methods to secrecy communication systems," Proposal 946, Fed. Telecommun. Lab., Nutley, NJ, Aug. 28, 1950.

[131] "A report on security of overseas transport," Project Hartwell, M.I.T., Cambridge, MA, Sept. 21, 1950 (ATI 205035, ATI 205036; not available from M.I.T.).

[132] "R.L.E.: 1946 + 20," M.I.T. Res. Lab. Electron., Cambridge, MA, May 1966.

[133] D. Lang, *An Inquiry into Enoughness*, New York: McGraw-Hill, 1965.

[134] "Problems of air defense," Project Charles, M.I.T., Cambridge, MA, Final Rep., Aug. 1, 1951 (ATI 139962; not available from M.I.T.).

[135] B. L. Basore, "Noise-like signals and their detection by correlation," M.I.T. Res. Lab. Electron. and Lincoln Lab., Tech. Rep. 7, May 26, 1952 (AD 004641).

[136] B. J. Pankowski, "Multiplexing a radio teletype system using a random carrier and correlation detection," M.I.T. Res. Lab. Electron. and Lincoln Lab., Tech. Rep. 5, May 16, 1952 (ATI 168857; not available from M.I.T.).

[137] "Engineering study and experimental investigation of secure directive radio communication systems," Sylvania Elec. Products, Buffalo, NY, Interim Eng. Rep., Contr. AF-33(616)-167. Aug. 5–Nov. 5, 1952 (AD 005243).

[138] W. B. Davenport, Jr., "NOMAC data transmission systems," presented at the RDB Symp. Inform. Theory Appl. Guided Missile Problems, California Inst. Technol., Pasadena, Feb. 2–3, 1953.

[139] P. E. Green, Jr., "Correlation detection using stored signals," M.I.T. Lincoln Lab., Tech. Rep. 33, Aug. 4, 1953 (AD 020524).

[140] B. M. Eisenstadt, P. L. Fleck, Jr., O. G. Selfridge, and C. A. Wagner, "Jamming tests on NOMAC systems," M.I.T. Lincoln Lab., Tech. Rep. 41, Sept. 25, 1953 (AD 020419).

[141] P. E. Green, Jr., "The Lincoln F9C radioteletype system," M.I.T. Lincoln Lab., Tech. Memo. 61, May 14, 1954 (not available from M.I.T.).

[142] P. E. Green, Jr., R. S. Berg, C. W. Bergman, and W. B. Smith, "Performance of the Lincoln F9C radioteletype system," M.I.T. Lincoln Lab., Tech. Rep. 88, Oct. 28, 1955 (AD 080345).

[143] N. Zierler, "Inverting the sum generator," M.I.T. Lincoln Lab., Group Rep. 34–48, Feb. 13, 1956 (AD 310397).

[144] B. M. Eisenstadt and B. Gold, "Autocorrelations for Boolean functions of

noiselike periodic sequences," *IRE Trans. Electron. Comput.*, vol. EC-10, pp. 383–388, Sept. 1961.

[145] R. C. Titsworth, "Correlation properties of cyclic sequences," Ph.D. dissertation, California Inst. Technol., Pasadena, 1962.

[146] J. M. Wozencraft, "Active filters," U.S. Patent 2 880 316, Mar. 31, 1959 (filed Mar. 21, 1955).

[147] M. L. Doelz and E. T. Heald, "A predicted wave radio teletype system," in *IRE Conv. Rec.*, New York, NY, Mar. 22–25, 1954, part 8, pp. 63–69.

[148] B. Goldberg, "Applications of statistical communications theory," presented at the Army Sci. Conf., West Point, NY, June 20–22, 1962 (AD 332048); republished in *IEEE Commun. Mag.*, vol. 19, pp. 26–33, July 1981.

[149] R. Price, "Statistical theory applied to communication through multipath disturbances," M.I.T. Res. Lab. Electron. Tech. Rep. 266 and M.I.T. Lincoln Lab. Tech. Rep. 34, Sept. 3, 1953 (AD 028497).

[150] R. Price, "Notes on ideal receivers for scatter multipath," M.I.T. Lincoln Lab., Group Rep. 34-39, May 12, 1955 (AD 224557).

[151] J. B. Wiesner and Y. W. Lee, "Experimental determination of system functions by the method of correlation," *Proc. IRE*, vol. 38, p. 205, Feb. 1950 (abstr.).

[152] D. G. Brennan, "On the maximum signal-to-noise ratio realizable from several noisy signals," *Proc. IRE*, vol. 43, p. 1530, Oct. 1955.

[153] G. L. Turin, "Communication through noisy, random-multipath channels," *IRE Conv. Rec.*, New York, NY, Mar. 19–22, 1956, part 4, pp. 154–166.

[154] P. Monsen, "Fading channel communications," *IEEE Commun. Mag.*, vol. 18, pp. 16–25, Jan. 1980.

[155] R. Price and P. E. Green, Jr., "An anti-multipath communication system," M.I.T. Lincoln Lab., Tech. Memo. 65, Nov. 9, 1956 (not available from M.I.T.).

[156] _____, "A communication technique for multipath channels," *Proc. IRE*, vol. 46, pp. 555–570, Mar. 1958.

[157] _____, "Anti-multipath receiving system," U.S. Patent 2 982 853, May 2, 1961 (filed July 2, 1956).

[158] M. G. Nicholson, Jr., "Time delay apparatus," U.S. Patent 2 401 094, May 28, 1946 (filed June 23, 1944).

[159] "Lincoln F9C-A radio teletype system," Sylvania Electron. Defense Lab., Mountain View, CA, Instruction Manual EDL-B8, Dec. 21, 1956.

[160] D. E. Sunstein and B. Steinberg, "Communication technique for multipath distortion," U.S. Patent 3 168 699, Feb. 2, 1965 (filed June 10, 1959).

[161] B. Goldberg, R. L. Heyd, and D. Pochmerski, "Stored ionosphere," *Proc. IEEE Int. Conf. Commun.*, Boulder, CO, June 1965, pp. 619–622.

[162] J. R. Pierce and A. L. Hopper, "Nonsynchronous time division with holding and with random sampling," *Proc. IRE*, vol. 40, pp. 1079–1088, Sept. 1952.

[163] A. R. Eckler, "The construction of missile guidance codes resistant to random interference," *Bell Syst. Tech. J.*, vol. 39, pp. 973–994, July 1960.

[164] A. Corneretto, "Spread spectrum com system uses modified PPM," *Electron. Design*, June 21, 1961.

[165] R. Lowrie, "A secure digital command link," *IRE Trans. Space Electron. Telem.*, vol. SET-6, pp. 103–114, Sept.–Dec. 1960.

[166] R. M. Jaffe and E. Rechtin, "Design and performance of phase-lock loops

capable of near optimum performance over a wide range of input signal and noise levels," Jet Propulsion Lab., Pasadena, CA, Progress Rep. 20-243, Dec. 1954; see also *IRE Trans. Inform. Theory*, IT-1, pp. 103–114, Mar. 1955.

[167] E. Rechtin, "An annotated history of CODORAC: 1953–1958," Jet Propulsion Lab., Pasadena, CA, Rep. 20-120, Contr. DA-04-495-Ord 18, Aug. 4, 1958 (AD 301248).

[168] Corporal Bimonthly Summary Rep. 37a (July 1–Sept. 1, 1953), Jet Propulsion Lab., Pasadena, CA, Oct. 1, 1953.

[169] W. F. Sampson, "Transistor pseudonoise generator," Jet Propulsion Lab., Pasadena, CA, Memo, 20-100, Dec. 7, 1954 (AD 056175).

[170] S. W. Golomb, *Shift Register Sequences*, San Francisco, CA: Holden-Day, 1967.

[171] B. L. Scott and L. R. Welch, "An investigation of iterative Boolean sequences," Jet Propulsion Lab., Pasadena, CA, Sect. Rep. 8-543, Nov. 1, 1955.

[172] Jupiter Bimonthly Summary No. 6 (Mar. 15–May 15, 1957), Jet Propulsion Lab., Pasadena, CA, June 1, 1957.

[173] W. K. Victor and M. H. Brockman, "The application of linear servo theory to the design of AGC loops," *Proc. IRE*, vol. 48, pp. 234–238, Feb. 1960.

[174] W. B. Davenport, Jr., "Signal-to-noise ratios in bandpass limiters," *J. Appl. Phys.*, vol. 24, pp. 720–727, June 1953.

[175] J. C. Springett and M. K. Simon, "An analysis of the phase coherent-incoherent output of the bandpass limiter," *IEEE Trans. Commun. Technol.*, CT 19, pp. 42–49, Feb. 1971.

[176] B. J. DuWaldt, "Survey of radio communications securing techniques," Space Technol. Lab., Los Angeles, CA, Tech. Rep. TR-59-0000-00789, Aug. 31, 1959 (AD 358618).

[177] S. W. Golomb, "Sequences with randomness properties," Glenn L. Martin Co., Baltimore, MD, Terminal Progress Rep., Contract Req. No. 639498, June 1955.

[178] _____, "Remarks on orthogonal sequences," Glenn L. Martin Co., Baltimore, MD, Interdepartment communication, July 28, 1954.

[179] N. Zierler, "Two pseudo-random digit generators," M.I.T. Lincoln Lab., Group Rep. 34-24, July 27, 1954.

[180] _____, "Several binary sequence generators," M.I.T. Lincoln Lab., Tech. Rep. 95, Sept. 12, 1955 (AD 089135).

[181] _____, "Linear recurring sequences," *J. SIAM*, vol. 7, pp. 31–48, Mar. 1959.

[182] E. N. Gilbert, "Quasi-random binary sequences," Bell Tel. Lab., unpublished memo., Nov. 27, 1953.

[183] D. A. Huffman, "Synthesis of linear sequential coding networks," presented at the 3rd London Symp. Inform. Theory, Sept. 12–16, 1955; published in *Information Theory*, C. Cherry, ed., New York: Academic, 1956.

[184] T. G. Birdsall and M. P. Ristenbatt, "Introduction to linear shift-register generated sequences," Univ. Michigan Res. Inst., Ann Arbor, Tech. Rep. 90, Oct. 1958 (AD 225380).

[185] M. Ward, "The arithmetical theory of linear recurring sequences," *Trans. Amer. Math. Soc.*, vol. 35, pp. 600–628, July 1933.

[186] M. Hall, "An isomorphism between linear recurring sequences and algebraic rings," *Trans. Amer. Math. Soc.*, vol. 44, pp. 196–218, Sept. 1938.

[187] J. Singer, "A theorem in finite projective geometry and some applications to number theory," *Trans. Amer. Math. Soc.*, vol. 43, pp. 377–385, May 1938.

[188] R. J. Turyn, "On Singer's parametrization and related matters," Appl. Res. Lab., Sylvania Electron. Syst., Waltham, MA, Eng. Note 197, Nov. 10, 1960.

[189] R. Gold, "Optimal binary sequences for spread spectrum multiplexing," *IEEE Trans. Inform. Theory*, IT-13, pp. 619–621, Oct. 1967.

[190] _____, "Maximal recursive sequences with 3-valued cross-correlation functions," *IEEE Trans. Inform. Theory*, IT-14, pp. 151–156, Jan. 1968.

[191] T. B. Whiteley and D. J. Adrian, "Random FM autocorrelation fuze system," U.S. Patent 4 220 952, Sept. 2, 1980 (filed Feb. 17, 1956).

[192] J. P. Costas and L. C. Widmann, "Data transmission system," U.S. Patent 3 337 803, Aug. 22, 1967 (filed Jan. 9, 1962).

[193] "Reliable tactical communications," General Electric Res. Lab., Schenectady, NY, Final Rep., Contr. DA-36-039sc-42693, Mar. 2, 1954 (AD 30344).

[194] E. J. Groth, "Notes on MUTNS, a hybrid navigation system," Conf. 6733 on Modern Navigation Systems, Univ. Michigan, Ann Arbor, Summer 1967 (Univ. Michigan Library Call No. VK 145.M624, 1967).

[195] E. J. Groth *et al.*, "Navigation, guidance, and control system for drone aircraft," Motorola, Final Eng. Rep., Contr. DA36-039sc78020, June 6, 1961 (AD 329101, AD 329102, AD 329103).

[196] L. S. Schwartz, "Wide-bandwidth communications," *Space/Aeronautics*, pp. 84–89, Dec. 1963.

[197] J. M. Wozencraft, "Reliable radio teletype coding," U.S. Patent 3 896 381, July 22, 1975 (filed Nov. 2, 1960).

[198] A. Kohlenberg, S. M. Sussman, and D. Van Meter, "Matched filter communication systems," U.S. Patent 3 876 941, Apr. 8, 1975 (filed June 23, 1961).

[199] E. Guillemin, "Matched filter communication systems," U.S. Patent 3 936 749, Feb. 3, 1976 (filed June 23, 1961).

[200] S. M. Sussman, "A matched filter communication system for multipath channels," *IRE Trans. Inform. Theory*, IT-6, pp. 367–373, June 1960.

[201] W. G. Ehrich, "Common channel multipath receiver," U.S. Patent 3 293 551, Dec. 20, 1966 (filed Dec. 24, 1963).

[202] J. J. Spilker, Jr., "Nonperiodic energy communication system capable of operating at low signal-to-noise ratios," U.S. Patent 3 638 121, Jan. 25, 1972 (filed Dec. 20, 1960).

[203] H. G. Lindner, "Communication security method and system," U.S. Patent 4 184 117, Jan. 25, 1980 (filed Apr. 16, 1956).

[204] J. W. Craig and R. Price, "A secure voice communication system," *Trans. Electron. Warfare Symp.*, 1959.

[205] J. W. Craig, Jr., "An experimental NOMAC voice communication system," M.I.T. Lincoln Lab., Rep. 34G-0007, Aug. 29, 1960 (AD 319610).

[206] R. M. Fano, "Patent disclosure," Nov. 14, 1951, disclosed orally to R. F. Schreitmueller, P. E. Green, Jr., and W. B. Davenport, Jr., Oct. 8, 1951.

[207] D. J. Gray, "A new method of teletype modulation," M.I.T. Lincoln Lab., Tech. Rep. 9, Sept. 22, 1952 (AD 000928).

[208] R. M. Fano, "Anti-multipath communication system," U.S. Patent 2 982 852, May 2, 1961 (filed Nov. 21, 1956).

[209] L. A. deRosa, M. Rogoff, W. B. Davenport, Jr., and P. E. Green, Jr., in "1981 Pioneer Award," *IEEE Trans. Aerosp. Electron. Syst.*, AES-18, pp. 153–160, Jan. 1982.

[210] R. C. Dixon, ed., *Spread Spectrum Techniques*, New York: IEEE Press, 1976.

[211] A. M. Semenov and A. A. Sikarev, *Shirokopolosnaya Radiosvyazy* [Wideband

Radio Communications]. Moscow, USSR: Voyenizdat, 1970.

[212] A. A. Kharkevich, "The transmission of signals by modulated noise," *Telecommunications (USSR)*, vol. 11, no. 11, pp. 43–47, 1957.

[213] F. H. Lange, *Korrelationselektronik* [Correlation Electronics], Berlin, Germany: VEB Verlag Technik, 1959.

[214] J. P. Costas, "Poisson, Shannon, and the radio amateur," *Proc. IRE*, vol. 47, pp. 2058–2068, Dec. 1959.

[215] ———, "Author's comment," *Proc. IRE*, vol. 48, p. 1911, Nov. 1960 (see also "Information capacity of fading channels under conditions of intense interference," *Proc. IEEE*, vol. 51, pp. 451–161, Mar. 1963).

[216] P. Bello, "Demodulation of a phase-modulated noise carrier," *IRE Trans. Inform. Theory*, vol. IT-7, pp. 19–27, Jan. 1961.

[217] H. Magnuski, "Wideband channel for emergency communication," in *IRE Int. Conv. Rec.*, New York, NY, Mar. 20–23, 1961, part 8, pp. 80–84.

[218] R. A. Marolf, "200 Mbit/s pseudo random sequence generator for very wide band secure communication systems," in *Proc. NEC*, Chicago, IL, 1963, vol. 19, pp. 183–187.

[219] A. I. Alekseyev, "Optimum noise immunity of noise like signals," (in Russian), presented at the 19th All-Union Conf. Popov Society, May 1963; trans. in *Telecommun. Radio Eng. (USSR)*, pt. 2, vol. 19, pp. 79–83, Aug. 1965.

[220] M. K. Razmakhnin, "Wideband communication systems" (in Russian), *Zarubezhnaya Radioelektronika* [*Foreign Radio Electron.*], no. 8, pp. 3–29, 1965.

[221] N. T. Petrovich and M. K. Razmakhnin, *Sistemy Svyazi s Shumopodobnymi Signalami* [*Communication Systems with Noise-Like Signals*], Moscow: Sovetskoye Radio, 1969.

[222] A. I. Alekseyev, A. G. Sheremet'yev, G. I. Tuzov, and B. I. Glazov, *The Theory and Application of Pseudorandom Signals* (in Russian), Moscow: 1969.

[223] Yu. B. Okunev and L. A. Yakovlev, *Shirokopolosnye Sistemy Svyazi a Sostavnymi Signalami* [*Wideband Systems of Communication with Composite Signals*], Moscow: Svyaz, 1968.

[224] L. S. Gutkin, V. B. Pestryakov, and V. N. Tipugin, *Radioupravleniye* [*Radio Control*], Moscow: 1970.

[225] M. D. Fagen, ed., *A History of Engineering and Science in the Bell System, National Service in War and Peace (1925–1975)*, Bell Lab., Murray Hill, NJ, 1978, pp. 296–317.

[226] R. Price, "Coding for greater effectiveness against jamming in NOMAC systems," M.I.T. Lincoln Lab., Tech. Rep. 78, 30 Mar. 1955 (not available from M.I.T.).

[227] R. H. Barker, "Synchronising arrangements for pulse code systems," U.S. Patent 2 721 318, Oct. 18, 1955 (filed in U.S. Feb. 16, 1953; in Great Britain, Feb. 25, 1952).

[228] S. O. Rice, "Communication in the presence of noise; probability of error for two encoding schemes," *Bell System. Tech J.*, Jan. 1950.

[229] R. M. Fano, "Signal-to-noise ratio in correlation detectors," M.I.T. Res. Lab. Electron., Tech. Rep. 186, Feb. 19, 1951 (PB 110543, ATI 103043).

[230] Y. W. Lee and J. B. Wiesner, "Correlation functions and communication applications," *Electronics*, June 1950, pp. 86–92.

[231] A. Hodges, *Alan Turing: The Enigma*, New York: Simon & Schuster, 1983.

Chapter 3

BASIC CONCEPTS AND SYSTEM MODELS

This chapter introduces the basic design approach for anti-jam communications and presents two simple idealized examples of such systems, specifically coherent direct-sequence spread, binary phase-shift-keying (DS/BPSK), and non-coherent frequency-hopped binary frequency-shift-keying (FH/BFSK) systems opposed by various jammer waveforms. Ideal signal parameter acquisition and synchronization is assumed throughout this chapter. Ground rules for system performance analysis are specified by stating assumptions regarding jammers and anti-jam systems along with definitions of the fundamental system parameters used throughout this three-volume work. Because coding and interleaving are extremely important in anti-jam system design, the impact of these techniques is illustrated with examples. The purpose of this chapter is to exemplify basic concepts with simple examples which serve as an introduction to the more advanced material in Volume I, Chapter 4, and Volume II, Chapters 1 and 2. The material in this and the previously mentioned chapters is based in part on the works of Jacobs and Viterbi [1], Houston [2], Scholtz [3], Viterbi [4], Clark and Cain [5], and Pickholtz et al. [6].

3.1 DESIGN APPROACH FOR ANTI-JAM SYSTEMS

How can a receiver overcome the effects of intentional jamming, particularly when the jammer has much more power than the transmitted signal? Classical communication theoretic investigations of the additive white Gaussian noise channel suggest the answer. White Gaussian noise is a mathematical model which has infinite power, spread uniformly over all frequencies. Effective communication is possible with this interfering noise

135

of infinite power because only the finite power noise components in the "signal coordinates" can do any harm. Thus, as long as the noise components in the signal coordinates are not too large, reliable communication over an additive white Gaussian noise channel is achievable. For a typical narrowband signal, this simply means that only the noise in the signal bandwidth can degrade performance. This classical theory suggests the following design approach to combatting intentional jamming:

SELECT SIGNAL COORDINATES SUCH THAT THE JAMMER CANNOT ACHIEVE LARGE JAMMER-TO-SIGNAL POWER RATIO IN THESE COORDINATES.

If many signal coordinates are available to a communication link and only a small subset of them are used at any time, and if *the jammer cannot determine the subset in use*, then the jammer is forced to jam *all coordinates* with little power in each coordinate or to jam a *few coordinates* with more power in each of the jammed coordinates. Naturally, the more signal coordinates available, the better the protection against jamming. For signals of bandwidth W and duration T the number of coordinates is roughly [7]

$$N \cong \begin{cases} 2WT & \text{coherent signals} \\ WT & \text{non-coherent signals.} \end{cases} \tag{3.1}$$

T typically represents the time to send a basic symbol. For fixed T, to make N large, W is commonly made large by one of two techniques:

· Direct Sequence Spreading (DS)
· Frequency Hopping (FH).

The signals resulting from these basic forms are referred to as spread-spectrum signals. Various hybrids of these two spreading techniques and other spread-spectrum signals are possible, but their performance does not significantly differ from that of these two basic ones. Throughout this work, the stored reference systems are assumed and in this chapter two simple examples are studied.

The assumption that the jammer does not know which subset of the many possible signal coordinates the signal uses at any given time is generally achieved by having identical synchronized pseudorandom (PN) sequence generators at both the transmitter and receiver. These continuously running PN generators are used by the transmitter and receiver to choose signal coordinates that continuously change in time. This PN generator pair must use a common key without which they would not work in unison. Throughout this book, we assume that the PN generator key is available to the transmitter and intended receiver, but *not* to the jammer. This, in fact, is the only information not available to the jammer. Our basic assumption concerning the jammer is:

> THE JAMMER HAS COMPLETE KNOWLEDGE OF THE SPREAD-SPECTRUM SYSTEM DESIGN EXCEPT HE DOES NOT HAVE THE KEY TO THE PSEUDORANDOM SEQUENCE GENERATORS.

Indeed, the jammer can have identical copies available (stolen perhaps) of the spread-spectrum transmitter and receiver, and certainly can try various keys; but without knowledge of the actual key used by the spread-spectrum system, the jammer cannot generate the same pseudorandom numbers used by the spread-spectrum transmitter and receiver. Chapter 5 of this volume reviews the properties and design of pseudorandom number generators.

3.2 MODELS AND FUNDAMENTAL PARAMETERS

The basic system is shown in Figure 3.1 where the following system parameters are fixed:

W_{ss} = total spread-spectrum signal bandwidth available

R_b = data rate in bits per second

$$\left. \begin{array}{l} S = \text{signal power} \\ J = \text{jammer power} \end{array} \right\} \text{at input to the intended receiver} \tag{3.2}$$

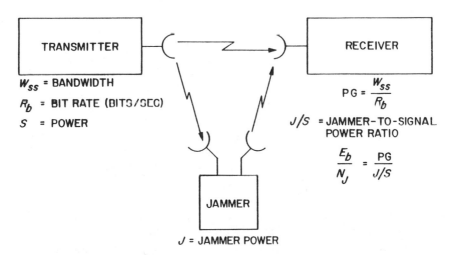

Figure 3.1. System overview.

The total spread-spectrum signal bandwidth available, W_{ss}, may not necessarily be contiguous; and it may be that a spread-spectrum system does not use all of this available bandwidth. Nevertheless, we define W_{ss} as the total available signal bandwidth and its definition is independent of how it is used. The data rate, R_b, is the uncoded information rate to be transmitted over the spread-spectrum communication link; this definition is independent of the use of coding. The signal power S and jammer power J are the time-averaged powers at the intended receiver. If, for example, the jammer transmits a pulsed signal where high peak power is achieved over short time intervals, the time-averaged power is still J. In practice, this relationship between peak power and average power may not always apply.

Regardless of the signal and jammer waveforms, an equivalent bit energy-to-jammer noise ratio is defined as

$$\frac{E_b}{N_J} = \frac{W_{ss}S}{R_bJ} \tag{3.3}$$

where we define processing gain (PG)

$$PG = \frac{W_{ss}}{R_b} \tag{3.4}$$

and

$$\frac{J}{S} = \text{jammer-to-signal power ratio.} \tag{3.5}$$

In decibels (dB), the bit energy-to-jammer noise ratio is

$$\frac{E_b}{N_J}(dB) = [PG]_{dB} - [J/S]_{dB} \tag{3.6}$$

where

$$[PG]_{dB} = 10\log_{10}\left(\frac{W_{ss}}{R_b}\right)$$

$$[J/S]_{dB} = 10\log_{10}\left(\frac{J}{S}\right). \tag{3.7}$$

In this work, bit error bounds are derived as a function of this E_b/N_J for various spread-spectrum systems and various types of jamming waveforms.

Note again that all of these definitions are independent of the type of spread-spectrum system being used, including the use of coding. The purpose of defining all of these basic parameters, independent of specific system details, is so that the performance of different systems for the same set of overall system parameter values can be compared. The particular definition of processing gain given in (3.4) may not agree with the definition of processing gain as the ratio of spread bandwidth to unspread signal bandwidth. This latter definition depends on both the modulation and coding technique used and is not as fundamental as the definition (3.4) used throughout this work.

For most of the performance evaluations in this work, it is assumed that the jammer limits performance, and therefore the effects of receiver noise in the channel can be ignored. Thus, the performance results are generally based on the assumption that the jammer power is much larger than the noise in the receiver system. An exception to this is considered when we examine both non-uniform fading channels and multiple access channels where other user signals are taken into the analysis.

3.3 JAMMER WAVEFORMS

There are an infinite number of possible jamming waveforms that could be considered. A class of jamming waveforms is selected to illustrate the basic spread-spectrum communication concepts and includes (to a good approximation) the worst types of jammer to the spread-spectrum systems of interest. *There is no single jamming waveform that is worst for all spread-spectrum systems and there is no single spread-spectrum system that is best against all jamming waveforms.* For most of this work the following types of possible jammers are considered.

3.3.1 Broadband and Partial-Band Noise Jammers

A broadband noise jammer spreads Gaussian noise of total power J evenly over the total frequency range of the spread bandwidth W_{ss} as shown in Figure 3.2(a). This results in an equivalent single-sided noise power spectral density

$$N_J - \frac{J}{W_{ss}}. \tag{3.8}$$

Since the signal energy per bit is ST_b where $T_b = 1/R_b$,

$$E_b = \frac{S}{R_b}. \tag{3.9}$$

Thus, in this case, the general definition (3.3) is exactly the bit energy-to-jammer noise ratio.

The broadband noise jammer is a brute force jammer that does not exploit any knowledge of the anti-jam communication system except its spread bandwidth W_{ss}. The resulting bit error probability of the anti-jam system is the same as that with additive white Gaussian noise of one-sided spectral density equal to N_J. We refer to the performance with this jammer as the *baseline performance*. In the following sections, we will illustrate how much worse the performance can become with other jamming waveforms of the same power J. Coding and interleaving, however, can be used to recover most of the performance loss as a result of these other jammers and to reduce the jammer's effectiveness to that of the baseline broadband noise

(a) BROADBAND

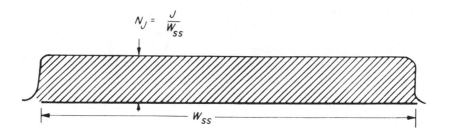

(b) PARTIAL-BAND:

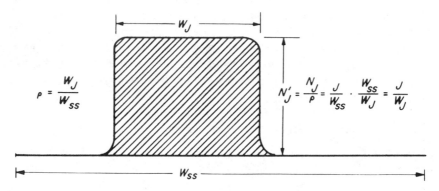

Figure 3.2.   Noise jammer frequency distribution.

jammer case. *An effective anti-jam communication system is one that gives performance close to or better than the baseline performance, regardless of the type of jammer waveform used.*

A partial-band noise jammer, as shown in Figure 3.2(b), spreads noise of total power J evenly over some frequency range of bandwidth W_J, which is a subset of the total spread bandwidth W_{ss}. We define ρ as the ratio

$$\rho = \frac{W_J}{W_{ss}} \leq 1 \tag{3.10}$$

which is the fraction of the total spread-spectrum band that has noise of power spectral density

$$\frac{J}{W_J} = \frac{J}{W_{ss}} \cdot \frac{W_{ss}}{W_J}$$

$$= N_J/\rho. \tag{3.11}$$

3.3.2 CW and Multitone Jammers

A CW jammer has the form

$$J(t) = \sqrt{2J} \cos[\omega t + \theta] \qquad (3.12)$$

while multitone jammers using N_t equal power tones can be described by

$$J(t) = \sum_{l=1}^{N_t} \sqrt{2J/N_t} \cos[\omega_l t + \theta_l]. \qquad (3.13)$$

These are shown in the frequency domain in Figures 3.3(a) and 3.3(b). All phases are assumed to be independent and uniformly distributed over $[0, 2\pi]$.

3.3.3 Pulse Jammer

Pulse jamming occurs when a jammer transmits with power

$$J_{\text{peak}} = \frac{J}{\rho} \qquad (3.14)$$

for a fraction ρ of the time, and nothing for the remaining fraction $1 - \rho$ of the time. During the pulse interval, noise or tones can be transmitted while the time-averaged power is J.

3.3.4 Arbitrary Jammer Power Distributions

A natural generalization of the broadband and partial-band noise jammers given above is a Gaussian noise jammer with an arbitrary power spectral density of total power J. For most cases of interest, the power spectral

(a) CW:

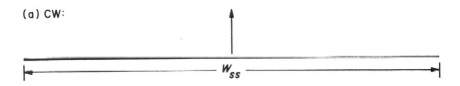

(b) MULTITONE:

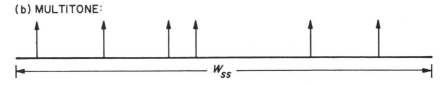

Figure 3.3. Tone jammer frequency distribution.

density that has the worst impact on the communicator has the simple partial-band form shown in Figure 3.2(b) for some corresponding worst case value of ρ. We illustrate this point in Volume II, Chapters 1 and 2. An exception to this occurs when the channel is non-uniform in the sense that the actual signal-to-noise ratio varies across the spread-spectrum band. This will be discussed in Volume II, Chapter 2, when modeling the high frequency (HF) channel that uses the skywave reflections off the ionosphere.

The pulse jammer discussed above can be generalized to include an arbitrary power distribution in time, keeping average power equal to J. Again, the worst case jammer is usually one that pulses on and off with a worst case value of ρ, the fraction of time the jammer is on with full peak power.

3.3.5 Repeat-Back Jammers

A repeat-back jammer first estimates parameters from the intercepted spread-spectrum signal and then transmits jamming waveforms that use this information. Such jammers are primarily effective against FH systems when the hop rate is slow enough for the repeat-back jammer to respond within the hop duration. *This type of jammer can be neutralized by increasing the hop rate or independently hopping each tone in the case of MFSK signals.* Here such jammers are sometimes referred to as "frequency-following jammers." An analysis of the average repeat-back jammer power reaching the receiver's data detection circuits is given in Scholtz [3].

In this work, the impact of repeat-back jammers is not discussed except to caution against low hop rates in an FH spread system. Generally the effectiveness of a repeat-back jammer depends on the hop rate and the distances between the transmitter, receiver, and jammer.

3.4 UNCODED DIRECT-SEQUENCE SPREAD BINARY PHASE-SHIFT-KEYING

To illustrate some basic concepts consider the example of an uncoded coherent, direct-sequence spread, binary phase-shift-keying (DS/BPSK) system. This type of anti-jam system is covered in greater detail in Volume II, Chapter 1.

Ordinary BPSK signals have the form

$$s(t) = \sqrt{2S} \, \sin[\omega_0 t + d_n \pi/2];$$

$$nT_b \le t < (n + 1)T_b, \quad n = \text{integer}. \tag{3.15}$$

Here T_b is the data bit time and $\{d_n\}$ is the sequence of independent data bits where

$$d_n = \begin{cases} 1, & \text{with probability } \tfrac{1}{2} \\ -1, & \text{with probability } \tfrac{1}{2}. \end{cases} \tag{3.16}$$

Equation (3.15) can also be expressed in the form

$$s(t) = d_n\sqrt{2S}\,\cos\omega_0 t;$$

$$nT_b \le t < (n+1)T_b, \quad n = \text{integer}. \tag{3.17}$$

Hence, we can view BPSK as phase modulation or amplitude modulation. Ordinary BPSK signals have a $(\sin^2 x)/x^2$ shaped power spectrum with one-sided first null bandwidth equal to $1/T_b$.

Direct sequence spreading of this BPSK signal is done with a pseudorandom (PN) binary sequence $\{c_k\}$ whose elements have values ± 1 and are generated by the PN sequence generator N times faster than the data rate. Thus, the time T_c of a PN binary symbol referred to as a "chip" is

$$T_c = \frac{T_b}{N}. \tag{3.18}$$

The PN chip rate might be several megabits per second while the data rate might be a few bits per second. The direct sequence spread-spectrum signal has the form

$$x(t) = \sqrt{2S}\,\sin[\omega_0 t + d_n c_{nN+k}\pi/2]$$

$$= d_n c_{nN+k}\sqrt{2S}\,\cos\omega_0 t;$$

$$nT_b + kT_c < t < nT_b + (k+1)T_c$$

$$k = 0, 1, 2, \ldots, N-1$$

$$n = \text{integer}. \tag{3.19}$$

This signal is similar to ordinary BPSK except that the apparent data rate is N times faster, resulting in a signal spectrum N times wider. Here the processing gain is simply

$$\text{PG} = \frac{W_{ss}}{R_b} = N, \tag{3.20}$$

where W_{ss} is again the direct sequence spread signal bandwidth.

Defining the data function

$$d(t) = d_n, \qquad nT_b \le t < (n+1)T_b \tag{3.21}$$

for all integers n, and PN function

$$c(t) = c_k, \qquad kT_c \le t < (k+1)T_c \tag{3.22}$$

for all integers k, the direct sequence spread BPSK signal can be expressed as

$$x(t) = \sqrt{2S}\,\sin[\omega_0 t + c(t)d(t)\pi/2]$$

$$= c(t)d(t)\sqrt{2S}\,\cos\omega_0 t. \tag{3.23}$$

Figure 3.4 illustrates $d(t)$, $c(t)$, and $c(t)d(t)$ for $N = 6$. Figure 3.5(a) displays the normal form of the DS/BPSK modulator and Figure 3.5(b) shows a more convenient model for analysis. Note that in this latter form,

$$x(t) = c(t)s(t), \tag{3.24}$$

where

$$s(t) = d(t)\sqrt{2S}\,\cos\omega_0 t \tag{3.25}$$

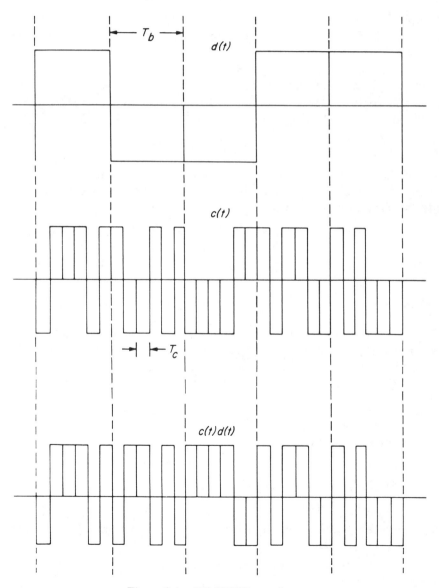

Figure 3.4. DS/BPSK waveforms.

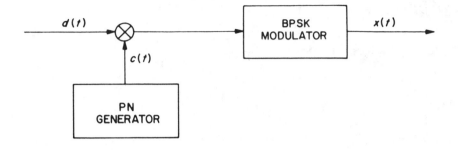

(a) Normal Form

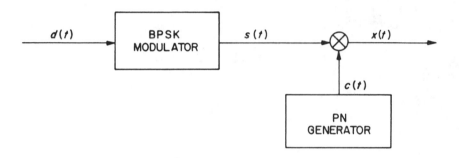

(b) Convenient Representation

Figure 3.5.   DS/BPSK modulation.

is the ordinary BPSK signal. Also, since

$$c^2(t) = 1 \text{ for all } t, \qquad (3.26)$$

it follows that

$$c(t)x(t) = s(t). \qquad (3.27)$$

Assuming the intended receiver has a PN sequence generator that is synchronized to the one in the transmitter, $c(t)$ is available to both the transmitter and receiver.

3.4.1 Constant Power Broadband Noise Jammer

Suppose a jammer is transmitting a signal $J(t)$ with constant power J in the system shown in Figure 3.6. An ideal coherent BPSK demodulator is assumed to follow the received signal $y(t)$ multiplied by the PN sequence

$c(t)$. Here, the channel output is[1]

$$y(t) = x(t) + J(t) \tag{3.28}$$

which is multiplied by the PN sequence $c(t)$ to obtain

$$
\begin{aligned}
r(t) &= c(t)y(t) \\
&= c(t)x(t) + c(t)J(t) \\
&= s(t) + c(t)J(t).
\end{aligned} \tag{3.29}
$$

This is the ordinary BPSK signal imbedded in some additive noise given by $c(t)J(t)$. The BPSK detector output which is optimum for broadband interference [8] is

$$r = d\sqrt{E_b} + n \tag{3.30}$$

where d is the data bit for this T_b second interval. $E_b = ST_b$ is the bit energy, and n is the equivalent noise component given by

$$n = \sqrt{\frac{2}{T_b}} \int_0^{T_b} c(t)J(t)\cos \omega_0 t\, dt. \tag{3.31}$$

The typical BPSK decision rule is

$$\hat{d} = \begin{cases} 1, & \text{if } r > 0 \\ -1, & \text{if } r \le 0. \end{cases} \tag{3.32}$$

Hence, the bit error probability is

$$
\begin{aligned}
P_b &= \Pr\{r > 0 | d = -1\} \\
&= \Pr\left\{n > \sqrt{E_b}\right\}
\end{aligned} \tag{3.33}
$$

assuming, without loss of generality, that $d = -1$.

The noise term depends upon many PN chips, viz.,

$$n = \sum_{k=0}^{N-1} c_k \sqrt{\frac{2}{T_b}} \int_{kT_c}^{(k+1)T_c} J(t)\cos \omega_0 t\, dt. \tag{3.34}$$

Assuming the jammer transmits broadband Gaussian noise of one-sided power spectral density given by (3.8), the terms

$$n_k = \sqrt{\frac{2}{T_c}} \int_{kT_c}^{(k+1)T_c} J(t)\cos \omega_0 t\, dt \tag{3.35}$$

are independent zero-mean Gaussian random variables with variance $N_J/2$. Thus, n defined by (3.34) and rewritten as

$$n = \sum_{k=0}^{N-1} c_k \sqrt{\frac{T_c}{T_b}}\, n_k \tag{3.36}$$

---

[1]Receiver noise which is typically modeled as additive white Gaussian noise is ignored here since jammer interference is assumed to dominate. In Section 3.9 we discuss this assumption.

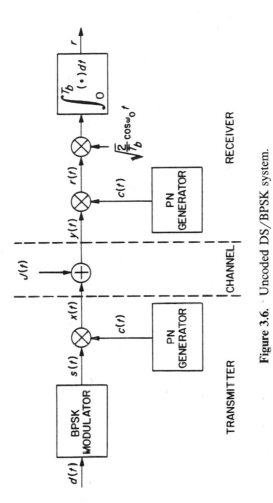

Figure 3.6. Uncoded DS/BPSK system.

is a zero-mean Gaussian random variable with variance $N_J/2$. For a continuous broadband noise jammer of constant power J the uncoded bit error probability is

$$P_b = Q\left(\sqrt{\frac{2E_b}{N_J}}\right) \tag{3.37}$$

where

$$Q(x) \triangleq \int_x^\infty \frac{1}{\sqrt{2\pi}} e^{-t^2/2}\, dt \tag{3.38}$$

is the Gaussian probability integral [8] and from (3.3) and (3.20),

$$\begin{aligned}
\frac{E_b}{N_J} &= \frac{\text{PG}}{(J/S)} \\
&= \frac{(W_{ss}/R_b)}{(J/S)} \\
&= \frac{N}{(J/S)}. \tag{3.39}
\end{aligned}$$

For large N, the above bit error probability also applies to most constant power jammer waveforms. This is based both on the assumption that the PN sequence $\{c_k\}$ is approximated as an independent binary sequence and the application of the Central Limit Theorem [9]. The details of this assumption are discussed in Volume I, Chapter 5 and Volume II, Chapter 1. Thus, this result *seems* to say that, regardless of the type of jammer, the performance of DS/BPSK is essentially the same as for the baseline jammer, namely the broadband Gaussian noise jammer. This is not true, as shown in the next example.

3.4.2 Pulse Jammer

Suppose that the jammer transmits broadband noise as in the previous section, but for only a fraction of the time with larger power. In particular, let ρ be the fraction of time the jammer is "on" and N_J/ρ be the jammer power spectral density where N_J is given by (3.8). Note that the time-averaged jammer power is assumed to be J, although the actual power during a jamming pulse duration is J/ρ. Also assume that the jammer pulse duration is greater than T_b, the data bit time, and that a particular transmitted data bit either encounters a channel with jammer "on" with probability ρ or jammer "off" with probability $1 - \rho$. We ignore the cases where the jammer might be "on" only a fraction of the transmitted data bit time.

We model the pulse jammer case as shown in Figure 3.6 with the detector output now given as

$$r = d\sqrt{E_b} + Zn \tag{3.40}$$

where n is the jammer noise term which is a Gaussian random variable with zero mean and variance $N_J/(2\rho)$. Here Z is a random variable independent of n that has probability,

$$\Pr\{Z = 1\} = \rho$$
$$\Pr\{Z = 0\} = 1 - \rho. \tag{3.41}$$

The random variable Z specifies whether or not the jammer is "on" during a particular T_b data bit time interval when one BPSK signal is transmitted.

The bit error probability is then

$$
\begin{aligned}
P_b &= \Pr\left\{ Zn > \sqrt{E_b} \right\} \\
&= \Pr\left\{ Zn > \sqrt{E_b} \,\middle|\, Z = 1 \right\} \Pr\{Z = 1\} \\
&\quad + \Pr\left\{ Zn > \sqrt{E_b} \,\middle|\, Z = 0 \right\} \Pr\{Z = 0\} \\
&= \Pr\left\{ n > \sqrt{E_b} \right\} \rho \\
&= \rho Q\left(\sqrt{\frac{2E_b}{N_J}\rho} \right)
\end{aligned}
\tag{3.42}
$$

where with jammer pulse "off"

$$\Pr\left\{ Zn > \sqrt{E_b} \,\middle|\, Z = 0 \right\} = 0. \tag{3.43}$$

This ignores the effects of noise and assumes the jammer signal "on" error term is much larger than the jammer signal "off" error term.

Figure 3.7 shows values of P_b versus $E_b/N_J = N/(J/S)$ for various values of ρ, the fraction of time the jammer pulse is on with power J/ρ. Note that the value of ρ that maximizes the bit error probability, P_b, decreases with increasing values of E_b/N_J. To see this analytically, differentiate (3.42) with respect to ρ to find that the value of ρ that maximizes P_b is

$$
\rho^* = \begin{cases} \dfrac{0.709}{E_b/N_J}, & E_b/N_J > 0.709 \\[2mm] 1, & E_b/N_J \le 0.709 \end{cases}
\tag{3.44}
$$

which results in the maximum bit error probability

$$
\begin{aligned}
P_b^* &= \max_{0 \le \rho \le 1} \rho Q\left(\sqrt{\frac{2E_b}{N_J}\rho} \right) \\
&= \rho^* Q\left(\sqrt{\frac{2E_b}{N_J}\rho^*} \right) \\
&= \begin{cases} \dfrac{0.083}{E_b/N_J}, & E_b/N_J > 0.709 \\[3mm] Q\left(\sqrt{\dfrac{2E_b}{N_J}} \right), & E_b/N_J \le 0.709. \end{cases}
\end{aligned}
\tag{3.45}
$$

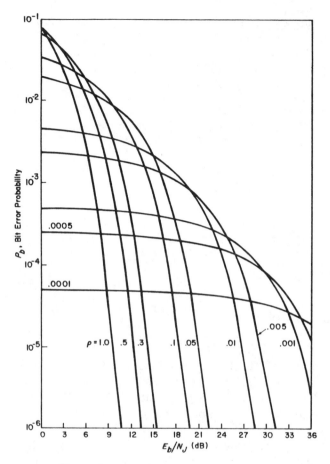

Figure 3.7.   DS/BPSK pulse jammer.

Figure 3.8 illustrates the huge difference between a constant power jammer ($\rho = 1$) and the worst case pulse jammer with $\rho = \rho^*$ for uncoded DS/BPSK systems. Note that at a bit error probability of 10^{-6} there is almost a 40 dB difference in E_b/N_J. Here for the same fixed average power J, the jammer can do considerably more harm to an uncoded DS/BPSK anti-jam system with pulse jamming than with constant power jamming. This can be explained by noting in Figure 3.8 the rapidly dropping bit error probability curve for $\rho = 1$. A small decrease in E_b/N_J results in a large increase in P_b. The jammer can vary the value of E_b/N_J with pulse jamming, resulting in the net average bit error probability skewed toward the high error probabilities associated with the small values of E_b/N_J that occur during a pulse. In general, if the jammer were to vary its power over several values while maintaining time-averaged power J, the resulting bit error probability would be the weighted sum of the bit error probabilities at each of the possible jammer power levels. In this sum, the largest error

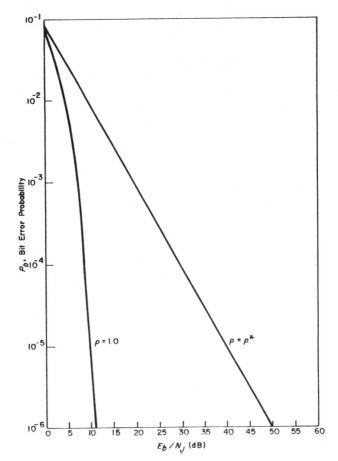

Figure 3.8. Constant power and worst case pulse jammer for DS/BPSK.

probabilities tend to dominate. We show in Volume II, Chapter 1 that the worst jammer power distribution is the "on" and "off" pulse jammer considered here.

In practice, the jammer may not be able to achieve small values of ρ or high peak power. Also when a jammer pulse length is shorter than a data bit time this analysis does not apply. In general, however, (3.45) represents an upper bound on the bit error probability.

3.5 CODED DIRECT-SEQUENCE SPREAD BINARY PHASE-SHIFT-KEYING

The impact of pulse jamming can be neutralized with coding techniques where the use of coding not only provides the usual coding gain but also forces the worst pulse jammer to be the constant power jammer. Thus, for example, at 10^{-6} bit error probability, the total coding gain against the worst pulse jammer is 40 dB as a result of pulse jammer neutralization plus the usual coding gain. This is a significant gain compared to the coding gain achieved in the usual additive white Gaussian noise channel.

Reducing the data rate or expanding the signal bandwidth is usually associated with coding since some form of redundancy is required. *For spread-spectrum signals it is not necessary to reduce the data rate or increase the signal bandwidth in order to use coding techniques.* To illustrate this consider a simple constraint length $K = 2$, rate $R = \frac{1}{2}$ bits per coded symbol convolutional code. For each data bit the encoder generates two coded bits. Let $\{d_n\}$ be the data bit sequence as before. For the k-th transmission time interval, the two coded bits are

$$\mathbf{a}_k = (a_{k1}, a_{k2}) \tag{3.46}$$

where

$$a_{k1} = d_k$$

$$a_{k2} = \begin{cases} 1; & d_k \neq d_{k-1} \\ -1; & d_k = d_{k-1}. \end{cases}$$

Suppose that T_b is the data bit time. Then each coded bit must be transmitted in $T_s = T_b/2$ seconds. Here T_s is the time of each coded symbol. Defining

$$a(t) = \begin{cases} a_{k1}, & kT_b \leq t < (k + \frac{1}{2})T_b \\ a_{k2}, & (k + \frac{1}{2})T_b \leq t < (k + 1)T_b \end{cases}$$

$$k = \text{integer} \tag{3.47}$$

shown in Figure 3.9 is the uncoded waveform $d(t)$, coded waveform $a(t)$, and PN waveform $c(t)$ for $N = 6$. With ordinary BPSK modulation, the coded waveform will have twice the bandwidth of the uncoded waveform. However, when we multiply the PN waveform with the coded and uncoded waveform, we obtain $c(t)d(t)$ and $c(t)a(t)$ shown in Figure 3.10. At this point, the resulting spread-spectrum signals have the same bandwidth. Also shown here is the convolutional code diagram when $\{1, -1\}$ binary symbols are converted to $\{1, 0\}$ binary symbols.

To illustrate the impact of coding, consider the simplest of all coding techniques, the repeat code. This is a code of rate $R = 1/m$ bits per coded symbol for some integer m. For each data bit d this code is to simply transmit m bits

$$\mathbf{a} = (a_1, a_2, a_3, \ldots, a_m) \tag{3.48}$$

where

$$a_i = d; \quad i = 1, 2, \ldots, m.$$

Thus, each coded bit a_i is transmitted within $T_s = T_b/m$ seconds. As we noted above, as long as $m < N$, the bandwidth of the direct sequence spread signal is unchanged by the use of this code.

The coded DS/BPSK system is sketched in Figure 3.11. Data bit d is encoded into a sequence of m coded bits each identical to d in value. The coded bits are scrambled in time by the interleaver and then BPSK modulated and direct sequence spread by the PN sequence $c(t)$. The transmitted DS/BPSK signal is given by (3.24) where $s(t)$ is the ordinary coded BPSK signal. The channel output is again $y(t)$ of (3.28) which after

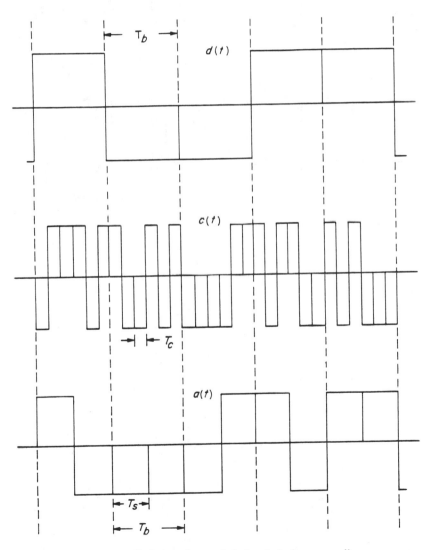

Figure 3.9. Coded and uncoded signals before spreading.

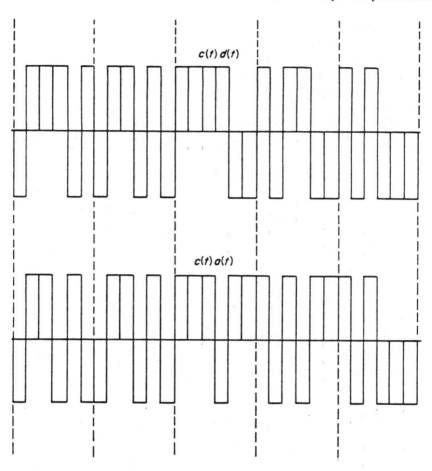

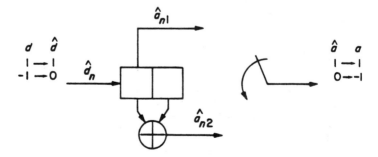

Figure 3.10. Coded and uncoded signals after spreading.

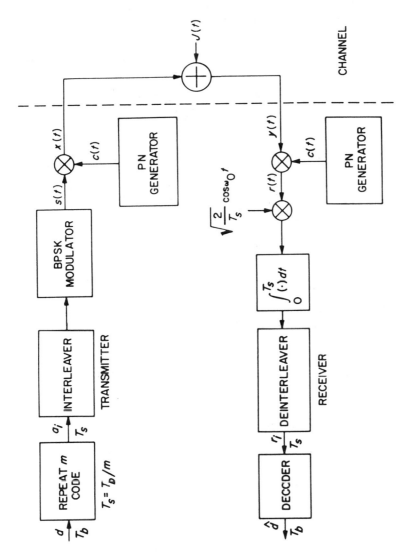

Figure 3.11. Repeat coded DS/BPSK system.

multiplication by $c(t)$ becomes $r(t)$ of (3.29). The detector for this signal consists of a correlator or matched filter with outputs, after deinterleaving, given by

$$r_i = a_i\sqrt{E_b/m} + Z_i n_i;$$
$$i = 1, 2, \ldots, m \tag{3.49}$$

where $n_1, n_2, \ldots, n_m$ are independent zero-mean Gaussian random variables with variance $N_J/(2\rho)$. Here again, ρ is the fraction of time the pulse jammer is on and Z_i indicates the jammer state where

$$Z_i = \begin{cases} 1, & \text{jammer on during } a_i \text{ transmission} \\ 0, & \text{jammer off during } a_i \text{ transmission} \end{cases} \tag{3.50}$$

and

$$\Pr\{Z_i = 1\} = \rho$$
$$\Pr\{Z_i = 0\} = 1 - \rho. \tag{3.51}$$

3.5.1 Interleaver and Deinterleaver

The purpose of the interleaver and deinterleaver is to scramble the time sequence of the coded symbols before transmission and unscramble the received signals after transmission so that the impact of the pulse jammer is independent in each of the coded symbols. For ideal interleaving and deinterleaving this results in $Z_1, Z_2, Z_3, \ldots, Z_m$ becoming independent random variables. In Appendix 3A, various forms of interleaver-deinterleaver pairs are discussed.

Suppose for the moment we did not have an interleaver and deinterleaver in the coded DS/BPSK system of Figure 3.11. Then the outputs of the channel have the form

$$r_i = d\sqrt{E_b/m} + Z n_i;$$
$$i = 1, 2, \ldots, m \tag{3.52}$$

where

$$a_i = d; \qquad i = 1, 2, \ldots, m$$

and since there is no interleaver and deinterleaver

$$Z_i = Z; \qquad i = 1, 2, \ldots, m. \tag{3.53}$$

This follows from the assumption that the jammer is either "on" or "off" during the whole transmission of a data bit. The interleaver-deinterleaver pair would normally scramble the coded bits over a large enough time span so that $Z_1, Z_2, \ldots, Z_m$ would come from widely spaced time segments and, thus, be independent of each other.

With no interleaver-deinterleaver pair the optimum decision rule is to form the sum

$$r = \sum_{i=1}^{m} r_i$$

$$= d\sqrt{mE_b} + Z \sum_{i=1}^{m} n_i \qquad (3.54)$$

and make the bit decision $\hat{d}$ in accordance with (3.32). The bit error probability is, thus,

$$P_b = \Pr\{ r > 0 | d = -1 \}$$

$$= \Pr\left\{ Z \sum_{i=1}^{m} n_i > \sqrt{mE_b} \right\}$$

$$= \Pr\left\{ Z \frac{1}{\sqrt{m}} \sum_{i=1}^{m} n_i > \sqrt{E_b} \right\}$$

$$= \rho \Pr\left\{ Z \frac{1}{\sqrt{m}} \sum_{i=1}^{m} n_i > \sqrt{E_b} \,\Big|\, Z = 1 \right\}$$

$$= \rho Q\left(\sqrt{\frac{2E_b}{N_J} \rho} \right). \qquad (3.55)$$

This bit error probability is identical to that of the uncoded DS/BPSK system (see (3.42)). Thus, without interleaving and deinterleaving, there is no difference between the uncoded and simple repeat-coded systems.

In general, coding techniques are ineffective without interleaving and deinterleaving. Burst error correcting codes [10] may be effective against some pulse jammers, but would be useless against others. For robust system designs, interleavers and deinterleavers must be used with coding techniques so that the effect of the pulse jammer is independent among coded symbol transmissions. This is also true for other types of coded spread-spectrum systems and jammers. For the remainder of this work, ideal interleaving and deinterleaving is assumed where the impact of jamming is independent among transmitted coded symbols. (See Appendix 3A for more details.)

3.5.2 Unknown Channel State

With ideal interleaving and deinterleaving the channel outputs are as given in (3.49) where again $Z_1, Z_2, \ldots, Z_m$ and $n_1, n_2, \ldots, n_m$ are all independent random variables. The decoder takes $r_1, r_2, \ldots, r_m$ and makes a decision $\hat{d} = 1$ or $\hat{d} = -1$. With pulse jamming there is the possibility that the decoder may have additional information about the values $Z_1, Z_2, \ldots, Z_m$ that might aid the decision rule. This might be possible with channel measurements. Here, however, assume the decoder has no knowledge of $Z_1, Z_2, \ldots, Z_m$ and consider two standard decision rules.

3.5.2.1 Soft Decision Decoder

The soft decision decoder computes

$$r = \sum_{i=1}^{m} r_i$$

$$= d\sqrt{mE_b} + \sum_{i=1}^{m} Z_i n_i \qquad (3.56)$$

and makes the decision in accordance with (3.32). The bit error probability

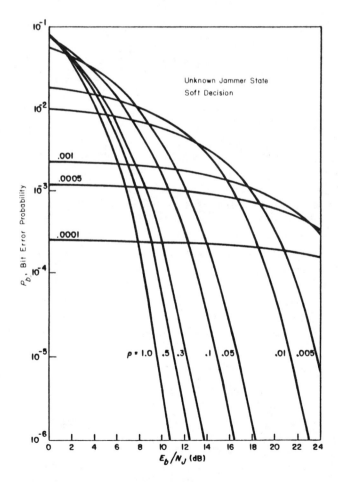

Figure 3.12. Repeat code $m = 5$ with unknown jammer state/soft decision.

is given by

$$P_b = \Pr\{\, r > 0 | d = -1 \,\}$$

$$= \Pr\left\{ \sum_{i=1}^{m} Z_i n_i > \sqrt{mE_b} \right\}. \tag{3.57}$$

To evaluate this, note that if k of the m coded symbols experience a pulse jammer then $\sum_{i=1}^{m} Z_i n_i$ is a sum of k independent Gaussian random variables each of variance $N_J/(2\rho)$. Using H_k to denote the condition of k pulse jammed symbols, we have

$$\Pr\left\{ \sum_{i=1}^{m} Z_i n_i > \sqrt{mE_b} \,\Big|\, H_k \right\} = Q\left(\sqrt{\frac{2mE_b}{kN_J}\rho} \,\right) \tag{3.58}$$

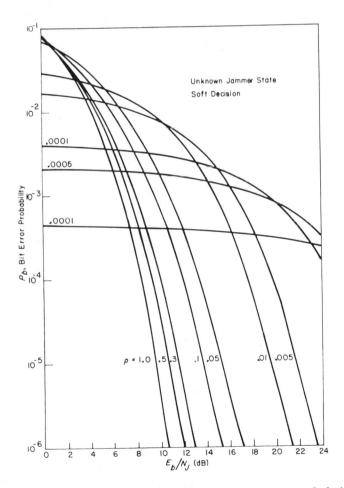

Figure 3.13. Repeat code $m = 9$ with unknown jammer state/soft decision.

where

$$\Pr\{H_k\} = \binom{m}{k}\rho^k(1-\rho)^{m-k};$$

$$k = 0, 1, \ldots, m. \qquad (3.59)$$

The bit error probability is, thus,

$$P_b = \sum_{k=0}^{m} \binom{m}{k}\rho^k(1-\rho)^{m-k}Q\left(\sqrt{\frac{2mE_b}{kN_J}}\rho\right). \qquad (3.60)$$

Figures 3.12 and 3.13 show P_b versus E_b/N_J for various values of ρ where $m = 5$ in Figure 3.12 and $m = 9$ for Figure 3.13. Comparing these with the uncoded case shown in Figure 3.7 there is only slight improvement with the repeat code. It is still clear that pulse jamming can cause much more degradation compared to constant power jamming ($\rho = 1$). Our conclusion is that *soft decision decoding with no jammer state knowledge is not very effective against pulse jammers.*

3.5.2.2 Hard Decision Decoder

The hard decision decoder makes a binary decision on each channel output symbol as follows:

$$\hat{d}_i = \begin{cases} 1, & r_i > 0 \\ -1, & r_i \le 0; \end{cases} \qquad (3.61)$$

$$i = 1, 2, \ldots, m.$$

The final decision is then given by,

$$\hat{d} = \begin{cases} 1, & \sum_{i=1}^{m} \hat{d}_i > 0 \\ -1, & \sum_{i=1}^{m} \hat{d}_i \le 0. \end{cases} \qquad (3.62)$$

For an odd integer m, the probability of error is the probability that $(m + 1)/2$ or more of the m symbol decisions are in error. The probability that a particular coded symbol decision is in error is given by

$$\varepsilon = \Pr\{r_i > 0 | d = -1\}$$

$$= \Pr\left\{Z_i n_i > \sqrt{E_b/m}\right\}$$

$$= \rho Q\left(\sqrt{\frac{2E_b}{mN_J}}\rho\right). \qquad (3.63)$$

The probability that more than half of the m coded symbol decisions are in error gives the overall bit error probability

$$P_b = \sum_{k=\frac{m+1}{2}}^{m} \binom{m}{k} \varepsilon^k (1 - \varepsilon)^{m-k}. \qquad (3.64)$$

Figures 3.14 to 3.16 show P_b as given by (3.64) versus E_b/N_J for various values of m. Here by increasing the value of m we can effectively combat pulse jamming. There is no loss in data rate nor any change in the spread-spectrum signal by increasing the number m of repeats of the data bit.

These results show that *against the worst pulse jammer the hard decision decoder does better than the soft decision decoder*. At first, this may seem like a contradiction of the fact that for the additive Gaussian noise channel, soft

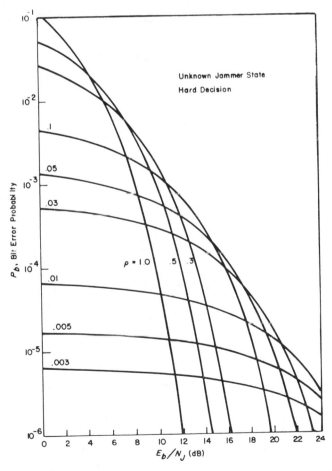

Figure 3.14. Repeat code $m = 3$ with unknown jammer state/hard decision.

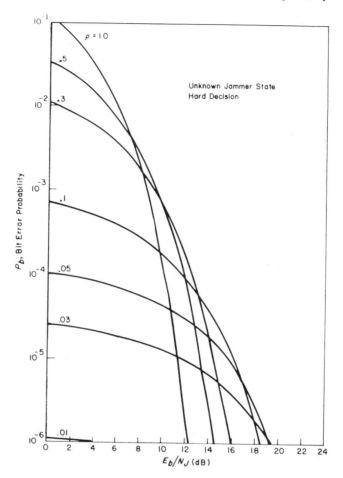

Figure 3.15. Repeat code $m = 5$ with unknown jammer state/hard decision.

decision decoders always do better than hard decision decoders. In our examples, the pulse jammer can do more harm to a soft decision decoder because it can place a lot more jammer noise power in the final decision statistics given by (3.56) through a single detector output. For the hard decision case, the amount of harm the pulse jammer can do to any given detector output is limited because of the hard decision. In general, decision rules that are optimum for the classical additive noise channel are not necessarily optimum over a channel with intentional jamming. Indeed, *there is no single decision rule that is optimum for all jamming signals.* A good decision rule is one that is robust in the sense that no jammer can degrade it very much. The soft decision rule can be degraded badly with high peak, low duty cycle pulses while the hard decision rule is less sensitive to pulse jamming. With coding, it is important to consider decision rules that limit the impact a single coded channel symbol output can have on the final

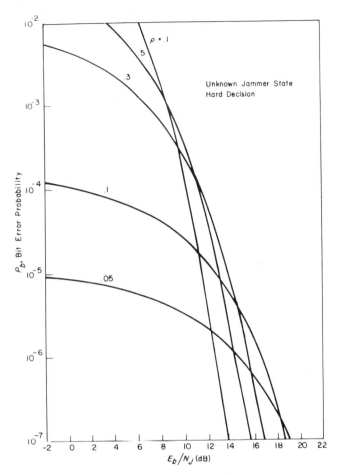

Figure 3.16. Repeat code $m = 9$ with unknown jammer state/hard decision.

decision. This usually means that some nonlinear function such as hard decisions or clipping must be used.

3.5.3 Known Channel State

Suppose that because of channel measurements, the decoder knows which of the m channel outputs $r_1, r_2, \ldots, r_m$ have a jammer term in them. This is equivalent to knowing the values of $Z_1, Z_2, \ldots, Z_m$ at the decoder. Naturally, when any $Z_i = 0$, then $r_i = d\sqrt{E_b/m}$ and d is known correctly. Hence, the only way an error can be made is when

$$Z_1 = Z_2 = \cdots = Z_m = 1. \qquad (3.65)$$

That is, an error can only occur when all m coded symbols encounter a jamming pulse. This occurs with probability

$$\Pr\{Z_1 = 1, Z_2 = 1, \ldots, Z_m = 1\} = \rho^m. \qquad (3.66)$$

3.5.3.1 Soft Decision Decoder

When $Z_1 = Z_2 = \cdots = Z_m = 1$, the soft decision decoder makes the decision

$$\hat{d} = \begin{cases} 1, & \sum_{i=1}^{m} r_i > 0 \\ -1, & \sum_{i=1}^{m} r_i \leq 0. \end{cases} \qquad (3.67)$$

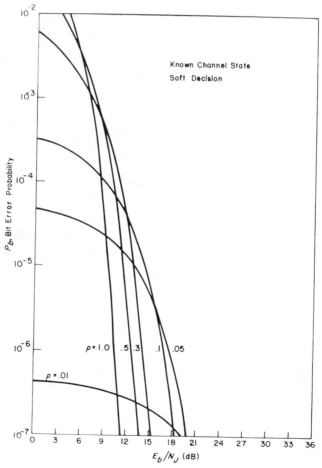

Figure 3.17. Repeat code $m = 3$ with known jammer state/soft decision.

Hence, the probability of a bit error is

$$P_b = \rho^m \Pr\left\{ \sum_{i=1}^{m} n_i > \sqrt{mE_b} \right\}$$

$$= \rho^m Q\left(\sqrt{\frac{2E_b}{N_J}\rho} \right). \tag{3.68}$$

Figures 3.17 to 3.19 show that for this case the impact of pulse jamming is effectively neutralized. For $m = 9$ and larger, for example, $\rho = 1$ gives the worst bit error probability for most values of interest. For $\rho = 1$, the repeat code gives the same performance as the uncoded DS/BPSK with constant power jamming. With more powerful codes, we can get coding gain in addition to neutralizing the degradation (40 dB at 10^{-6} bit error probability) as a result of worst case pulse jamming.

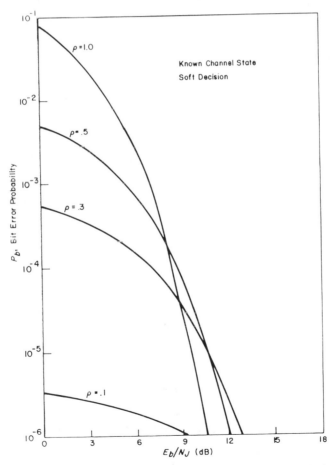

Figure 3.18.   Repeat code $m = 5$ with known jammer state/soft decision.

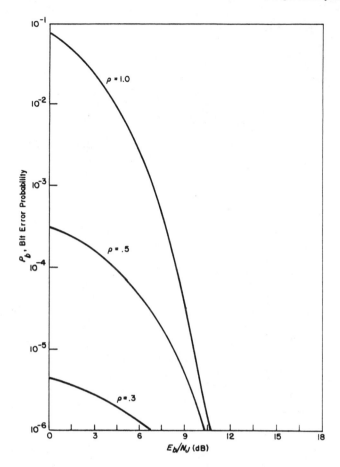

Figure 3.19. Repeat code $m = 9$ with known jammer state/soft decision.

3.5.3.2 Hard Decision Decoder

When $Z_1 = Z_2 = \cdots = Z_m = 1$, the hard decision decoder makes hard decisions on each coded symbol in accordance with (3.61), where a coded symbol error is

$$\varepsilon = \Pr\left\{ n_i > \sqrt{E_b/m} \right\}$$

$$= Q\left(\sqrt{\frac{2E_b}{mN_J}} \rho \right). \tag{3.69}$$

Applying the overall bit decision rule of (3.62) results in the bit error probability

$$P_b = \rho^m \sum_{k=\frac{m+1}{2}}^{m} \binom{m}{k} \varepsilon^k (1 - \varepsilon)^{m-k}. \tag{3.70}$$

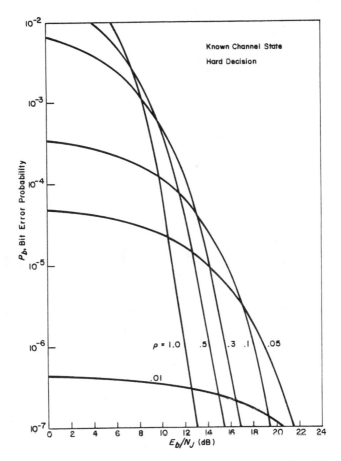

Figure 3.20.   Repeat code $m = 3$ with known jammer state/hard decision.

Figures 3.20–3.22 show this bit error probability for various values of m. As expected, the bit error probabilities are smaller than the unknown channel state case for hard decision detectors. With jammer state knowledge the soft decision case yields better performance than the hard decision case.

3.6 UNCODED FREQUENCY-HOPPED BINARY FREQUENCY-SHIFT-KEYING

Taking a basic modulation technique and changing the carrier frequency in some pseudorandom manner is the frequency-hopping approach to generating a spread-spectrum signal. The most common modulations used with frequency hopping are the M-ary frequency-shift-keying (MFSK) modulations together with non-coherent reception. This section illustrates some

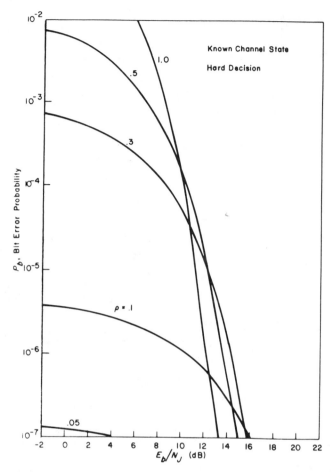

Figure 3.21. Repeat code $m = 5$ with known jammer state/hard decision.

additional basic concepts with the frequency-hopped binary frequency-shift-keying (FH/BFSK) spread-spectrum signals.

Ordinary BFSK signals have the form

$$s(t) = \sqrt{2S} \, \sin[\omega_0 t + d_n \Delta\omega t];$$

$$nT_b \leq t < (n + 1)T_b, \qquad n = \text{integer}. \qquad (3.71)$$

Here T_b is the data bit time and $\{d_n\}$ are the independent data bits where

$$d_n = \begin{cases} 1, & \text{with probability } \frac{1}{2} \\ -1, & \text{with probability } \frac{1}{2}. \end{cases} \qquad (3.72)$$

Typically we choose

$$\Delta\omega T_b = \pi \qquad (3.73)$$

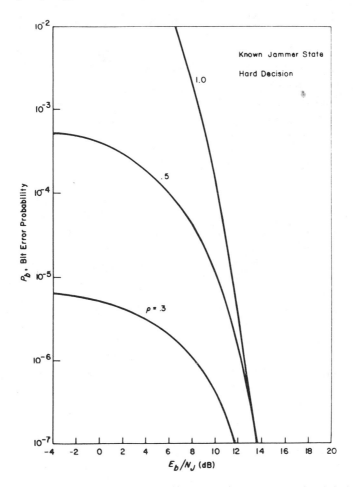

Figure 3.22. Repeat code $m = 9$ with known jammer state/hard decision.

so that the two possible transmitter tones are orthogonal for all relative phase shifts over the T_b second interval.

Frequency hopping of this BFSK signal is done with a pseudorandom binary sequence that is used to select a set of carrier frequency shifts resulting in the frequency-hopped signal

$$x(t) = \sqrt{2S} \, \sin[\omega_0 t + \omega_n t + d_n \Delta\omega t];$$

$$nT_b \leq t < (n + 1)T_b, \qquad n = \text{integer} \qquad (3.74)$$

where ω_n is the particular hop frequency chosen for the n-th transmission interval. Generally, if L pseudorandom binary symbols are used to select a frequency shift each T_b seconds, then there are at most 2^L distinct frequency shift values possible. The range of values taken by these frequency shifts defines the total spread-spectrum signal bandwidth W_{ss}. Although this total spread bandwidth need not be contiguous, it is assumed here that this band is a continuous frequency range.

Figure 3.23 illustrates the basic uncoded FH/BFSK system. For simplicity, assume the receiver's PN sequence generator is synchronized with that of the transmitter and, thus, the frequency dehop at the receiver removes the effects of the pseudorandom frequency shifts. A conventional non-coherent BFSK receiver follows the frequency dehop. Essentially, the transmitted signal is a conventional BFSK signal that has a shifting carrier frequency and the receiver has a conventional BFSK receiver that merely shifts its center frequency together with that of the transmitter.

The outputs of the energy detectors in Figure 3.23 are denoted e_+ and e_-. If there were no jamming signal present and if $d = 1$ were transmitted, these outputs would be $e_- = 0$ and $e_+ = ST_b$, the BFSK pulse energy. In general, the non-coherent decision rule based on the additive white Gaussian noise channel [8] is

$$\hat{d} = \begin{cases} 1, & e_+ > e_- \\ -1, & e_+ \le e_-. \end{cases} \tag{3.75}$$

During any T_b second interval, the transmitted signal is a tone of duration T_b seconds and has a $(\sin^2 x)/x^2$ spectrum of bandwidth roughly $2/T_b$ centered at frequency $\omega_0 + \omega_n$. The transmitted signal would then be one of two possible tones separated in frequency by $2\Delta\omega$. This "instantaneous bandwidth" is generally a small fraction of the total spread-spectrum signal bandwidth W_{ss}, which is primarily determined by the range of frequency shift values generated by the frequency hopping.

For each T_b second interval, the particular bit error probability is determined by the amount of jammer power in the "instantaneous bandwidth" of the signal that contributes to the energy terms e_+ and e_-. The overall bit error probability is then the average of these particular bit error probabilities where the average is taken over all frequency-hopped shifts.

3.6.1 Constant Power Broadband Noise Jammer

Assume that the jammer transmits broadband noise over the total spread-spectrum band with constant power J. Thus, during any T_b second interval, regardless of the carrier frequency shift, there will be an equivalent white Gaussian noise process in the "instantaneous bandwidth" of the transmitted signal. The one-sided noise spectral density is $N_J = J/W_{ss}$.

Since an equivalent white Gaussian noise process is encountered in all parts of the total spread-spectrum band, the bit error probability for the uncoded FH/BFSK system of Figure 3.23 is the same as that for conventional BFSK in white Gaussian noise, namely

$$P_b = \tfrac{1}{2}e^{-(E_b/2N_J)} \tag{3.76}$$

where E_b/N_J is still given by (3.39). This is the baseline performance of the FH/BFSK system.

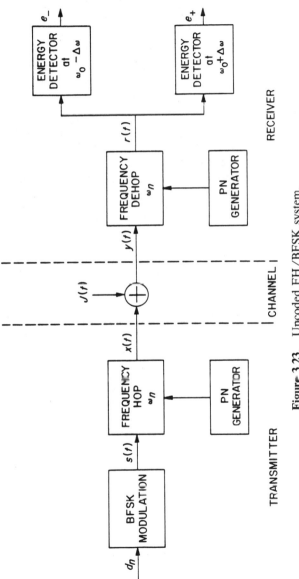

Figure 3.23. Uncoded FH/BFSK system.

3.6.2 Partial-Band Noise Jammer

Next, consider the impact of partial-band noise jamming where the jammer transmits noise over a fraction ρ of the total spread-spectrum signal band. Denoting the jammed frequency band by W_J, then ρ is given by (3.10) and in the jammed part of the band, the equivalent single-sided noise spectral density is given by (3.11).

Assume that W_J is large compared to the bandwidth of the unhopped BFSK signal and the effects of the signal hopping onto the edge of this band are negligible. That is, ignore the possibility that when a signal is sent it is frequency-hopped to the edge where only part of the instantaneous band of the signal is jammed. This assumes either a signal is hopped into the jammed band or not. In addition, the jammer is allowed to change the band it is jamming and so the transmitter and receiver never know *a priori* which frequency range is being jammed.

We again introduce the jammer state parameter Z for each T_b interval where now

$$Z = \begin{cases} 1, & \text{signal in jammed band} \\ 0, & \text{signal not in jammed band} \end{cases} \qquad (3.77)$$

with probability distribution as in (3.41). The bit error probability is then given by

$$
\begin{aligned}
P_b &= \Pr\{e_+ > e_- | d = -1\} \\
&= \Pr\{e_+ > e_- | d = -1, Z = 1\} \Pr\{Z = 1\} \\
&\quad + \Pr\{e_+ > e_- | d = -1, Z = 0\} \Pr\{Z = 0\} \\
&= \frac{\rho}{2} e^{-\rho(E_b/2N_J)}
\end{aligned}
\qquad (3.78)
$$

where there are no errors when the signal hops out of the jammed band.

Figure 3.24 illustrates the bit error probability for various values of ρ. The value of ρ that maximizes P_b is easily obtained by differentiation and found to be

$$
\rho^* = \begin{cases} \dfrac{2}{E_b/N_J}, & E_b/N_J > 2 \\ 1, & E_b/N_J \leq 2. \end{cases}
\qquad (3.79)
$$

This yields the maximum value of P_b given by

$$
P_b = \begin{cases} \dfrac{e^{-1}}{E_b/N_J}, & E_b/N_J > 2 \\ \frac{1}{2} e^{-(E_b/2N_J)}, & E_b/N_J \leq 2. \end{cases}
\qquad (3.80)
$$

Figure 3.25 shows this worst case value of the bit error probability. Here at 10^{-6} bit error probability there is a 40 dB difference between broadband

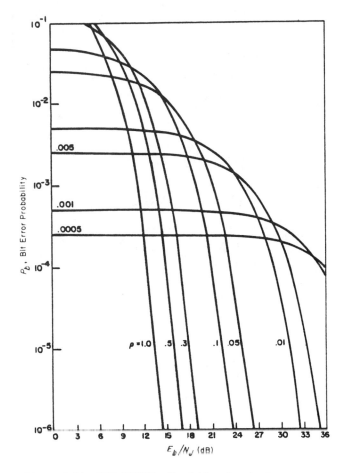

Figure 3.24.   FH/BFSK—Partial-band noise jammer.

noise jamming and the worst case partial-band noise jamming for the same jammer power.

The partial-band noise jammer effect on the uncoded FH/BFSK system is analogous to the pulse noise jammer effect on the uncoded DS/BPSK system of Section 3.4. In both systems, these jammers cause considerable degradation by concentrating more jammer power on a fraction of the transmitted uncoded symbols. This potentially large degradation is explained by the fact that the uncoded bit error probability varies dramatically with small changes in the effective bit energy-to-jammer noise ratio, E_b/N_J. Thus, the jammer can cause high error probabilities for a fraction of the transmitted bits resulting in a high average bit error probability.

For the uncoded FH/BFSK system, pulse noise jamming and partial-band noise jamming have the same effect on performance. These are essentially equivalent ways of concentrating more jammer power on some fraction of

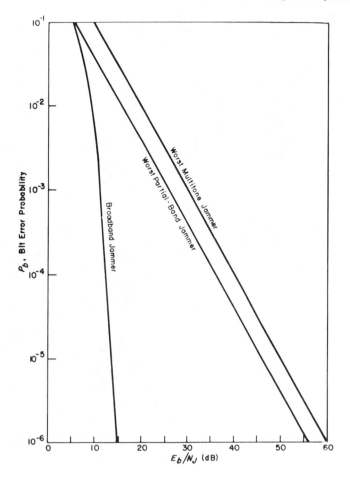

Figure 3.25.   FH/BFSK—Against jammers.

the uncoded transmitted symbols. Using pulse noise jamming or a combination of pulse noise and partial-band noise jamming would give the same results as we found for partial-band noise alone.

3.6.3 Multitone Jammer

Recall that each signal tone of T_b second duration has one-sided first null bandwidth $1/T_b$. For the total spread-spectrum signal bandwidth W_{ss}, there are

$$N = W_{ss}T_b$$

possible orthogonal tone positions. Each FH/BFSK signal would then use an adjacent pair of these tone positions to transmit one data bit. The pair of tone positions selected at any data bit time is determined by the PN sequence generator.

Consider a jammer that transmits many tones each of energy $S_J T_b$. With total power J there are at most

$$N_t = \frac{J}{S_J} \tag{3.81}$$

jammer tones randomly scattered across the band. The probability that any given signal tone position is jammed with a jammer tone is, thus,

$$\rho = \frac{N_t}{N}$$

$$= \frac{J}{S_J W_{ss} T_b}. \tag{3.82}$$

Here ρ is also the fraction of the signal tone positions that are jammed.

Assume that the jammer has exact knowledge of the N possible signal tone positions and places the N_t jamming tones in some subset of these N positions, where $N_t < N$ is always assumed.

During the transmission of a data bit, one of two possible adjacent tone positions is used by the transmitter. An error occurs if the detected energy in the alternate tone position not containing the transmitted signal tone is larger than the detected energy in the transmitted tone position. This can occur only if a jammer tone occurs in this alternative tone position. Here, ignore the smaller probability of a jammer tone in both positions and assume an error occurs if and only if a jammer tone with power $S_J \geq S$ occurs in the alternative tone position. Thus, the probability of a bit error is

$$P_b = \rho = \frac{J}{S_J W_{ss} T_b} \tag{3.83}$$

provided $S_J > S$. From the communicator's standpoint, the worst choice of S_J is $S_J = S$ resulting in the maximum bit error probability

$$P_b^* = \frac{J}{S W_{ss} T_b}$$

$$= \frac{1}{E_b/N_J}. \tag{3.84}$$

This bit error probability is slightly larger than the worst partial-band noise jammer performance; the results are essentially the same. Figure 3.25 shows the bit error probabilities for broadband noise jamming, worst partial-band noise jamming, and worst multitone jamming. Volume II, Chapter 2, will examine these cases in greater detail.

3.7 CODED FREQUENCY-HOPPED BINARY FREQUENCY-SHIFT-KEYING

Figure 3.25 illustrates the up to 45 dB of degradation at 10^{-6} bit error probability that a jammer can cause to an uncoded FH/BFSK system using the same average power J. As was done for the DS/BPSK system, we show next how a simple repeat code can effectively neutralize the degradation because of multitone jamming. This same result is also shown for worst case partial-band noise jamming in Volume II, Chapter 2.

Assume that m FH/BFSK tones are transmitted for each data bit. In particular assume the simple repeat m code where for each data bit, m identical BFSK tones are sent where each of these tones are hopped separately. Referring to these tones or codeword components as "chips," m chips make up a single data bit. The chip duration is

$$T_c = \frac{T_b}{m}. \qquad (3.85)$$

Requiring each of the chip tones to be orthogonal results in the total number of orthogonal chip tones to be

$$\begin{aligned} N_c &= W_{ss}T_c \\ &= W_{ss}T_b/m \end{aligned} \qquad (3.86)$$

which is m times smaller than the uncoded case. As before, assume the jammer sends multiple tones where the number of jammer tones is still given by (3.81). Again, choose $S_J = S$ so that the probability that a particular chip tone position is jammed is given by

$$\begin{aligned} \delta &= \frac{N_t}{N_c} \\ &= \frac{J/S}{W_{ss}T_b/m} \\ &= \frac{m}{E_b/N_J}. \end{aligned} \qquad (3.87)$$

After dehopping, the receiver is assumed to detect the energy in each of the two possible chip tone frequencies for every T_c second interval. The decoder adds up the chip energies for each of the two possible BFSK frequencies and makes a decision based on which of these has more total energy. In this case, an error is made only if a jammer tone occurs in all m of the chip tone frequencies corresponding to the BFSK frequency that was not transmitted. This occurs with probability

$$\begin{aligned} P_b &= \delta^m \\ &= \left(\frac{m}{E_b/N_J}\right)^m \end{aligned} \qquad (3.88)$$

since each chip is independently hopped.

This analysis ignored the effects of jamming tones occurring in the same frequencies as the transmitted chips. Also, it could have considered the cases where the jammer tone power S_J is larger than S so that fewer than m jammed tones could still cause an error. These more general cases are examined in detail in Volume II, Chapter 2.

The bit error probability given in (3.88) is plotted in Figure 3.26 for various values of m. The $m = 1$ case is the uncoded case considered in the previous section. Note that there exists a value of m that achieves a bit error probability close to the baseline case of broadband noise jamming.

The repeat m code is a simple code of rate $R = 1/m$ bits per coded bit. It is also referred to as *diversity of order m*. Diversity techniques are useful in combatting deep fades in a fading channel. For similar reasons, diversity is effective for multitone jamming and for worst case partial-band jamming. We shall see later, however, that there are more effective codes than simple diversity.

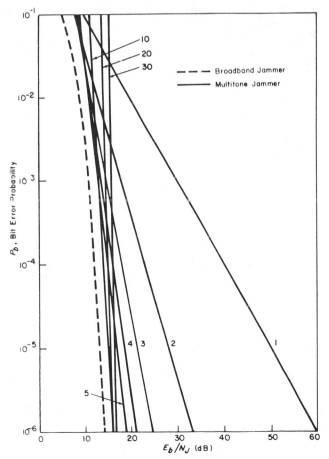

Figure 3.26. FH/BFSK with diversity—multitone jammer.

As with the DS/BPSK example, the use of coding here did not change the data rate or the total spread-spectrum bandwidth W_{ss}. Only the instantaneous bandwidth associated with each coded bit or chip became wider.

3.8 INTERLEAVER / HOP RATE TRADEOFF

In the coded FH/BFSK example considered above, the hop rate was increased from once every T_b seconds to once every $T_c = T_b/m$ seconds. Rather than increasing the hop rate by a factor of m, the same result can be achieved with an interleaver and deinterleaver. This is illustrated in Figure 3.27 with an example of $m = 3$.

Assume the frequency hop rate is fixed at once every T_b seconds, but $m = 3$ chip tones are transmitted during each hop. As illustrated in the frequency and time diagram of Figure 3.27, during each hop transmit chip tones corresponding to three different data bits. Thus, in three hop intervals of $3T_b$ seconds transmit nine chip tones where chip tones corresponding to the same data bit appear in different hop intervals. This ensures that the transmitted chips for each data bit are hopped independently. The performance achieved with this interleaver and deinterleaver is the same as the case where each transmitted chip is independently hopped at the rate of three times per T_b seconds.

3.9 RECEIVER NOISE FLOOR

In this chapter receiver noise was ignored. A more accurate model of the received signal is

$$y(t) = x(t) + J(t) + n(t) \tag{3.89}$$

where $x(t)$ is the transmitted signal, $J(t)$ is the jammer signal, and $n(t)$ is the receiver noise which is typically modeled as additive white Gaussian noise of single-sided spectral density N_0. With this receiver noise included in the analysis, the resulting bit error probability would have the form

$$P_b = f(E_b/N_J, E_b/N_0) \tag{3.90}$$

which is a function of the bit energy-to-noise ratio, E_b/N_0, as well as the effective bit energy-to-jammer power density ratio, E_b/N_J defined in (3.3).

The examples illustrated in this chapter assumed $N_0 = 0$ or $E_b/N_0 = \infty$, which is equivalent to assuming the jammer interference is much stronger than the receiver noise which can then be ignored. This resulted in bit error probability expressions of the form

$$P_b^* = f(E_b/N_J, \infty) \tag{3.91}$$

which were derived in this chapter for a few simple examples. Of course, if

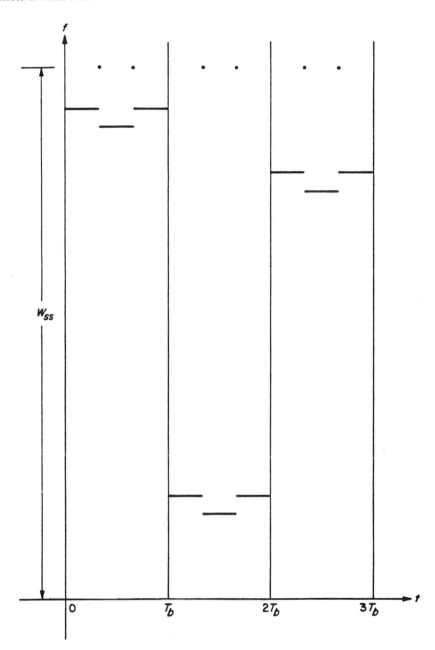

Figure 3.27. Interleaver for FH/BFSK.

the jammer is not transmitting ($N_J = 0$), then the impact of receiver noise must be considered since it is no longer negligible. Here,

$$P_b^{**} = f(\infty, E_b/N_0) \tag{3.92}$$

can be obtained using conventional performance analysis for the additive white Gaussian noise channel [8].

Figure 3.28 illustrates the typical relationship between P_b^*, P_b^{**}, and the exact bit error probability P_b. The overall analysis is considerably simplified by examining separately the performance as a result of jamming alone and then as a result of noise alone, since the exact bit error probability is approximately the larger of these two bit error probabilities.

The additive white Gaussian noise, $n(t)$, is sometimes referred to as the receiver noise floor implying that it sets the lower limit on the bit error probability when considering other types of channel interference. The

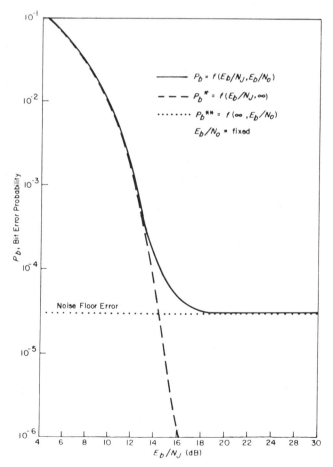

Figure 3.28.   Impact of receiver noise.

analysis presented here and in Volume I, Chapter 4, and Volume II, Chapters 1 and 2, ignores this receiver noise and considers the impact of the jammer alone. Some receiver noise floor, as illustrated in Figure 3.28, will always exist and must ultimately be considered. We are primarily concerned with the impact of jamming which is assumed to be the primary limitation on anti-jam systems.

3.10 DISCUSSION

In this chapter we have stated basic assumptions and defined key system parameters for anti-jam communication systems and jamming signals. Concepts such as coding, interleaving, and diversity were illustrated with examples. Even simple repeat codes (diversity) were shown to effectively neutralize the over 40 dB of degradation at 10^{-6} bit error probabilities that some jammers can cause in uncoded anti-jam systems.

Also illustrated was the important fact that, in designing a coded anti-jam communication system, one ought to be careful in choosing the decoding metric. There is no single metric that is optimum for all types of jammers. We have seen that the maximum-likelihood metrics for the additive white Gaussian noise channel can cause the anti-jam system to be vulnerable to certain jammers and that a hard decision metric can yield much better performance than a soft decision metric. Also, having a decision rule that uses side information about when a channel is jammed or not can improve performance significantly. A good decision rule and associated metric is one that is robust in the sense that no jammer can degrade it much more than the baseline jammer.

Except for these very simple examples, coded spread-spectrum systems under attack by various jamming signals are difficult to analyze. Analysis becomes even more difficult when unconventional decoding metrics are used. In the next chapter, a general approach to the analysis of spread-spectrum communication systems is presented. It provides a means for handling these more difficult cases as well as placing the basic concepts of this chapter on firmer ground.

3.11 REFERENCES

[1] A. J. Viterbi and I. M. Jacobs, "Advances in coding and modulation for noncoherent channels affected by fading, partial band, and multiple access interference," in *Advances in Communication Systems*, vol. 4, New York: Academic Press, 1975, pp. 279–308.

[2] S. W. Houston, "Modulation techniques for communication, Part I: Tone and noise jamming performance of spread spectrum M ~ ary FSK and 2,4 ~ ary DPSK waveforms," in *Proceedings of the IEEE National Aerospace and Electronics Conference* (NAECON '75), Dayton, Ohio, June 10–12, 1975, pp. 51–58.

[3] R. A. Scholtz, "The spread spectrum concept," *IEEE Trans. Commun.*, COM-25, pp. 748–755, August 1977.

[4] A. J. Viterbi, "Spread spectrum communication—Myths and realities," *IEEE Communications Magazine*, vol. 17, no. 3, pp. 11–18, May 1979.

[5] G. C. Clark, Jr., and J. B. Cain, *Error-Correction Coding for Digital Communications*, New York: Plenum Press, 1981.

[6] R. L. Pickholtz, D. L. Schilling, and L. B. Milstein, "Theory of spread-spectrum communications—A tutorial," *IEEE Trans. Commun.*, COM-30, no. 5, pp. 855–884, May 1982.

[7] H. J. Landau, and H. O. Pollak, "Prolate spheroidal wave functions, Fourier analysis, and uncertainty," *Bell Syst. Tech. J.*, Part II, vol. 40, 1961, pp. 65–84, Part III, in vol. 41, 1962, pp. 1295–1336.

[8] A. J. Viterbi and J. K. Omura, *Principles of Digital Communication and Coding*, New York: McGraw-Hill, 1979.

[9] A. Papoulis, *Probability, Random Variables, and Stochastic Processes*, New York: McGraw-Hill, 1965, Section 8-6.

[10] S. Lin and D. Costello, *Error Control Coding Fundamentals and Applications*, Englewood Cliffs, NJ: Prentice-Hall, 1983, chapter 14.

[11] J. L. Ramsey, "Realization of optimum interleavers," *IEEE Trans. Inform. Theory*, IT-6, pp. 338–345, May 1970.

[12] I. A. Richer, "A simple interleaver for use with Viterbi decoding," *IEEE Trans. Commun.*, COM-26, pp. 406–408, March 1978.

APPENDIX 3A. INTERLEAVING AND DEINTERLEAVING

The purpose of interleaving at the transmitter and deinterleaving at the receiver is to convert a channel with memory to one that is memoryless. In the pulse jammer example in Section 3.4.2, for example, it was assumed that the jammer pulse would jam several successive transmitted symbols resulting in a channel whose interference is not independent from symbol to symbol. By scrambling the order of channel symbols at the transmitter with interleaving and unscrambling it at the receiver with deinterleaving, an approximately memoryless channel is achieved where the impact of jamming or other channel disturbances is independent from transmitted symbol to symbol. Coding techniques are primarily designed for memoryless channels and, therefore, require interleaving and deinterleaving to be effective.

Interleaving is a form of time diversity which requires no knowledge of the channel memory other than its approximate length and is consequently very robust to changes in memory statistics. This is particularly important in a jamming environment where jammer waveforms may change. Since in all practical cases, memory decreases with time separation, if all symbols of a given codeword are transmitted at widely spaced intervals and the intervening spaces are filled similarly by symbols of other codewords, the statistical dependence between symbols of a given codeword can be effectively eliminated.

Figure 3A.1 illustrates an example of a block interleaver with depth $I = 5$ and interleaver span $N = 15$. Here the coded symbols are written into the interleaver memory filling it along columns while the transmitted symbols are read out of this memory along rows. Thus, if $x_1, x_2, x_3, x_4, \ldots$ are the coded symbols entering the interleaver, the inputs to the channel are $x_1, x_{16}, x_{31}, x_{46}, x_{61}, x_2, x_{17}, x_{32}, \ldots$. From this figure it is clear that the minimum separation is at least $I = 5$ for any two symbols entering the block interleaver within a separation of $N - 1 = 14$ symbols.

In general, with a block interleaver with depth I and span N, there are IN symbols of memory required. For synchronous transmission where one such memory is used to write in the symbols while another memory is used for reading out symbols into the channel, there is a total of $2IN$ symbols of memory required.

At the receiver, the deinterleaver simply performs the inverse operation where symbols are written into the deinterleaver by rows and read out by columns. Note that here a jamming pulse of b coded symbol duration where $b \leq I$ will result in these jammed symbols at the deinterleaver output separated by at least N symbols. Thus, if we had a block code of block-length $n \leq N$, there would be only one jammed symbol in a transmitted

Read in by columns: $x_1, x_2, x_3, \ldots$
Read out by rows: $x_1, x_{16}, x_{31}, \ldots$

Figure 3A.1. Block interleaver example.

codeword as a result of this jammer pulse. For convolutional codes a similar result is obtained if N is larger than a constraint length. Clearly I should be chosen to be larger than the channel memory and N should be larger than the code memory measured in number of coded symbols.

Synchronization of the block interleaver requires some sort of standard frame synchronization technique. This requires some extra overhead symbols which are inserted periodically.

Another type of interleaver is the convolutional interleaver proposed by Ramsey [11]. This is illustrated in Figure 3A.2 for the same parameters $I = 5$ and $N = 15$. Here the coded symbols are switched between $I = 5$ tapped shift registers of the interleaver bank. The zero-th element of this bank provides no storage (the symbols are transmitted immediately), while each successive element provides $j = 3$ symbols more storage than the

x_1	—	—	—	—
x_6	—	—	—	—
x_{11}	—	—	—	—
x_{16}	x_2	—	—	—
x_{21}	x_7	—	—	—
x_{26}	x_{12}	—	—	—
x_{31}	x_{17}	x_3	—	—
x_{36}	x_{22}	x_8	—	—
x_{41}	x_{27}	x_{13}	—	—
x_{46}	x_{32}	x_{18}	x_4	—
x_{51}	x_{37}	x_{23}	x_9	—
x_{56}	x_{42}	x_{28}	x_{14}	—
x_{61}	x_{47}	x_{33}	x_{19}	x_5
x_{66}	x_{52}	x_{38}	x_{24}	x_{10}
x_{71}	x_{57}	x_{43}	x_{29}	x_{15}
x_{76}	x_{62}	x_{48}	x_{34}	x_{20}
$\vdots$	$\vdots$	$\vdots$	$\vdots$	$\vdots$

$I = 5$
$j = 3$
$N = jI = 15$

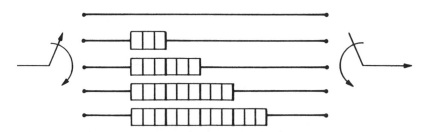

Figure 3A.2. Convolutional interleaver example.

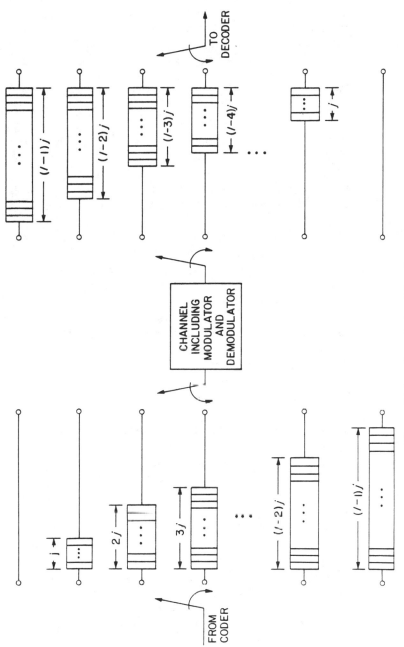

Figure 3A.3. Convolutional interleaver system.

preceding one. The input and output commutator switches move together from one register to the next. Again I is the minimum transmission separation provided for any two coded symbols entering the interleaver with a separation of less than $N = jI$ symbols.

In Figure 3A.2 the symbols entering the convolutional interleaver are $x_1, x_2, x_3, \ldots$. The interleaver outputs (coded channel input symbols) are shown by the rows of the diagram. Initially, there are some unused channel input slots until the shift registers in the interleaver fill up. After sixty channel symbols are transmitted the convolutional interleaver outputs are $x_{61}, x_{47}, x_{33}, x_{19}, x_5, x_{66}, x_{52}, \ldots$.

The convolutional deinterleaver inverts the action of the interleaver as shown in Figure 3A.3. Observables are fed in with each channel output going to a different shift register. Again, we have the property that a jamming pulse of b coded symbol duration where $b \leq I$ will result in these jammed symbols at the deinterleaver output being separated by at least N symbols.

From this it is clear that the convolutional interleaver requires roughly $IN/2$ symbols of memory compared to $2IN$ for the block interleaver memory for synchronous transmission. A fair comparison, however, must consider the fact that shift registers of varying lengths may be considerably more costly in terms of numbers of required integrated circuits than, for example, a random access memory (RAM) of larger size. Also, note that the deinterleaver memories for both approaches are often several times larger than the interleaver memory. This is because of the fact that channel output symbols are often quantized with more bits than the channel input symbols. For BPSK modulations, for example, each coded channel input symbol is one bit while the usual channel outputs are three bit quantized to eight levels.

With the deterministic interleaver and deinterleaver techniques described above, a jammer might cleverly pulse jam selected channel symbols in such a way that the deinterleaver outputs a burst of jammed symbols. This would dramatically degrade a coding system designed for independent channel disturbances. A pseudorandom interleaver/deinterleaver system can overcome this vulnerability.

A pseudorandom interleaver/deinterleaver is one where a pseudorandom sequence generator is used at the transmitter and receiver to pseudorandomly choose between many possible interleaver/deinterleaver structures. Typically for block interleaver techniques, coded symbols are written into a RAM and read out pseudorandomly. There may be several possible read out patterns, each stored in a read only memory (ROM). For each interleaved output block, a PN sequence can be used to select one of the read out patterns. (See Clark and Cain [5] and Richer [12].)

The main point of this discussion is that channels with memory can be converted into essentially memoryless channels at a cost of buffer storage and transmission delay. This cost can become prohibitive if the channel memory is very long compared to the transmission time per coded symbol.

Chapter 4

GENERAL ANALYSIS OF ANTI-JAM COMMUNICATION SYSTEMS

The evaluation of bit error probabilities for anti-jam communication systems is generally more difficult than for conventional communication systems. This is partly because of the fact that intentional jamming signals can be more varied than the additive Gaussian noise model which is typically assumed in conventional communication systems. In addition, most anti-jam communication systems use coding techniques where coding gains at 10^{-5} bit error rates are typically 30 dB to 60 dB compared to the 4 dB to 5 dB with the additive Gaussian noise channel. (This will be shown in greater detail in Volume II, Chapters 1 and 2). Even with conventional communication systems, the evaluation of exact coded bit error probabilities is usually difficult to do and, thus, easy-to-evaluate upper bounds are used [1], [2], [3].

The purpose of this chapter is to present a general expression for upper bounds on coded bit error probabilities which apply to all coded communication systems that use enough interleaving and deinterleaving so that the channel can be modeled as memoryless. We shall show that these error bounds are tight over ranges of interest by comparing them with exact bit error probabilities when they can be found. They are also applicable to arbitrary decoding decision metrics and a wide class of detectors. This generalization is necessary for anti-jam communication systems where there are no optimum detectors and maximum likelihood decoding decision rules are not known at the receiver. Here, we are generally interested in robust detectors that are effective with all types of jammers and easy to compute decoding metrics. Under intentional jamming, it is unrealistic to assume that the receiver has complete knowledge of the jamming signal's statistical characteristics.

With channel measurements, the receiver may have some side information which can aid the decoding process. For example, the receiver may have knowledge of when a jammer signal is present or not during the transmission of a coded symbol. Side information may also include time-varying channel parameter values such as propagation conditions at HF frequencies measured with channel sounder equipment or the number of other active transmitters in a spread-spectrum multiple access environment. Such side information is included in the general formulation of the expression for upper bounds to coded bit error probabilities.

In summary, because for anti-jam communication systems exact bit error probabilities are difficult or impossible to evaluate, we derive general upper bounds to these probabilities that are tight and can be numerically evaluated. These bounds allow for receivers that have incomplete knowledge of the channel using any of several types of detectors and arbitrary decoding metrics. Channel measurements that provide additional side information can be included in the decision metrics and the impact of such side information can be evaluated with these upper bounds. Ideal signal acquisition and synchronization is assumed here.

4.1 SYSTEM MODEL

Consider the general anti-jam communication system illustrated in Figure 4.1. Here the channel can include noise, jamming, and possibly channel distortions such as fading and dispersion. The modulator is assumed to be memoryless in that any given modulation waveform depends only on one input coded symbol. In addition to modulation, the coded symbols are interleaved and the signal is spread in bandwidth for anti-jam protection. At the receiver, despreading is done first, followed by detection. The detector can have many possible forms including hard decision, soft decision, threshold, possibility of erasures, and for MFSK modulation, an ordered list of the energy detector outputs.

As shown in Figure 4.1, channel measurements may provide the decoder with side information. Long-term measurements can include slowly varying parameters of the channel such as propagation conditions. Short-term measurements will usually be used to determine if a particular coded symbol experienced jamming during transmission.

The part of the system shown inside the dotted lines in Figure 4.1 is called the coding channel. It is the effective channel as seen by the encoder and decoder system. Here a general coded bit error bound will be derived which will serve as a basis for evaluating the performance of all such complex communication systems. The key feature of this approach is the decoupling of the coding aspects of the system from the remaining part of the communication system. Specifically, the cutoff rate parameter [4], [5], [6], [7]

$$R_0 \text{ bits/channel use} \tag{4.1}$$

is computed. It represents the practically achievable reliable data rate per coded symbol. This cutoff rate will be a function of the equivalent channel

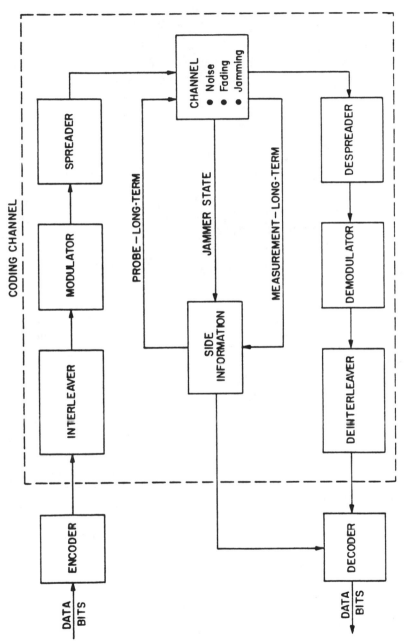

Figure 4.1. AJ system overview.

symbol energy-to-jammer noise ratio

$$\frac{E_s}{N_J} = R\left(\frac{E_b}{N_J}\right) \tag{4.2}$$

which is shown here to be directly related to the bit energy-to-jammer noise ratio given by (assuming jamming limits performance)

$$\frac{E_b}{N_J} = \frac{PG}{J/S}. \tag{4.3}$$

Here R is the code rate in data bits per channel symbol and the usual jamming parameters are

$$PG = \frac{W_{ss}}{R_b}, \quad \text{processing gain} \tag{4.4}$$

S = signal power

J = jammer power

W_{ss} = spread bandwidth

R_b = data rate in bits per second.

For conventional direct sequence (DS) coherent BPSK signals, E_s is the energy per coded symbol while for frequency-hopped (FH) non-coherent MFSK, E_s is the energy of each coded M-ary signal.

For any specific code the bound on the coded bit error probability will have the form

$$P_b \le B(R_0) \tag{4.5}$$

which is only a function of the cutoff rate R_0. Since the function $B(R_0)$ is unique for each code and the cutoff rate parameter, R_0, is independent of the code used, we are able to decouple the coding from the rest of the communication system. Thus, to evaluate various anti-jam communication systems, first compare them using the cutoff rate parameter. Codes can then be evaluated separately. This decoupling of the codes and the coding channels (shown enclosed by dotted lines in Figure 4.1) is possible at least for antipodal and orthogonal signals such as those commonly used in anti-jam communication systems.

The basic model for this analysis is shown in Figure 4.2. Here a coded symbol sequence of length N is denoted

$$\mathbf{x} = (x_1, x_2, \ldots, x_N) \tag{4.6}$$

with corresponding channel output sequence

$$\mathbf{y} = (y_1, y_2, \ldots, y_N). \tag{4.7}$$

In addition, a corresponding side information sequence is denoted

$$\mathbf{z} = (z_1, z_2, \ldots, z_N). \tag{4.8}$$

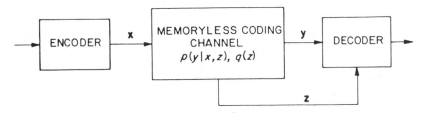

Figure 4.2.   General memoryless channel.

With adequate interleaving and deinterleaving, assume that the coding channel is memoryless. That is, assume that the channel probabilities satisfy

$$p_N(\mathbf{y}|\mathbf{x}, \mathbf{z}) = \prod_{n=1}^{N} p(y_n|x_n, z_n) \tag{4.9}$$

and

$$q_N(\mathbf{z}) = \prod_{n=1}^{N} q(z_n). \tag{4.10}$$

For any coded communication, the decoding process uses a metric of the form $m(y, x; z)$ if side information is available and $m(y, x)$ if it is not available. To simplify the decoding process, the metric is required to have an additivity property where, for sequences of symbols, the total metric of the sequences is the sum of the metrics of each channel input and output pair of the sequence. Thus, for $\mathbf{x}$, $\mathbf{y}$ and $\mathbf{z}$ given by (4.6), (4.7), and (4.8) the total metric is $m(y_1, x_1; z_1) + m(y_2, x_2; z_2) + \cdots + m(y_N, x_N; z_N)$ or $m(y_1, x_1) + m(y_2, x_2) + \cdots + m(y_N, x_N)$ depending on the availability of the side information sequence $\mathbf{z}$. The metric is used by the decoder to make decisions as to which sequence was transmitted given the corresponding channel output sequences. The additivity property is important in reducing the decoder processing complexity.

In general in a jamming environment the receiver would most likely have incomplete knowledge of the channel conditional probabilities[1] $p(y|x, z)$. Thus, the maximum-likelihood metric

$$m(y, x; z) = \log p(y|x, z) \tag{4.11}$$

when side information is available or

$$m(y, x) = \log p(y|x) \tag{4.12}$$

---

[1]In a jamming environment, it is unrealistic to assume the jamming signal statistical characteristics are known. This is also true of many other cases and, thus, the receiver design is often "mismatched" to the actual channel.

when no side information is available, is not necessarily used in the decoder. Instead, an arbitrary metric denoted $m(y, x; z)$ when side information is available and $m(y, x)$ when it is not available is assumed in the following. (The case where no side information is available can be regarded as a special case where the metric $m(y, x; z)$ does not depend on z.)

4.2 CODED BIT ERROR RATE BOUND

Assume for the moment that there are only two possible coded sequences of length N given by (4.6) and

$$\hat{\mathbf{x}} = (\hat{x}_1, \hat{x}_2, \ldots, \hat{x}_N). \tag{4.13}$$

When the channel output sequence is (4.7) and side information is (4.8) the decoder decides $\hat{\mathbf{x}}$ is the transmitted sequence if

$$\sum_{n=1}^{N} m(y_n, \hat{x}_n; z_n) \geq \sum_{n=1}^{N} m(y_n, x_n; z_n). \tag{4.14}$$

Otherwise, it decides $\mathbf{x}$ is the transmitted sequence. Assuming $\mathbf{x}$ is the actual transmitted coded sequence, the probability that the decoder incorrectly decides $\hat{\mathbf{x}}$ is given by

$$P(\mathbf{x} \rightarrow \hat{\mathbf{x}}) = \Pr\left\{ \sum_{n=1}^{N} m(y_n, \hat{x}_n; z_n) \geq \sum_{n=1}^{N} m(y_n, x_n; z_n) \Big| \mathbf{x} \right\}. \tag{4.15}$$

This is called a *pairwise error probability*.

Applying the Chernoff bound (see Appendix 4A) (4.15) has the following bound:

$$P(\mathbf{x} \rightarrow \hat{\mathbf{x}}) = \Pr\left\{ \sum_{n=1}^{N} \left[m(y_n, \hat{x}_n; z_n) - m(y_n, x_n; z_n) \right] \geq 0 \Big| \mathbf{x} \right\}$$

$$\leq E\left\{ \exp\left(\lambda \sum_{n=1}^{N} \left[m(y_n, \hat{x}_n; z_n) - m(y_n, x_n; z_n) \right] \right) \Big| \mathbf{x} \right\}$$

$$= E\left\{ \prod_{n=1}^{N} \exp\left(\lambda \left[m(y_n, \hat{x}_n; z_n) - m(y_n, x_n; z_n) \right] \right) \Big| \mathbf{x} \right\}$$

$$= \prod_{n=1}^{N} E\left\{ \exp\left(\lambda \left[m(y_n, \hat{x}_n; z_n) - m(y_n, x_n; z_n) \right] \right) \Big| \mathbf{x} \right\}$$

$$\tag{4.16}$$

for any $\lambda \geq 0$. For $\hat{x}_n = x_n$ it is clear that

$$E\left\{ \exp\left(\lambda \left[m(y_n, \hat{x}_n; z_n) - m(y_n, x_n; z_n) \right] \right) \Big| \mathbf{x} \right\} = 1. \tag{4.17}$$

Antipodal and orthogonal waveforms commonly used in anti-jam communication systems and all metrics of interest have the important property,

$$E\left\{\exp\left(\lambda\left[m(y_n, \hat{x}_n; z_n) - m(y_n, x_n; z_n)\right]\right)|\mathbf{x}\right\}\Big|_{\hat{x}_n \neq x_n} = D(\lambda) \quad (4.18)$$

where $D(\lambda)$ is independent of x_n and $\hat{x}_n$ as long as $\hat{x}_n \neq x_n$. Thus, the pairwise error probability is bounded by

$$P(\mathbf{x} \rightarrow \hat{\mathbf{x}}) \leq \left[D(\lambda)\right]^{w(\mathbf{x}, \hat{\mathbf{x}})} \quad (4.19)$$

where $w(\mathbf{x}, \hat{\mathbf{x}})$ is the number of places where $\hat{x}_n \neq x_n$, $n = 1, 2, \ldots, N$. This is sometimes called the Hamming distance [1]. Since (4.19) applies for all $\lambda \geq 0$, define

$$D = \min_{\lambda \geq 0} D(\lambda)$$

$$= \min_{\lambda \geq 0} E\left\{\exp(\lambda\left[m(y, \hat{x}; z) - m(y, x; z)\right])|\mathbf{x}\right\}\Big|_{\hat{x} \neq x} \quad (4.20)$$

which gives the final form of the pairwise error probability Chernoff bound,

$$P(\mathbf{x} \rightarrow \hat{\mathbf{x}}) \leq D^{w(\mathbf{x}, \hat{\mathbf{x}})}. \quad (4.21)$$

Note that the parameter D depends only on the coding channel and the choice of decoder metric.

In deriving the above results, we assumed a metric $m(y, x; z)$ which was arbitrary. For the special case of the maximum-likelihood metric of (4.11), (4.20) has the special form

$$D = E\left\{\sum_y \sqrt{p(y|x, z)p(y|\hat{x}, z)}\right\}\Bigg|_{\hat{x} \neq x} \quad (4.22a)$$

where the expectation is over the jammer state random variable z. When there is no jammer state information and the maximum-likelihood metric (4.12) is used, then (4.20) has the form

$$D = \sum_y \sqrt{p(y|x)p(y|\hat{x})}\Bigg|_{\hat{x} \neq x}. \quad (4.22b)$$

When (4.22) is used in the pairwise error bound (4.21), this is referred to as the Bhattacharyya bound [2]. In order to analyze realistic cases where the channel conditional probabilities are not known and, thus, receivers are not based on the maximum-likelihood metric (mismatched receivers), the general form of D is given in (4.20).

The pairwise error probability is the basis of general bit error bounds for coded communication systems. This is based on the union bound where the bit error probability is upper bounded by the sum of the probabilities of all ways a bit error can occur. For any two coded sequences $\mathbf{x}$ and $\hat{\mathbf{x}}$ let $a(\mathbf{x}, \hat{\mathbf{x}})$ denote the number of bit errors occurring when $\mathbf{x}$ is transmitted and $\hat{\mathbf{x}}$ is chosen by the decoder. If $p(\mathbf{x})$ is the probability of transmitting sequence $\mathbf{x}$ then the coded bit error bound has the form

$$
\begin{aligned}
P_b &\le \sum\sum_{\mathbf{x}, \hat{\mathbf{x}} \in \mathscr{C}} a(\mathbf{x}, \hat{\mathbf{x}}) p(\mathbf{x}) P(\mathbf{x} \to \hat{\mathbf{x}}) \\
&\le \sum\sum_{\mathbf{x}, \hat{\mathbf{x}} \in \mathscr{C}} a(\mathbf{x}, \hat{\mathbf{x}}) p(\mathbf{x}) D^{w(\mathbf{x}, \hat{\mathbf{x}})}
\end{aligned}
\tag{4.23}
$$

where $\mathscr{C}$ is the set of all coded sequences. Thus, we have the general form [2]

$$
P_b \le G(D) \tag{4.24}
$$

where $G(\cdot)$ is a function determined solely by the specific code whereas the parameter D depends only on the coding channel and the decoder metric.

The above coded bit error bound is based on two inequalities: the Chernoff bound (see Appendix 4A) and the union bound which has the general form

$$
\Pr\left\{ \bigcup_n A_n \right\} \le \sum_n \Pr\{A_n\}. \tag{4.25}
$$

In addition, in many cases of interest we can reduce the bound based on these two inequalities by a factor of one-half. In all cases where the maximum-likelihood metric is used, the factor of one-half applies [8], [9]. This is shown in Appendix 4B. In all examples given in this book we will introduce this factor of one-half whenever it can be used.

4.3 CUTOFF RATES

The parameter D is directly related to the coding channel cutoff rate R_0. In general, the channel capacity is the theoretical upper limit on data rates where arbitrarily small bit error probabilities can be achieved with coding [2]. From a practical viewpoint, it is difficult to obtain small bit error probabilities with data rates near the channel capacity. Most practical coding techniques operate near the smaller data rate R_0. Thus, R_0 is known as the practically achievable data rate with coding. In addition, R_0 is easier to evaluate than the capacity of the coding channel.

Next we examine the relationship between R_0 and the pairwise error bound for the two coded sequences of length N denoted $\mathbf{x}$ and $\hat{\mathbf{x}}$. Assuming the code symbol alphabet consists of M distinct symbols, randomly select the $2N$ symbols of $\mathbf{x}$ and $\hat{\mathbf{x}}$ where each symbol is independently selected with uniform probability. The average of the pairwise error bound is then

$$\overline{P(\mathbf{x} \to \hat{\mathbf{x}})} \le \overline{D^{w(\mathbf{x}, \hat{\mathbf{x}})}}$$

$$= \overline{\prod_{n=1}^{N} D^{w(x_n, \hat{x}_n)}}$$

$$= \prod_{n=1}^{N} \overline{D^{w(x_n, \hat{x}_n)}}$$

$$= \left\{ \frac{1 + (M - 1)D}{M} \right\}^{N} \tag{4.26}$$

where the overbar indicates the average over the coded sequences.

The cutoff rate for this case is defined as

$$R_0 = \log_2 M - \log_2 \{1 + (M - 1)D\} \text{ bits/symbol} \tag{4.27}$$

so that the average pairwise error probability has the form

$$\overline{P(\mathbf{x} \to \hat{\mathbf{x}})} \le 2^{-NR_0}. \tag{4.28}$$

This definition of the cutoff rate is the natural generalization of the usual definition where the maximum-likelihood metric is assumed. Since from (4.20) D is defined for an arbitrary decoding metric, (4.27) likewise has been generalized to apply to arbitrary decoding metrics. For the special case where the maximum-likelihood metric is used and D is given by (4.22), we have the usual definition of R_0.

Note from (4.27) that there is a one-to-one relationship between R_0 and D where D can be expressed as

$$D = \frac{M2^{-R_0} - 1}{M - 1} \tag{4.29}$$

and for a specific code the bit error bound has the form

$$P_b \le G(D)$$

$$= G\left(\frac{M2^{-R_0} - 1}{M - 1} \right)$$

$$= B(R_0) \tag{4.30}$$

where $B(\cdot)$ is a function determined solely by the specific code. In this final form of the coded bit error bound, the decoupling of the coding technique given by $B(\cdot)$ and the coding channel characterized by R_0, the cutoff rate is apparent.

In several cases of interest considered in this book, the parameter D is independent of the channel input alphabet size M whereas the cutoff rate given by (4.27) depends on both D and M. In these commonly occurring cases, the cutoff rate for an arbitrary code alphabet size M can be obtained from the cutoff rate for the binary alphabet special case where $M = 2$.

4.4 CONVENTIONAL COHERENT BPSK

For the usual coherent BPSK modulation with the additive white Gaussian noise channel the coding channel model is shown in Figure 4.3(a) where the additive noise component n is a zero-mean Gaussian random variable with variance

$$E\{n^2\} = \frac{N_0}{2}. \tag{4.31}$$

Here N_0 is the single-sided power spectral density of the additive white Gaussian noise process.

The quantizer is necessary when decoding with a digital processor. Here a "0" coded symbol results in a cosine waveform of energy E_s while a "1" coded symbol is a negative cosine waveform of energy E_s. If there is no quantizer in Figure 4.3(a) then the conventional maximum-likelihood metric is

$$m(y, x) = yx. \tag{4.32}$$

This is referred to as a soft decision channel. If x is the transmitted symbol then

$$y = x + n \tag{4.33}$$

and

$$\begin{aligned} D(\lambda) &= E\{ e^{\lambda[y(\hat{x}-x)]} | x \}\big|_{\hat{x} \neq x} \\ &= E\{ e^{\lambda(x+n)(\hat{x}-x)} \}\big|_{\hat{x} \neq x} \\ &= e^{-2\lambda E_s + \lambda^2 E_s N_0} \end{aligned} \tag{4.34}$$

or

$$\begin{aligned} D &= \min_{\lambda \geq 0} \left\{ e^{-2\lambda E_s + \lambda^2 E_s N_0} \right\} \\ &= e^{-E_s/N_0} \end{aligned} \tag{4.35}$$

where the minimum occurs for $\lambda = 1/N_0$.

Suppose next a hard decision channel sketched in Figure 4.3(b) is used where the quantizer forces a decision on each transmitted coded symbol. This results in the Binary Symmetric Channel (BSC) where the coded symbol error probability is

$$\begin{aligned} \varepsilon &= \Pr\left\{ n \geq \sqrt{E_s} \right\} \\ &= \Pr\left\{ \frac{n}{\sqrt{N_0/2}} \geq \sqrt{\frac{2E_s}{N_0}} \right\} \\ &= Q\left(\sqrt{\frac{2E_s}{N_0}} \right) \end{aligned} \tag{4.36}$$

where

$$Q(x) \triangleq \int_x^\infty \frac{1}{\sqrt{2\pi}} e^{-t^2/2} \, dt. \tag{4.37}$$

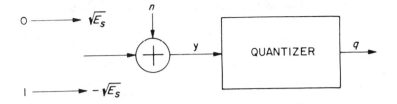

(a) General BPSK Coding Channel

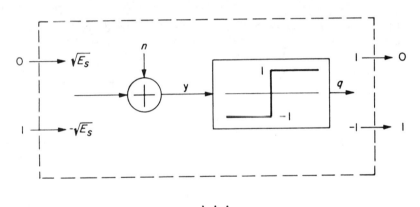

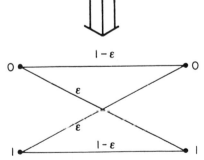

(b) Hard Decision BPSK Coding Channel

Figure 4.3. BPSK coding channels.

For the hard decision channel the maximum-likelihood metric is

$$m(y, x) = \begin{cases} 1, & y = x \\ 0, & y \neq x. \end{cases} \tag{4.38}$$

Hence,

$$D(\lambda) = E\left\{ e^{\lambda[m(y,\hat{x}) - m(y,x)]} \big| x \right\}\Big|_{\hat{x} \neq x}$$
$$= \varepsilon e^{\lambda} + (1 - \varepsilon)e^{-\lambda} \tag{4.39}$$

and

$$D = \min_{\lambda \geq 0}\left\{ \varepsilon e^{\lambda} + (1 - \varepsilon)e^{-\lambda} \right\}$$
$$= \sqrt{4\varepsilon(1 - \varepsilon)}. \tag{4.40}$$

In general for binary symbols ($M = 2$),

$$R_0 = 1 - \log_2(1 + D) \text{ bits/symbol.} \tag{4.41}$$

Figure 4.4 shows R_0 versus E_s/N_0 for the soft and hard decision detectors. Note that there is roughly a 2 dB difference for most values of E_s/N_0.

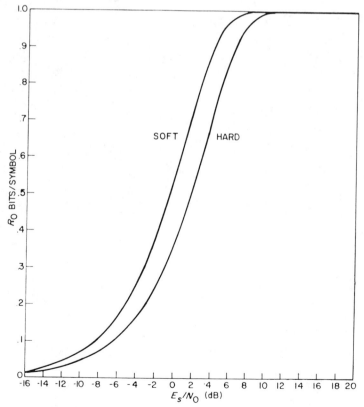

Figure 4.4. R_0 for hard and soft decision detectors.

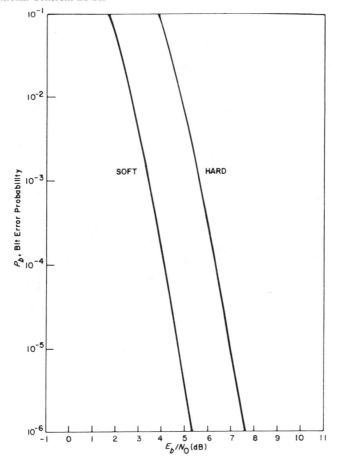

Figure 4.5a. $K = 7$, $R = 1/2$ binary convolutional code—BPSK.

The most commonly used code for coherent BPSK and QPSK modulations is the constraint length $K = 7$ rate $R = 1/2$ bits per coded bit convolutional code found by Odenwalder [10]. This code has the bit error bound[2]

$$P_b \leq \tfrac{1}{2}[36D^{10} + 211D^{12} + 1404D^{14} + 11633D^{16} + \cdots]. \quad (4.42)$$

Another common code is the $K = 7$, $R = 1/3$ convolutional code also found by Odenwalder where the bit error bound is given by

$$P_b \leq \tfrac{1}{2}[7D^{15} + 8D^{16} + 22D^{17} + 44D^{18} + \cdots]. \quad (4.43)$$

Figure 4.5 shows these two bit error bounds for the hard and soft decision

---

[2] The factor of one-half applies to all maximum-likelihood metrics considered here.

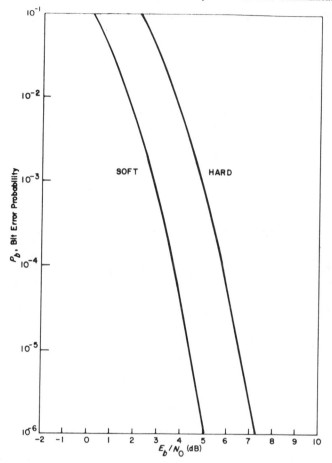

Figure 4.5b. $K = 7$, $R = 1/3$ binary convolutional code—BPSK.

detectors where the maximum-likelihood Viterbi decoders [2] are used for these convolutional codes.

Note that, regardless of the code used, the same difference in E_b/N_0 between hard and soft decision detectors occurs as seen in Figure 4.4 for R_0 versus E_s/N_0. The difference in E_s/N_0 for fixed value of R_0 directly translates to the difference in E_b/N_0 for the corresponding bound on bit error probability where E_s/N_0 and E_b/N_0 are related by (4.2) with N_J replaced by N_0.

One can view the uncoded case as a special case of coding where the code rate is $R = 1$. For this special case, for any transmitted sequence the maximum-likelihood receiver bases its decision on a symbol-by-symbol basis. That is, the optimum decision rule is to decide which particular symbol in a sequence was transmitted based only on the corresponding output symbol. This is the form of all memoryless channels with no coding since each channel output symbol depends only on the corresponding channel input symbol and all channel input symbols are independent of each other. With coding, of course, there are a restricted number of channel input sequences, thus introducing dependence among coded channel input sequences.

For the uncoded case, the symbol-by-symbol optimum decision rule is identical to the hard decision rule shown in Figure 4.3(b) where the exact bit error probability is

$$P_b = Q\left(\sqrt{\frac{2E_b}{N_0}}\right). \tag{4.44}$$

Note that these decisions assume the soft decision metric of (4.32). For the corresponding result using the general bound

$$P_b \le \tfrac{1}{2}D$$
$$= \tfrac{1}{2}e^{-E_b/N_0}. \tag{4.45}$$

Figure 4.6 shows both the exact bit error probability and this special case of

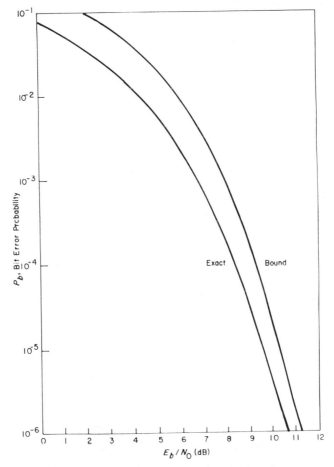

Figure 4.6. BPSK bit error probability.

the general coded bit error bound. For most values of E_b/N_0, there is approximately a 1 dB difference between the exact uncoded bit error probability and the bound.

4.5 DS/BPSK AND PULSE JAMMING

The previous results for the white Gaussian noise channel also apply to the case where there is continuous jamming of a direct sequence (DS) spread BPSK signal. The only difference is that N_J replaces N_0 in E_b/N_0 given by (4.3). Jamming, however, can take many forms. To illustrate the application of the general bounds presented in this chapter, consider the following example.

Suppose there is a pulse jammer with average power J/ρ for a fraction ρ of the time and zero power for a fraction $1 - \rho$ of the time. Assume that when the jammer is on, the channel is like an additive white Gaussian noise channel with power spectral density (one sided)

$$N_J' = \frac{J}{\rho W_{ss}} = \frac{N_J}{\rho} \tag{4.46}$$

where W_{ss} is the spread-spectrum signal bandwidth. During the transmission of a coded symbol define the jammer state random variable Z where

$$\Pr(Z = 1) = \rho$$

and

$$\Pr(Z = 0) = 1 - \rho. \tag{4.47}$$

This is the jammer state side information that may be available at the receiver where $Z = 1$ indicates a jammer is transmitting during a coded symbol time while $Z = 0$ indicates there is no jammer signal. With z available at the receiver, the metric we consider is

$$m(y, x; z) = c(z) yx \tag{4.48}$$

which is a weighted correlation metric.

Next compute the parameter

$$
\begin{aligned}
D(\lambda) &= E\{ e^{\lambda[c(Z)y(\hat{x}-x)]} |x\}|_{\hat{x} \neq x} \\
&= \rho E\{ e^{\lambda c(1)(x+n)(\hat{x}-x)} \}|_{\hat{x} \neq x} + (1 - \rho) e^{\lambda c(0)x(\hat{x}-x)}|_{\hat{x} \neq x} \\
&= \rho \exp\{ -2\lambda c(1) E_s + \lambda^2 c^2(1) E_s N_J/\rho \} \\
&\quad + (1 - \rho) \exp\{ -2\lambda c(0) E_s \}.
\end{aligned}
\tag{4.49}
$$

If the receiver has jammer state side information (knowledge of z) then the metric can have $c(0)$ as large as possible to make the second term in (4.49) negligibly small. Also without loss of generality normalize $c(1) = 1$ to obtain

$$D(\lambda) = \rho \exp\{ -2\lambda E_s + \lambda^2 E_s N_J/\rho \} \tag{4.50}$$

and

$$D = \min_{\lambda \geq 0} \left[\rho \exp\left\{ -2\lambda E_s + \lambda^2 E_s N_J/\rho \right\} \right]$$

$$= \rho e^{-\rho(E_s/N_J)}. \tag{4.51}$$

Suppose the receiver has no side information. Then the metric is independent of z, or equivalently has weighting $c(1) = c(0)$ which we can normalize to unity. Then, from (4.49)

$$D(\lambda) = \rho \exp\left\{ -2\lambda E_s + \lambda^2 E_s N_J/\rho \right\} + (1 - \rho)\exp\left\{ -2\lambda E_s \right\}$$

$$= e^{-2\lambda E_s} \left[\rho e^{\lambda^2 E_s N_J/\rho} + 1 - \rho \right] \tag{4.52}$$

and D is obtained by minimizing $D(\lambda)$ over $\lambda \geq 0$.

Both (4.50) and (4.52) correspond to soft decision detectors. The only difference between these two is the availability to the decoder of side information concerning the jammer state. The difference is shown in Figures 4.7 and 4.8 where $R_0 = 1 - \log_2(1 + D)$ is sketched versus E_s/N_J for various values of ρ. Clearly having jammer state knowledge helps improve the overall bit error probability. The special case where $\rho = 1$ coincides with the conventional soft decision curve shown in Figure 4.4. Similar results can be obtained for hard decision detectors. A complete discussion of DS/BPSK systems is covered in Chapter 1 of Volume II.

4.6 TRANSLATION OF CODED ERROR BOUNDS

The numerical evaluation of R_0 versus E_s/N_J is straightforward for most coding channels encountered in anti-jam communication systems. This cutoff rate parameter is independent of the specific code used, but can now be used to evaluate the coded bit error probability of any code whose standard bit error probability curve is available. For example, Figure 4.5 shows the coded bit error bounds for two codes commonly used over the additive white Gaussian noise channel. The basic modulation here is coherent BPSK or QPSK. Such curves are typically available in textbooks and published papers. For the standard additive white Gaussian noise channel, the standard cutoff rates are shown in Figure 4.4 for coherent BPSK modulation. The soft decision case in Figure 4.4 is the $\rho = 1$ case in Figures 4.7 and 4.8.

Suppose we now want to evaluate the coded bit error bound for the constraint length $K = 7$ rate $R = 1/2$ convolutional code used in a DS/BPSK anti-jam system where there is a pulse jammer with $\rho = .05$. Also, suppose the detector used was a soft decision detector and no jammer state information is available. Here we assume the metric used is simply that of (4.32). Figure 4.8 shows the R_0 versus E_s/N_J for this case. We can now translate the standard curve for this code shown in Figure 4.5 to determine the bit error bound with this same code for the new pulse jammed channel described above.

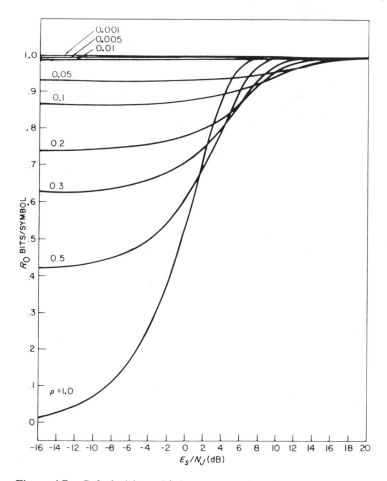

Figure 4.7. Soft decision with jammer state knowledge—DS/BPSK.

Note that for 10^{-6} coded bit error probability, the standard curve in Figure 4.5a shows a required

$$E_b/N_0 = 5.3 \text{ dB} \tag{4.53}$$

or since $E_s = E_b/2$ for $R = 1/2$,

$$E_s/N_0 = 2.3 \text{ dB}. \tag{4.54}$$

Next, for this choice of E_s/N_0, from Figure 4.4 we have that the cutoff rate required is (also see $\rho = 1$ curve in Figure 4.8)

$$R_0 = .76 \text{ bits/symbol}. \tag{4.55}$$

Recall that the coded bit error bound can always be expressed as a function of R_0 as shown in (4.5) or, more specifically for this code, by (4.42) where D is given by (4.29). Thus, if we have another coding channel with the same value of the cutoff rate parameter R_0, then the bit error bound is also the

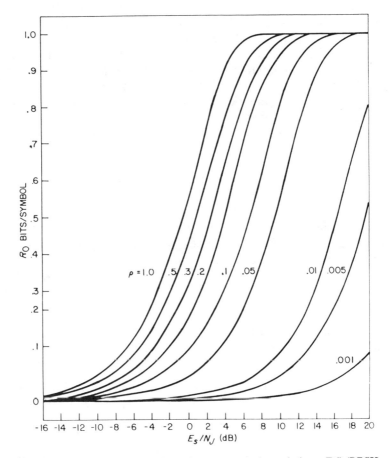

Figure 4.8. Soft decision with no jammer state knowledge—DS/BPSK.

same, which in this case is 10^{-6}. For our new coded channel, we determine from Figure 4.8 that for R_0 given by (4.55) and $\rho = .05$, the new required E_s/N_J is

$$E_s/N_J = 12.2 \text{ dB} \qquad (4.56)$$

or

$$E_b/N_J = 15.2 \text{ dB}. \qquad (4.57)$$

Thus, in this pulse jamming example, we require 15.2 dB of E_b/N_J defined by (4.3) to achieve 10^{-6} coded bit error probability with the given convolutional code. By continuing this translation for several bit error probabilities, we obtain the complete coded bit error bound for the case of interest by translating the standard coded bit error bound curve.

The translation of standard coded bit error bounds to obtain corresponding bit error bounds for different coding channels applies to all binary input coding channels. It also generalizes to M-ary input coding channels of the kind that arise in most spread-spectrum systems. This will be shown next with frequency-hopped non-coherent MFSK signals with partial-band jamming. For anti-jam communication systems, where jamming and receiver structures can take many possible forms, the general approach of first obtaining numerically computable values of the cutoff rate parameter and then translating standard error rate curves serves as a useful general approach to the analysis of anti-jam communication systems.

4.7 CONVENTIONAL NON-COHERENT MFSK

Conventional MFSK modulation consists of transmitting one of M symbols where the symbol is represented by one of M non-coherently orthogonal frequency tones of duration T_s seconds and energy E_s. We assume the M tones are separated in frequency by

$$\Delta f = \frac{1}{T_s}. \tag{4.58}$$

For MFSK modulation with the additive white Gaussian noise channel, the optimum non-coherent detector has the form shown in Figure 4.9. The detector output consists of

$$y = (e_1, e_2, \ldots, e_M) \tag{4.59}$$

where e_i is the normalized energy of the received signal at the i-th frequency. That is, e_i is the normalized energy out of the i-th frequency detector.

We denote the M modulation symbols by $\{1, 2, \ldots, M\}$ and a general input symbol into the modulator by x. Assuming $x = i$, which corresponds to the i-th frequency tone being transmitted, $e_1, e_2, \ldots, e_M$ are independent random variables where $r_i = \sqrt{e_i}$ has probability density functions [2],

$$p(r_i | x = i) = r_i e^{-r_i^2/2} e^{-E_s/N_0} I_0\left(\sqrt{\frac{2E_s}{N_0}}\, r_i\right) \tag{4.60}$$

for $r_i \geq 0$ where $I_0(\cdot)$ is the modified Bessel function of the first kind and zero-th order (this is a Rician density function) and

$$p(r_j | x = i) = r_j e^{-r_j^2/2} \tag{4.61}$$

for $r_j \geq 0$ and $j \neq i$ (these are Rayleigh density functions).

4.7.1 Uncoded

With no coding, the maximum-likelihood decision rule is: given channel output y given by (4.59), choose the transmitted symbol corresponding to the largest value of e_i; $i = 1, 2, \ldots, M$. This is equivalent to using a metric

$$m(y, x) = e_x. \tag{4.62}$$

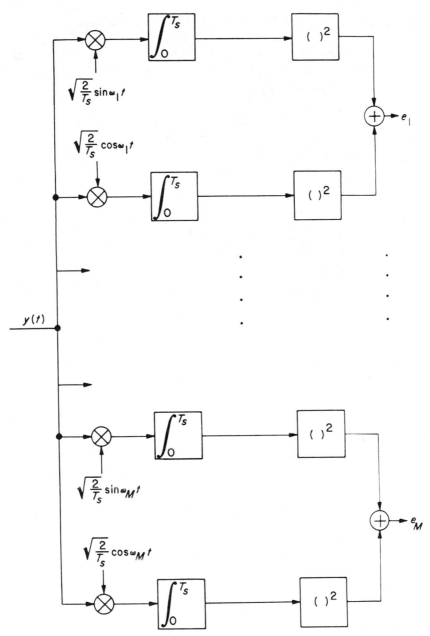

Figure 4.9. Non-coherent MFSK detector.

Note that, for the uncoded case, the decisions are made on a symbol-by-symbol basis without regard to sequences. Thus, we do not require the additivity property of metrics for the uncoded case. For this uncoded case, the probability of making a symbol decision error is [11]

$$P_s = \Pr\{\hat{x} \neq x\}$$

$$= \sum_{l=1}^{M-1} \binom{M-1}{l}(-1)^{l+1}\frac{1}{l+1}e^{-l(E_s/N_0)/(l+1)}. \qquad (4.63)$$

When $M = 2^K$, we can associate K data bits to each modulation symbol. Denoting E_b as the energy per data bit we have

$$E_s = KE_b \qquad (4.64)$$

and the bit error probability [10]

$$P_b = \frac{\frac{1}{2}M}{M-1}P_s$$

$$= \frac{2^{K-1}}{2^K-1}\sum_{l=1}^{2^K-1}\binom{2^K-1}{l}(-1)^{l+1}\frac{1}{l+1}e^{-lK(E_b/N_0)/(l+1)}. \qquad (4.65)$$

For the special case of binary symbols (BFSK), (4.65) becomes

$$P_b = \frac{1}{2}e^{-E_b/2N_0}. \qquad (4.66)$$

Applying the general coded error bound to the special case of uncoded non-coherent MFSK with metric $m(y, x) = e_x$ yields

$$D(\lambda) = E\{e^{\lambda[e_{\hat{x}}-e_x]}|x\}|_{\hat{x} \neq x}$$

$$= E\{e^{\lambda e_{\hat{x}}}|x\}|_{\hat{x} \neq x}E\{e^{-\lambda e_x}|x\} \qquad (4.67)$$

where

$$E\{e^{-\lambda e_x}|x\} = \int_0^\infty e^{-\lambda t^2}te^{-t^2/2}e^{-E_s/N_0}I_0\left(\sqrt{\frac{2E_s}{N_0}}\,t\right)dt$$

$$= \frac{1}{1+2\lambda}e^{-2\lambda(E_s/N_0)/(1+2\lambda)} \qquad (4.68)$$

and when $\hat{x} \neq x$,

$$E\{e^{\lambda e_{\hat{x}}}|x\}|_{\hat{x} \neq x} = \int_0^\infty e^{\lambda t^2}te^{-t^2/2}\,dt$$

$$= \frac{1}{1-2\lambda} \qquad (4.69)$$

provided $2\lambda < 1$. Since λ is a free variable, substituting λ in place of 2λ results in the form,

$$D = \min_{0 \leq \lambda \leq 1}\left\{\frac{1}{1-\lambda^2}e^{-\lambda(E_s/N_0)/(1+\lambda)}\right\}. \qquad (4.70)$$

The pairwise error probability is, thus,

$$P(x \to \hat{x}) \leq D \qquad (4.71)$$

yielding the symbol error probability

$$P_s = \frac{1}{M} \sum_{i=1}^{M} \Pr\{ e_i \le e_j \text{ for some } j \ne i | x = i \}$$

$$\le \tfrac{1}{2}(M-1)D. \tag{4.72}$$

This bound is based on the union bound with the factor of one-half applying. (See Appendix 4B). The bit error probability is then given by

$$P_b = \frac{\tfrac{1}{2}M}{M-1} P_s$$

$$\le \tfrac{1}{4}MD. \tag{4.73}$$

Figure 4.10 compares (4.65) and (4.73) for $M = 2, 4, 8, 16,$ and 32. There is approximately a 1 dB difference between the exact bit error probability

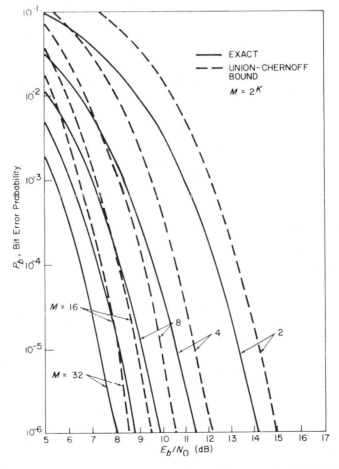

Figure 4.10. Exact versus union-Chernoff bound for uncoded MFSK bit error probability.

and the general coded bit error rate bound applied to this special case of no coding. Recall that the bound in (4.73) is based on the union bound (4.25) and the Chernoff bound (4.16). The 1 dB difference between the exact bit error probabilities and the bounds is due primarily to the Chernoff bound. For the uncoded case, we can derive the exact pairwise error probability which is given by

$$\Pr\{e_i \leq e_j | x = i\} = \tfrac{1}{2} e^{-E_s/2N_0} \tag{4.74}$$

for $j \neq i$. This is just the two signal case equivalent to the BFSK bit error probability. The symbol error union bound is, thus,

$$P_s \leq \sum_{j \neq i} \Pr\{e_i \leq e_j | x = i\}$$

$$= (M - 1)\tfrac{1}{2} e^{-E_s/2N_0} \tag{4.75}$$

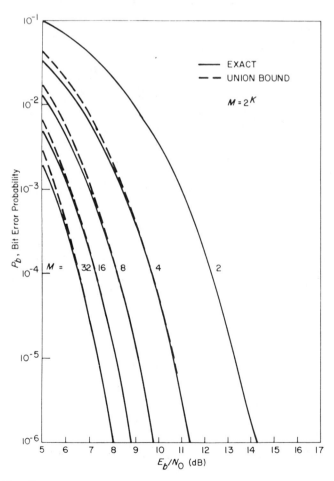

Figure 4.11. Exact versus union bound for uncoded MFSK bit error probability.

and the corresponding uncoded bit error bound using only the union bound is

$$P_b \leq \tfrac{1}{4} M e^{-E_s/2N_0}. \tag{4.76}$$

Figure 4.11 shows this bound and the exact bit error probability for $M = 2$, 4, 8, 16, and 32.

For the special case of no coding, the Chernoff bound was not required since an exact expression for the pairwise error probability is available. For the coded cases, however, the Chernoff bound for the pairwise error probabilities is needed since easy-to-evaluate exact pairwise error probability expressions are difficult to obtain.

4.7.2 Coded

With the use of coding, each coded MFSK symbol has an associated M-ary signal of energy denoted E_c and of duration T_c. With no coding these parameters become E_s and T_s as in the previous section.

The maximum-likelihood decision rule with coding requires a metric that includes a zero-th order Bessel function. This is because in order to have the additivity property, the maximum-likelihood metric is always proportional to the logarithm of the channel conditional probability which in this case has the form given in (4.60). Without coding where only symbol-by-symbol decisions are made, the maximum-likelihood metric can be any monotone function of the channel conditional probability such as (4.62).

For high signal-to-noise ratios the optimum metric can be approximated by (4.62). Although this is not the maximum-likelihood metric, it is the most commonly used metric for coded non-coherent MFSK systems. Volume II, Chapter 2 shows that it is the maximum-likelihood metric when there is Rayleigh fading in the channel. The choice of this metric results in the parameter D given by (4.70) with $E_c = RE_b$ where R is the code rate in bits per coded M-ary symbol.

For the binary alphabet ($M = 2$), the binary convolutional codes found by Odenwalder are given by (4.42) for rate $R = 1/2$ and by (4.43) for rate $R = 1/3$. Thus, for non-coherent BFSK with the metric given by (4.62), we would merely use D given by (4.70) with $E_c = RE_b$. For $M = 4$, Trumpis [12] has found an optimum constraint length $K = 7$ convolutional code of rate $R = 1$ bit per coded 4-ary symbol whose performance is given by the bound

$$P_b \leq \tfrac{1}{2} [7D^7 + 39D^8 + 104D^9 + 352D^{10} + 1187D^{11} + \cdots]. \tag{4.77}$$

Trumpis also found an optimum convolutional code for alphabet $M = 8$ and constraint length $K = 7$ with rate $R = 1$ bit per 8-ary symbol whose performance is given by the bound

$$P_b \leq \tfrac{1}{2} [D^7 + 4D^8 + 8D^9 + 49D^{10} + 92D^{11} + \cdots]. \tag{4.78}$$

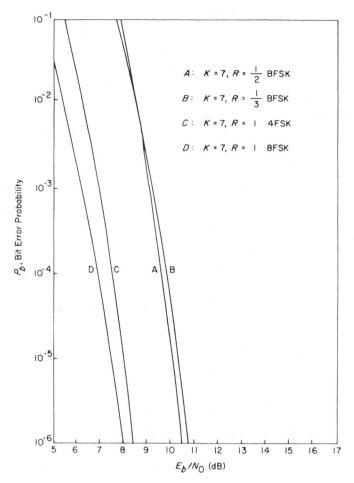

Figure 4.12. Convolutional code bit error bounds—MFSK.

Figure 4.12 shows all these codes for the non-coherent MFSK modulation in the conventional white Gaussian noise channel.

The metric given by (4.62) results in a non-negative real number. Since this metric is not quantized, it is referred to as a soft decision metric analogous to the soft decision metric for coherent BPSK modulation discussed earlier in Section 4.4. Also, the detector of Figure 4.9 is referred to as a soft decision detector since the detector output $y = (e_1, e_2, \ldots, e_M)$ is unquantized. If the detector is forced to make a decision as if there was no coding the result is the hard decision detector with outputs $1, 2, \ldots, M$ where the detector output is the integer i if $e_i > e_j$ for all $j \neq i$.

Essentially the hard decision detector makes the usual uncoded non-coherent MFSK decision and results in a symmetric coding channel shown in Figure 4.13 for $M = 4$ and $M = 8$.

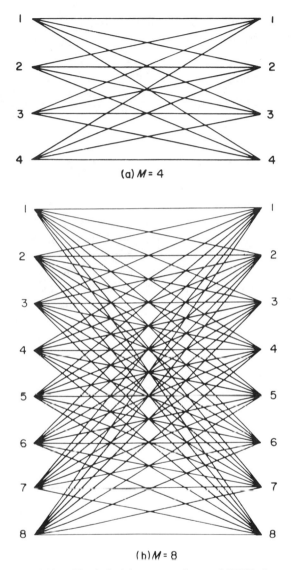

(a) $M = 4$

(b) $M = 8$

Figure 4.13.   Hard decision non-coherent MFSK channels.

If the output of the hard decision detector is given as y, then the hard decision coding channel for fixed M is characterized by

$$p(y|x) = \begin{cases} 1 - P_s, & y = x \\ \dfrac{P_s}{M - 1}, & y \neq x \end{cases} \tag{4.79}$$

where P_s is the symbol error probability given by (4.63). For this hard decision channel the maximum-likelihood metric is given by (4.38) and the parameter $D(\lambda)$ is

$$D(\lambda) = E\left\{ e^{\lambda[m(y,\hat{x})-m(y,x)]}|x \right\}|_{\hat{x} \neq x}$$

$$= (1 - P_s)e^{-\lambda} + \frac{P_s}{M-1}e^{\lambda} + \frac{P_s}{M-1}(M-2). \quad (4.80)$$

Thus,

$$D = \sqrt{\frac{4P_s(1-P_s)}{M-1}} + \left(\frac{M-2}{M-1}\right)P_s. \quad (4.81)$$

Note that for $M = 2$, D becomes identical to (4.40) with ε replaced by P_s; i.e., the crossover probability of the BSC.

For both hard and soft decision detectors the cutoff rate is given by (4.27). Figure 4.14 shows the cutoff rates for both the hard and soft decision detectors with the additive white Gaussian noise channel. The coded bit error bounds for the hard decision detector can be obtained for the convolutional codes discussed above by shifting the bit error bounds shown in Figure 4.12 for the soft decision detectors by the amount of difference between hard and soft decision detectors shown in Figure 4.14 for R_0 versus E_s/N_0. This is the basic motivation for evaluating the easy-to-compute cutoff rate R_0 associated with various channels and detectors.

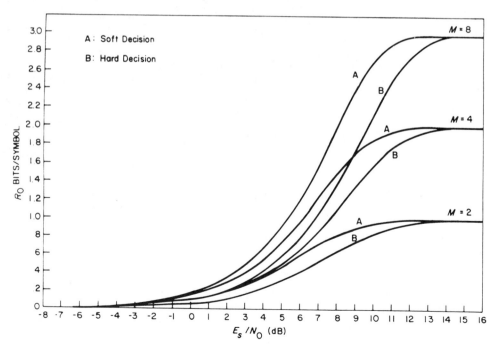

Figure 4.14.   Cutoff rates for MFSK.

4.8 FH/MFSK AND PARTIAL-BAND JAMMING

The previous section also applies to FH/MFSK with broadband noise jamming where $N_0 = N_J$ and E_b/N_J is given by (4.3). Here the case of partial-band noise jamming is examined to illustrate our general coded error bound evaluation.

Assume a jammer with power J that transmits Gaussian noise with constant power spectral density over a total bandwidth W_J. This jammed bandwidth is some subset of the total FH/MFSK signal bandwidth W_{ss} and it is assumed that the transmitter and receiver do not know this jammed bandwidth before transmission. Indeed, the jammed subset of total bandwidth W_J may be changed randomly by the jammer to prevent the transmitter and receiver from knowing what frequencies will be jammed. The fraction of the jammed band is denoted

$$\rho = \frac{W_J}{W_{ss}}. \tag{4.82}$$

Thus, defining $N_J = J/W_{ss}$, the true noise power spectral density is

$$N_J' = \frac{N_J}{\rho} = \frac{J}{W_J} \tag{4.83}$$

for W_J Hz and zero for the rest of the spread bandwidth.

Assuming that an MFSK signal is transmitted during the time interval of T_s seconds, the hop time interval is denoted T_h seconds, then

$$T_h = dT_s \tag{4.84}$$

where d is a fraction or a positive integer. Thus, there are d MFSK symbols transmitted during each hop interval.[3] Assume that each hop is independent of other hops and equally likely to be in any part of the total spread-spectrum signal frequency band of W_{ss} Hz. Thus, the probability of transmitting an FH/MFSK symbol in a jammed frequency band is given by ρ.

Throughout this analysis, also assume that

$$\frac{M}{T_s} \ll W_J \tag{4.85}$$

so that during any hop interval the whole set of M possible tones is either totally in the jammed band or not. This ignores the unlikely cases where the M possible tones straddle the edge of the jammed band leaving only some of the M tone positions with jamming noise present. When (4.85) holds, this edge effect can be ignored.

---

[3]Later in the text, we refer to the case where $d > 1$ as slow frequency hopping (SFH) and $d \le 1$ as fast frequency hopping (FFH).

As always, ideal interleaving is assumed so that the coding channel is memoryless (see Volume I, Chapter 3, Section 3.8). In this FH/MFSK example, it means that each transmitted M-ary symbol is jammed with probability ρ and not jammed with probability $1 - \rho$ independent of other transmitted symbols. As before, define the jammer state random variable Z where

$$P\{Z = 1\} = \rho$$

and

$$P\{Z = 0\} = 1 - \rho. \tag{4.86}$$

This is the jammer state information that may be available at the receiver where $Z = 1$ indicates that the transmitted symbol hopped into the jammed band while $Z = 0$ indicates that it hopped outside the jammed band.

With z available at the receiver, the metric we consider is

$$m(y, x; z) = c(z)e_x \tag{4.87}$$

which is a weighted version of the energy detector output corresponding to input x.

For the above metric the parameter (4.18) has the form

$$D(\lambda) = E\left\{ e^{\lambda c(Z)[e_{\hat{x}} - e_x]}|x \right\}|_{\hat{x} \neq x}$$

$$= \rho E\left\{ e^{\lambda c(1)[e_{\hat{x}} - e_x]}|x, Z = 1 \right\}|_{\hat{x} \neq x}$$

$$+ (1 - \rho)E\left\{ e^{\lambda c(0)[e_{\hat{x}} - e_x]}|Z = 0 \right\}|_{\hat{x} \neq x}. \tag{4.88}$$

From (4.67), (4.68), and (4.69) we have

$$E\left\{ e^{\lambda c(1)[e_{\hat{x}} - e_x]}|x, Z = 1 \right\}|_{\hat{x} \neq x} = \frac{1}{1 - (2\lambda c(1))^2} e^{-2\lambda c(1)\rho(E_s/N_J)/[1 + 2\lambda c(1)]}$$

$$\tag{4.89}$$

for $0 < 2\lambda c(1) < 1$ and

$$E\left\{ e^{\lambda c(0)[e_{\hat{x}} - e_x]}|x, Z = 0 \right\}|_{\hat{x} \neq x} = e^{-2\lambda c(0)\rho(E_s/N_J)}. \tag{4.90}$$

Thus,

$$D(\lambda) = \frac{\rho}{1 - (2\lambda c(1))^2} e^{-2\lambda c(1)\rho(E_s/N_J)/[1 + 2\lambda c(1)]}$$

$$+ (1 - \rho)e^{-2\lambda c(0)\rho(E_s/N_J)}. \tag{4.91}$$

When side information is available, the metric can be chosen with $c(0)$ large enough so that the second term in (4.91) is negligible and $c(1) = 1/2$ chosen for normalization. Then, the parameter D becomes

$$D = \min_{0 \leq \lambda \leq 1} \left\{ \frac{\rho}{1 - \lambda^2} e^{-\lambda\rho(E_s/N_J)/(1 + \lambda)} \right\}. \tag{4.92}$$

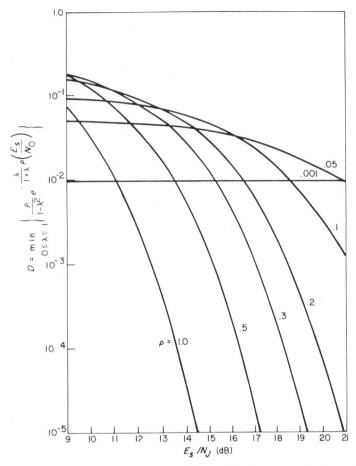

Figure 4.15a. Parameter D for soft decision with jammer state knowledge—FH/MFSK.

With no side information, $c(1) = c(0)$ which can be normalized to $1/2$ to get

$$D = \min_{0 \le \lambda \le 1} \left\{ \frac{\rho}{1 - \lambda^2} e^{-\lambda \rho (E_s/N_J)/(1+\lambda)} + (1 - \rho) e^{-\lambda \rho (E_s/N_J)} \right\}. \quad (4.93)$$

Figures 4.15a and 4.15b show D given by (4.92) and (4.93), respectively, for various values of ρ. Next, in Figures 4.16 through 4.18, the corresponding cutoff rates given by (4.27) for $M = 2$, 4, and 8 are shown. The special case of $\rho = 1.0$ is the broadband noise jamming case discussed in the previous section.[4] There D is given by (4.70) which is the same as (4.92) and

---

[4] The broadband noise jammer case is the same as having an additive white Gaussian noise channel with $N_0 = N_J$.

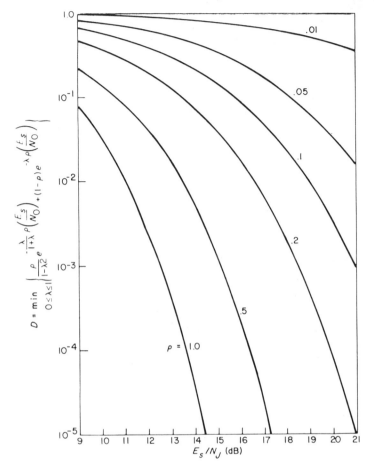

Figure 4.15b. Parameter D for soft decision with no jammer state knowledge—FH/MFSK.

(4.93) for $\rho = 1.0$. Thus, in Figures 4.15 through 4.18, we can compare the broadband noise jammer ($\rho = 1.0$) with various partial-band noise jammers. For any code, the bit error bounds for all these cases can be directly compared using these figures.

As an example, suppose we assume the receiver has no jammer state knowledge and the jammer is a partial-band noise jammer which jams a fraction

$$\rho = .05 \tag{4.94}$$

of the total spread-spectrum band. Assume the conventional non-coherent MFSK metric of (4.62).

With $M = 8$ and using the error bound on the Trumpis code given by (4.78), what is the E_b/N_J given by (4.3) required to achieve 10^{-5} bit error

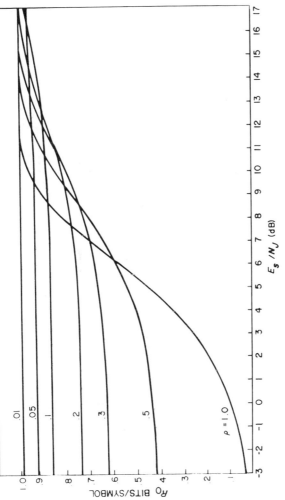

Figure 4.16a. Cutoff rate of FH/BFSK for soft decision with jammer state knowledge.

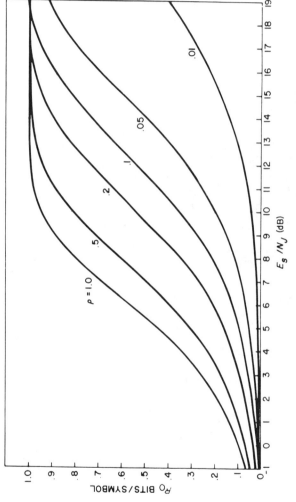

Figure 4.16b. Cutoff rate of FH/BFSK for soft decision with no jammer state knowledge.

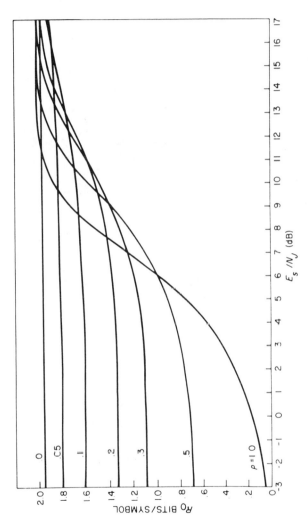

Figure 4.17a. Cutoff rate of FH/4FSK for soft decision with jammer state knowledge.

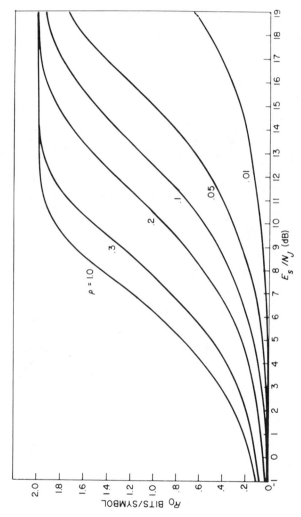

Figure 4.17b. Cutoff rate of FH/4FSK for soft decision with no jammer state knowledge.

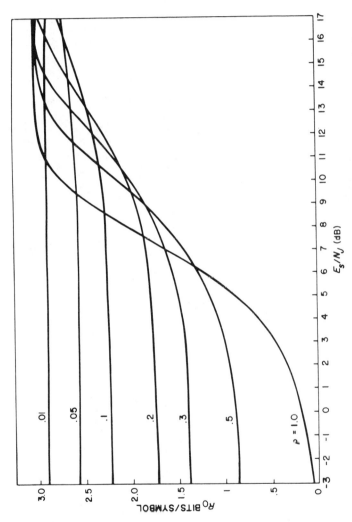

Figure 4.18a. Cutoff rate of FH/8FSK for soft decision with jammer state knowledge.

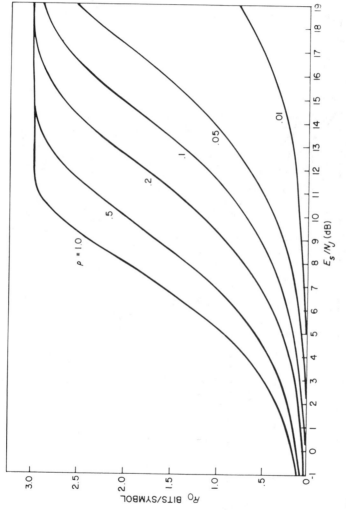

Figure 4.18b. Cutoff rate of FH/8FSK for soft decision with no jammer state knowledge.

probability? From the curve labelled D in Figure 4.12, the required E_b/N_0 for the additive white Gaussian noise channel is

$$E_b/N_0 = 7.5 \text{ dB}. \qquad (4.95)$$

Here, since $R = 1$, $E_s = E_b$ so this is also the value of E_s/N_0, the coded symbol energy-to-noise ratio. Figure 4.14 shows that the cutoff rate is

$$R_0 = 1.8 \text{ bits/symbol}. \qquad (4.96)$$

This is also shown in the $\rho = 1.0$ curve in Figure 4.18b since broadband noise jamming and the additive white Gaussian noise channels are the same. To achieve the same value of R_0 and, thus, the same 10^{-5} bit error probability for $\rho = 0.05$ we require

$$E_s/N_J = E_b/N_J = 16.8 \text{ dB}. \qquad (4.97)$$

This is determined from Figure 4.18b. By repeating this procedure for several values of the bit error probability, the bit error probability curve for the Trumpis code with an additive white Gaussian noise channel can be translated into the corresponding bit error probability curve for a partial-band noise jammer with $\rho = .05$.

4.9 DIVERSITY FOR FH/MFSK

In the FH/MFSK example with partial-band noise jamming for each transmitted M-ary symbol, the channel output signal-to-noise ratio is low with probability ρ and high with probability $1 - \rho$. This is similar to a fading channel except that here there are only two discrete fade levels.[5] Since diversity techniques are useful in fading channels, one would expect that they would be useful against partial-band noise jamming as well. This is indeed true and will be shown next.

Define diversity of order m as the case where each M-ary symbol is transmitted m times over the channels described in the previous sections. Thus, if E_c is the energy for each transmission then each M-ary symbol requires a total energy of

$$E_s = mE_c. \qquad (4.98)$$

The transmitted symbol for each single use of the channel is now referred to as a "chip." Thus, with diversity of order m, each M-ary symbol consists of the transmission of m chips.

Each transmitted chip consists of a frequency-hopped MFSK tone over the partial-band noise jammed channel described in the previous section. If symbol $x = i$ is to be transmitted, then the i-th MFSK tone is transmitted

---

[5]We take the high signal-to-noise ratio to be infinite for one level.

m times where each tone is independently hopped (ideal interleaving assumption is used for slower hop rates) resulting in independent jamming conditions for each chip or transmitted tone. The output corresponding to symbol $x = i$ consists of the m chip outputs

$$y_k = (e_{k1}, e_{k2}, \ldots, e_{kM})$$
$$k = 1, 2, \ldots, m \tag{4.99}$$

with the channel input $x_k = x$ for $k = 1, 2, \ldots, m$ where e_{kj} is the detected energy at the j-th tone frequency during the k-th chip interval or k-th use of the channel. Thus, when symbols $x_k = x$, $k = 1, 2, \ldots, m$ are transmitted the m channel outputs are

$$\mathbf{y} = (y_1, y_2, \ldots, y_m). \tag{4.100}$$

For the conventional additive white Gaussian noise channel, the natural choice for the metric is

$$m(\mathbf{y}, x) = e_{1x} + e_{2x} + \cdots + e_{mx} \tag{4.101}$$

for $x = 1, 2, \ldots, M$. That is, the decision is based on the non-coherent combining (or sum) of the energies of each of the m chips corresponding to each of the MFSK tones. With the partial-band jammer and jammer state knowledge available a more general metric is

$$m(\mathbf{y}, x; \mathbf{z}) = c(z_1)e_{1x} + c(z_2)e_{2x} + \cdots + c(z_m)e_{mx} \tag{4.102}$$

where

$$\mathbf{z} = (z_1, z_2, \ldots, z_m) \tag{4.103}$$

represents the jammer state information for the m uses of the channel.

A new extended channel is obtained where each input symbol results in m uses of the channel considered in previous sections and where the energy per symbol is m times the energy per chip. To emphasize this we use the notation $D(\lambda; m)$ in place of our earlier parameter $D(\lambda)$. That is, define

$$D(\lambda; m) = E\left\{ e^{\lambda[m(\mathbf{y}, \hat{x}; \mathbf{Z}) - m(\mathbf{y}, x; \mathbf{Z})]} | x \right\}\Big|_{\hat{x} \neq x}$$

$$= E\left\{ \prod_{j=1}^{m} e^{\lambda c(Z_j)[e_{j\hat{x}} - e_{jx}]} | x \right\}\Big|_{\hat{x} \neq x}$$

$$= [D(\lambda)]^m \tag{4.104}$$

where $D(\lambda)$ is given by (4.91) where E_c replaces E_s. Next, let $c(0)$ be arbitrarily large, normalize $c(1) = 1/2$ and define

$$D(m) = \min_{0 \leq \lambda \leq 1} D(\lambda; m)$$

$$= D^m \tag{4.105}$$

where D is given by (4.92) with E_c replacing E_s. If jammer state knowledge is not available, then choose $c(0) = c(1) = 1/2$ and D above is given by

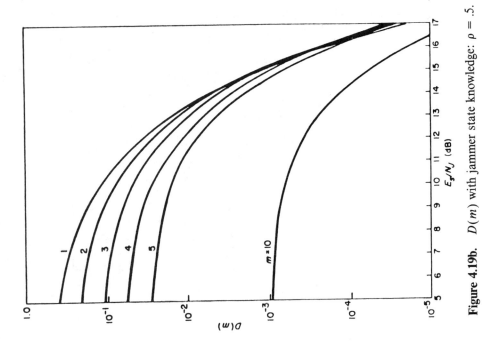

Figure 4.19b. $D(m)$ with jammer state knowledge: $\rho = .5$.

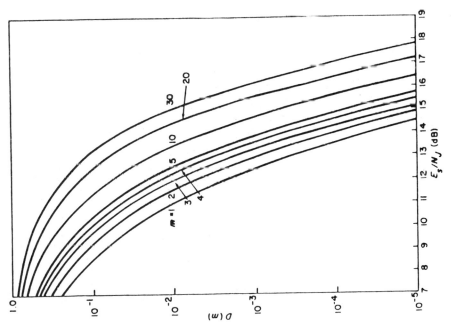

Figure 4.19a. $D(m)$ with jammer state knowledge: $\rho = 1.0$.

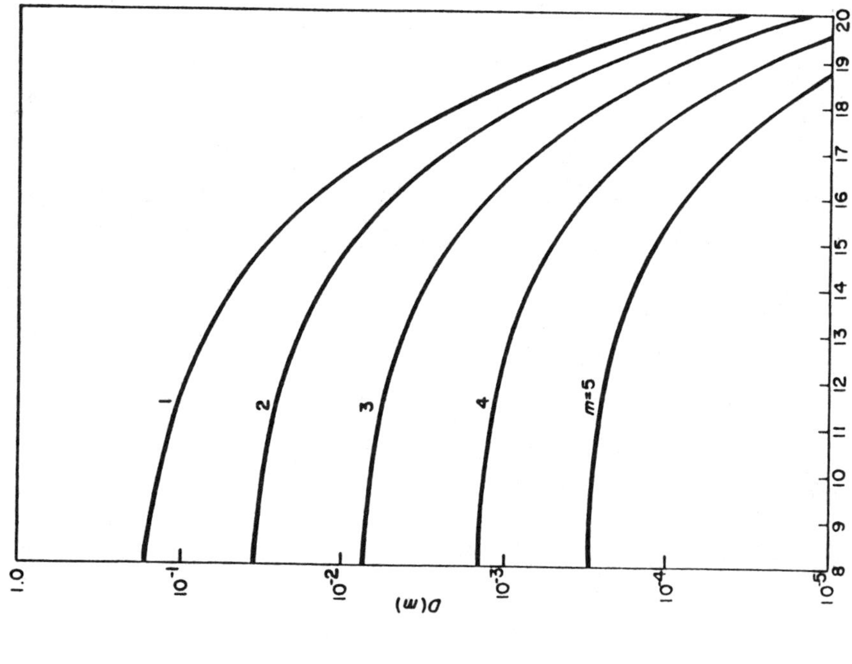

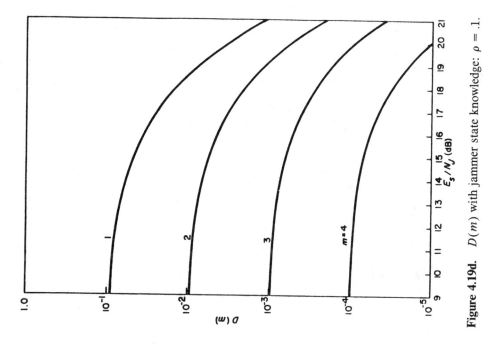

Figure 4.19d. $D(m)$ with jammer state knowledge: $\rho = .1$.

Figure 4.19c. $D(m)$ with jammer state knowledge: $\rho = .2$.

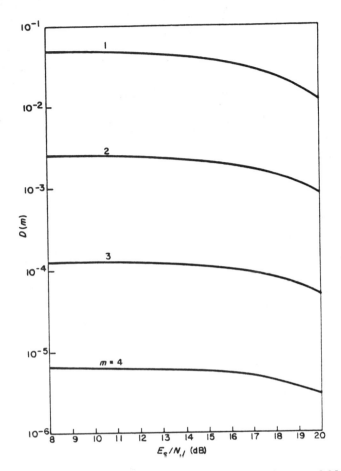

Figure 4.19e. $D(m)$ with jammer state knowledge: $\rho = 0.05$.

(4.93) again with E_c for E_s. Figure 4.19 shows $D(m)$ versus E_s/N_0 for the jammer state knowledge case while Figure 4.20 shows $D(m)$ when there is no jammer state knowledge. The special case of $m = 1$ corresponds to Figure 4.15.

With m diversity, the channel parameter D defined earlier is now replaced by $D(m)$ given in (4.105). The cutoff rate for this extended channel with diversity is denoted by $R_0(m)$ and given by

$$R_0(m) = \log_2 M - \log_2\{1 + (M - 1)D(m)\} \text{ bits/symbol.}$$

$$(4.106)$$

Thus, if an M-ary alphabet code characterized by $G(\cdot)$ or $B(\cdot)$ is used, the coded bit error bound for this code with m diversity of the MFSK signals is given by

$$P_b \leq G(D(m))$$
$$= B(R_0(m)).$$

$$(4.107)$$

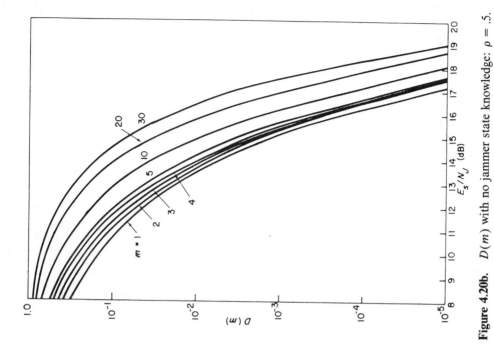

Figure 4.20b. $D(m)$ with no jammer state knowledge: $\rho = .5$.

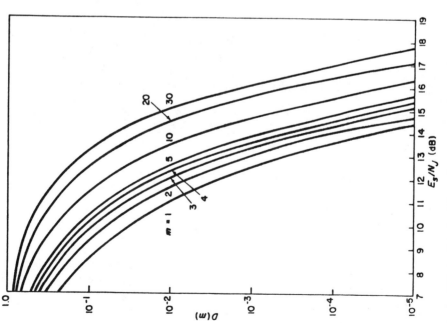

Figure 4.20a. $D(m)$ with no jammer state knowledge: $\rho = 1.0$.

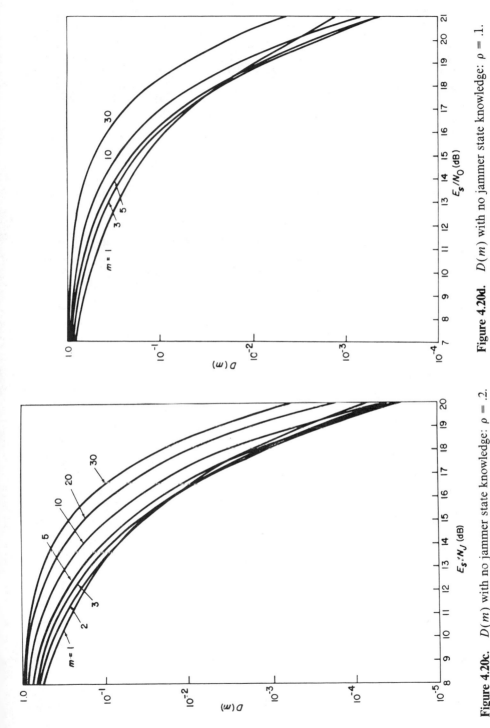

Figure 4.20d. $D(m)$ with no jammer state knowledge: $\rho = .1$.

Figure 4.20c. $D(m)$ with no jammer state knowledge: $\rho = .2$.

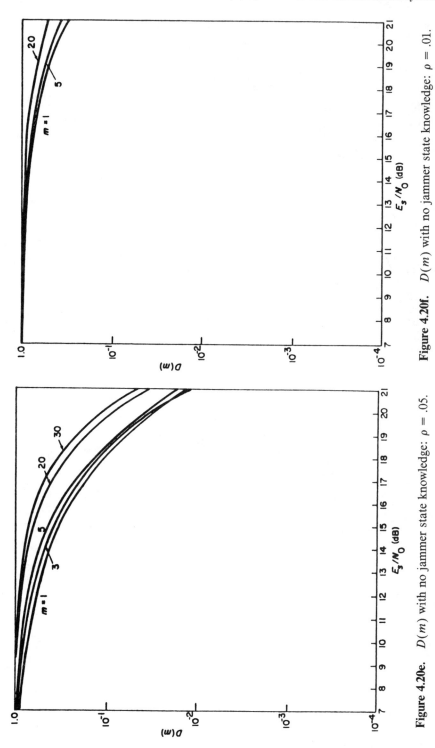

Figure 4.20f. $D(m)$ with no jammer state knowledge: $\rho = .01$.

Figure 4.20e. $D(m)$ with no jammer state knowledge: $\rho = .05$.

The special uncoded case results in the symbol error bound

$$P_s \leq \tfrac{1}{2}(M-1)D(m) \qquad (4.108)$$

and the bit error bound

$$P_b \leq \tfrac{1}{4}MD(m) \qquad (4.109)$$

where the energy per bit E_b is related to the symbol energy E_s and chip energy E_c by

$$E_s = KE_b = mE_c \qquad (4.110)$$

when $M = 2^K$.

Note that diversity of order m can also be viewed as a special case of coding where the code rate is K/m bits per chip. This simple repeat m times code can be quite effective against partial-band jamming as seen in Figure 4.19 where jammer state knowledge is assumed.

4.10 CONCATENATION OF CODES

Although in principle it is possible to achieve arbitrarily small coded bit error probabilities as long as the data rate is less than the channel capacity or cutoff rate, practical applications are limited by the processing complexity and speed required in channel decoding. To achieve small error rates in practice, Forney [13] proposed using a concatenation of codes where two layers of coding are used. This is illustrated in Figure 4.21 where the coding channel is the same as shown enclosed in dotted lines in Figure 4.1. The encoder and decoder of Figure 4.1 now become the "inner" encoder and decoder in Figure 4.21. To create a memoryless super channel interleaving and deinterleaving of the symbols that are encoded and decoded by the inner code is often required.

4.10.1 Binary Super Channel

Typically the inputs to the inner encoder and the inner decoder outputs are binary symbols. With interleaving and deinterleaving, the resulting super channel becomes a memoryless binary symmetric channel (BSC) with crossover probability ϵ. Here ϵ is the coded bit error probability for the inner code with the coding channel. Previously examples were presented where this bit error probability has a bound,

$$\epsilon \leq B(R_0) \qquad (4.111)$$

where R_0 is determined by the coding channel (see Figure 4.1) and given as a function of E_c/N_J where now E_c is the energy of each coded symbol of the inner code. If we denote the energy per super channel binary symbol by E_s then

$$E_c = RE_s \qquad (4.112)$$

where R is the inner code rate in bits per symbol.

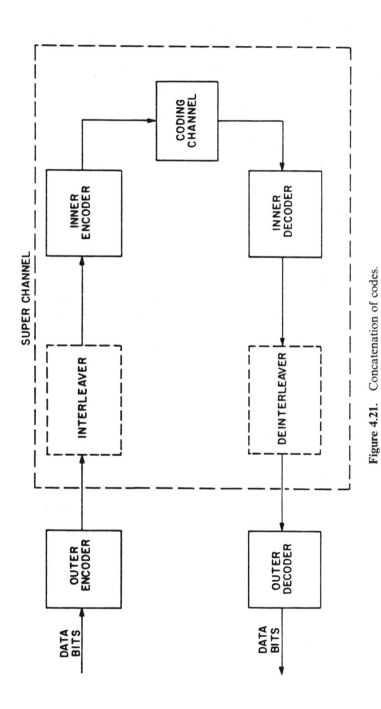

Figure 4.21.　Concatenation of codes.

The BSC super channel such as sketched in Figure 4.21 has a parameter like D which is denoted as D_s. Thus,

$$D_s = \sqrt{4\varepsilon(1 - \varepsilon)} \qquad (4.113)$$

and the cutoff rate is

$$R_s = 1 - \log_2(1 + D_s) \qquad (4.114)$$

which is measured in bits per super channel bit. Next, consider an outer code characterized by $B_s(\cdot)$ with the resulting final bit error bound

$$P_b \leq B_s(R_s). \qquad (4.115)$$

Note that if R_s is the outer code rate then the overall rate measure in data bits per coding channel symbol is $R_s R$ bits/coding channel symbol and the energy per bit is given by

$$E_b = R_s^{-1} E_s$$
$$= (R_s R)^{-1} E_c. \qquad (4.116)$$

As an example, suppose that with 8-ary FSK there is diversity of order $m = 5$. The diversity can be viewed as a special case of a repeat 5 times inner code with the bit error probability which has the bound (see (4.109))

$$\varepsilon \leq \tfrac{1}{4} M D(5) \qquad (4.117)$$

where (see (4.105))

$$D(5) = D^5 \qquad (4.118)$$

with D given by (4.92) or (4.93). Here the data rate is

$$R - \tfrac{3}{5} \text{ bits/coding channel 8 ary symbol.} \qquad (4.119)$$

Suppose the outer code is the optimum rate $R = 1/2$, $K = 7$ binary convolutional code. Then the final bit error bound is (see (4.42))

$$P_b \leq \tfrac{1}{2}\left[36 D_s^{10} + 211 D_s^{12} + 1404 D_s^{14} + 11633 D_s^{16} + \cdots\right] \qquad (4.120)$$

where

$$D_s = \sqrt{4\varepsilon(1 - \varepsilon)} \qquad (4.121)$$

with ε bounded by (4.117). Here the concatenation of codes gives an overall code rate of

$$\tfrac{1}{2} \cdot \tfrac{3}{5} = \tfrac{3}{10} \text{ bits/coded channel 8-ary symbol} \qquad (4.122)$$

with

$$\frac{E_b}{N_0} = \left(\frac{10}{3}\right)\frac{E_c}{N_0} \qquad (4.123)$$

where E_c is the energy per coded 8-ary FSK symbol in the coding channel.

4.10.2 *M*-ary Super Channel

In the previous example, with the inner code being the diversity or repeat 5 times code, the inputs to the inner encoder are 8-ary symbols. Each 8-ary symbol was taken to be equivalent to three bits each in deriving the bound on the BSC super channel crossover probability given by (4.117). The inputs and outputs of the inner encoder and decoder respectively could also be taken to be 8-ary symbols with the symbol error probability bounded by (4.108).

In general, we can have a super channel with M-ary input and output symbols where in all cases of interest these M-ary channels are symmetric with the super channel probabilities of (4.79) where y and x are now M-ary outputs and inputs respectively of the super channel and P_s is the symbol error probability of the inner code. In this case, parameter D_s is given by (4.81) and

$$R_s = \log_2 M - \log_2\left[1 + (M - 1)D_s\right]. \tag{4.124}$$

Thus, for $M = 4$ or $M = 8$ we can have outer codes found by Trumpis with bit error bounds given by (4.77) and (4.78) respectively where the super channel is characterized by D_s and R_s given by (4.81) and (4.124).

Another class of outer codes to consider are the Reed-Solomon codes which are discussed in the next section.

4.10.3 Reed-Solomon Outer Codes

Figure 4.22 shows the most popular form of concatenation of codes. Basically, this approach is to create a super channel as in Figure 4.21, where the super channel begins at the input to a convolutional encoder and ends at the Viterbi decoder output. The convolutional encoder with the Viterbi decoder that forms part of the super channel is the "inner code." The inner code reduces the error probabilities of the first coding channel (see Figure 4.1) which consists of the modulator, radio channel, and demodulator. This then forms a new coding channel for another "outer code" which can further reduce the error rate.

In general, it is difficult to analytically obtain the bit error statistics of the super channel for a specific convolutional code. In the following example, some simulation results assuming an ideal additive white Gaussian noise radio channel [14] are shown. These simulations show that burst lengths out of Viterbi decoders can be modeled by a geometric probability distribution.

Figures 4.23 and 4.24 show how decoded bit errors out of a Viterbi decoder tend to occur in bursts of various lengths with BPSK or QPSK modulation with three bit quantization. Generally, the decoded bits are error free for a while and then when a decoding bit error occurs the errors occur in a burst or string of length L_b with a probability distribution that is geometric [14],

$$\Pr\{L_b = m\} = p(1 - p)^{m-1}; \qquad m = 1, 2, \ldots \tag{4.125}$$

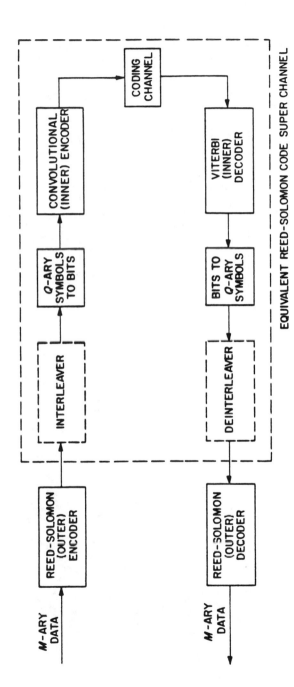

Figure 4.22. Reed-Solomon outer code.

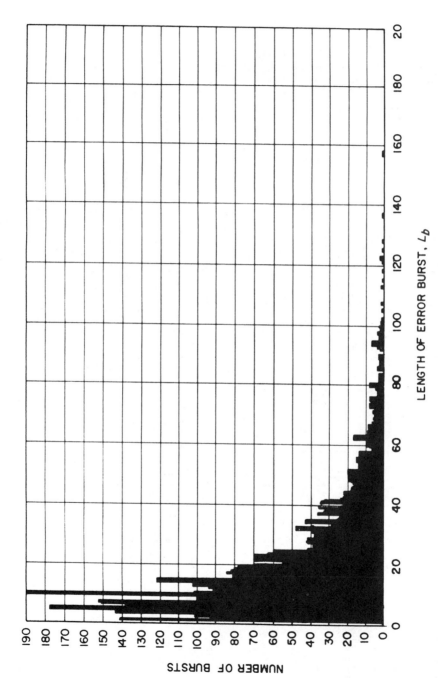

Figure 4.23. Histogram of burst lengths; Viterbi decoded constraint length 7, rate 1/2 convolutional code; $E_b/N_0 = 1.0$ dB (reprinted from [14]).

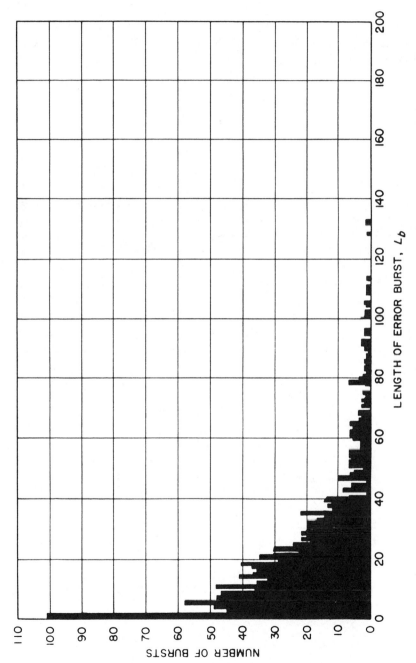

Figure 4.24. Histogram of burst lengths; Viterbi decoded constrain: length 10, rate 1/3 convolutional code; $E_b/N_0 = 0.75$ dB (reprinted from [14]).

where

$$p = \frac{1}{\overline{L}_b} \qquad (4.126)$$

and $\overline{L}_b$ is the average burst length in data bits. The waiting time, W, between bursts has the empirical distribution

$$\Pr\{W = n\} = q(1 - q)^{n-K-1}, \qquad n = K + 1, K + 2, \ldots \quad (4.127)$$

where

$$q = \frac{1}{\overline{W} - K + 2}. \qquad (4.128)$$

In (4.128), $\overline{W}$ is the average waiting time and K is again the convolutional code constraint length.

By choosing a Reed-Solomon (RS) outer code [15], we can take advantage of the bursty nature of the bit errors out of the inner code decoder. RS codes use higher order Q-ary symbols, where typically $Q = 2^m$ for some integer m. In most applications $m = 8$ ($Q = 256$). By taking $m = 8$ bits or 1 byte to form a single Q-ary symbol, a burst of errors in the 8 bits results in only one Q-ary symbol error which tends to reduce the impact of bit error bursts. In Figure 4.22 the conversion from Q-ary symbols to bits and back again for the coding channel of the RS outer code is shown. In addition, to avoid bursts of Q-ary symbol errors interleavers and deinterleavers may be used to provide a memoryless (non-bursty) Q-ary coding channel for the RS code.

As mentioned above, the most commonly employed RS code has $m = 8$ ($Q = 256$) where each symbol is an 8-bit byte and a block length of $N = Q - 1 = 255$ Q-ary symbols. To be able to correct up to $t = 16$ Q-ary symbols the number of data Q-ary symbols must be $K = Q - 1 - 2t = 223$ resulting in a $(255, 223)$ block code using 256-ary symbols that can correct up to $t = 16$ symbols. This is equivalent to a binary $(2040, 1784)$ block code[6] that can correct up to $t = 128$ bits in error as long as these bits are confined to at most 16 Q-ary symbols where $Q = 256$.

Figures 4.25 and 4.26 show the performance of the concatenation system with no interleaving for two convolutional inner codes with Viterbi decoding. Ideal interleaving is assumed in Figure 4.27. These curves show the currently most powerful (non-sequential) coding technique available. Sequential decoding of convolutional codes with large constraint lengths can also achieve similar performance but with the possibility of losing data caused by buffer overflows [1], [2].

For the above examples where a convolutional code with the Viterbi decoder forms the inner code, it is difficult to determine conditional

---

[6]Each 1784 data bits are encoded into a codeword of 2040 coded bits.

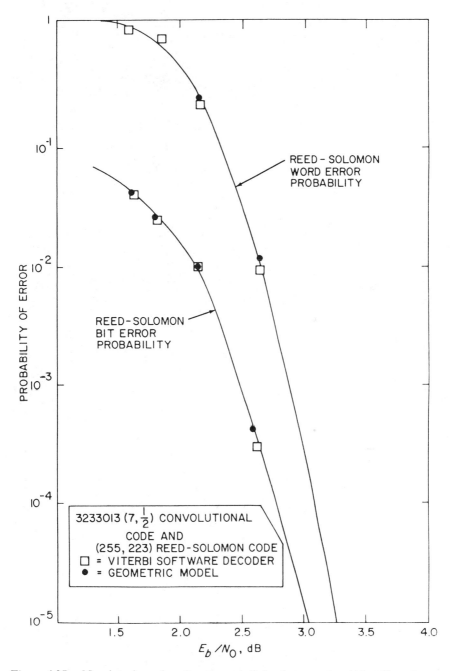

Figure 4.25. Non-interleaved performance statistics for concatenated coding scheme assuming no system losses; (7, 1/2) convolutional code (reprinted from [14]).

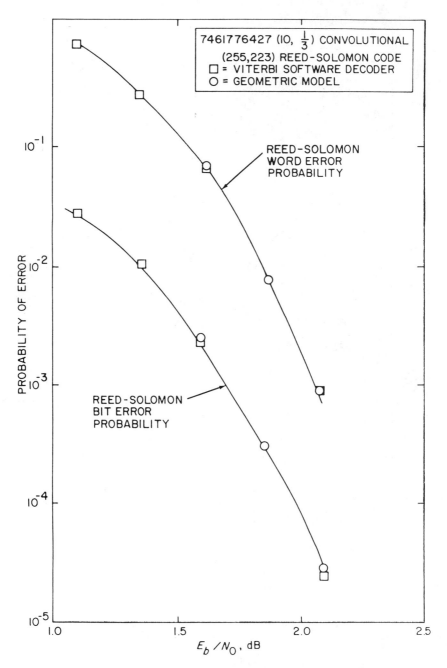

Figure 4.26. Non-interleaved performance statistics for concatenated coding scheme assuming no system losses; $(10, 1/3)$ convolutional code (reprinted from [14]).

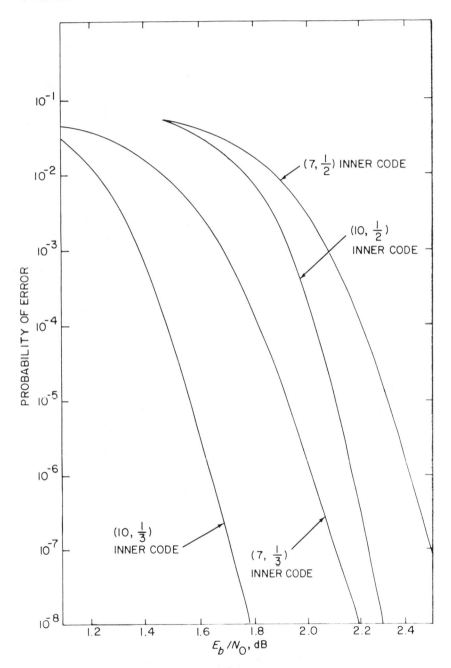

Figure 4.27. Comparison of concatenated channel decoder bit error rates for several convolutional inner codes and a Reed-Solomon $(255, 223)$ outer code with ideal interleaving assuming no system losses (reprinted from [14]).

probabilities of the super channel formed. A natural approximation is to assume a symmetric Q-ary super channel characterized by the single parameter P_s, the symbol error probability. For this model the super channel has conditional probabilities given by (4.79) with parameter D_s given by (4.81) and R_s given by (4.124). This type of symmetric Q-ary super channel occurs for many of the spread-spectrum systems considered earlier. For example, in a symmetric M-ary channel of the type considered earlier, collect L symbols to form Q-ary symbols for the outer channel where

$$M^L = Q. \tag{4.129}$$

For the binary case with $M = 2$, the choice of $L = 8$ gives the $Q = 256$ alphabet for the RS code. Similarly $M = 4$ and $L = 4$ also yields $Q = 256$. For these Q-ary symmetric channels the decoded bit error probability of the RS code P_b is directly a function of the symbol error probability of the inner code denoted P_s. That is,

$$P_b = F(P_s). \tag{4.130}$$

In Appendix 4C, a table showing this relationship for various RS codes is presented.

4.11 SUMMARY OF BIT ERROR BOUNDS

We conclude this chaper with tables of some basic expressions for bit error bounds of direct sequence spread coherent BPSK systems with pulse jamming and frequency-hopped spread non-coherent MFSK systems with partial-band noise jamming. These expressions are based on the discussion in this chapter. There are many other detector types such as detectors with various quantizations, clipping, and rank order lists which have not been included here. The basic analysis approach presented in this chapter, however, can be applied to derive easy to evaluate bit error bounds for these more complex detectors. Some of these will be discussed in Volume II, Chapters 1 and 2.

Recall that regardless of the use of coding the energy per bit-to-jammer noise ratio, E_b/N_J, is defined by (4.3). With time-continuous jamming of DS/BPSK systems, the bit error bound has the same form as for the additive white Gaussian noise channel. Similarly, with broadband noise jamming of FH/MFSK the bit error bound has the same form as for the additive white Gaussian noise channel.

4.11.1 DS/BPSK with Pulse Jamming

Table 4.1 presents the various expressions for the channel parameter D for DS/BPSK systems with pulse jamming. Here ρ is the fraction of time the system is being jammed where continuous jamming is the special case of $\rho = 1$.

Table 4.1.
DS/BPSK with pulse jamming

	Jammer State Information		Metric		$P_b \le G(D)$; $E_c = RE_b$
	Yes	No	Hard	Soft	$R_0 = 1 - \log_2(1 + D)$ bits/symbol
Continuous Jamming	X	X		X	$D = \exp(-E_c/N_J)$
	X	X	X		$D = \sqrt{4\varepsilon(1 - \varepsilon)}$; $\varepsilon = Q\left(\sqrt{2E_c/N_J}\right)$
Pulse Jamming	X			X	$D = \rho \exp(-\rho E_c/N_J)$
	X		X		$D = \rho\sqrt{4\varepsilon(1 - \varepsilon)}$; $\varepsilon = Q\left(\sqrt{2\rho E_c/N_J}\right)$
		X		X	$D = \min\limits_{\lambda \ge 0}\{\exp(-2\lambda E_c)[\rho \exp(\lambda^2 E_c N_J/\rho) + (1 - \rho)]\}$
		X	X		$D = \sqrt{4\varepsilon(1 - \varepsilon)}$; $\varepsilon = \rho Q\left(\sqrt{2\rho E_c/N_J}\right)$

For soft decision detectors the metric has the form

$$m(y, x; z) = c(z)yx \qquad (4.131)$$

while hard decision detectors have the metric

$$m(y, x; z) = \begin{cases} c(z), & y = x \\ 0, & y \ne x. \end{cases} \qquad (4.132)$$

When there is jamming state information

$$c(0) \gg c(1). \qquad (4.133)$$

That is, the unjammed metric values are given much more weight than the jammed metric values. With no jammer state information

$$c(0) = c(1). \qquad (4.134)$$

With continuous jamming ($\rho = 1$), the jammer always exists and all metrics are weighted the same.

4.11.2 FH / MFSK with Partial-Band Noise Jamming

Table 4.2 gives the expressions for parameter D for FH/MFSK systems with partial-band noise jamming where ρ is the fraction of the spread-spectrum band being jammed. Here $\rho = 1$ corresponds to broadband jamming.

The soft decision metric is

$$m(y, x; z) = c(z)e_x \qquad (4.135)$$

while the hard decision metric is

$$m(y, x; z) = \begin{cases} c(z), & y = x \\ 0, & y \ne x. \end{cases} \qquad (4.136)$$

As with DS/BPSK $c(z)$ has the form (4.133) when there is jammer state

Table 4.2.

FH/MFSK with partial-band noise jamming

$$P_b \leq G(D^m); \quad E_c = RE_b/m$$

$$R_0 = \log_2 M - \log_2\{1 + (M-1)D^m\} \text{ bits/symbol}$$

	Jammer State Information		Metric		D
	Yes	No	Hard	Soft	
Broad-band	X	X		X	$D = \min\limits_{0\leq\lambda\leq1}\left\{\dfrac{1}{1-\lambda^2}\exp\left[-\dfrac{\lambda}{1+\lambda}\left(\dfrac{E_c}{N_J}\right)\right]\right\}$
	X	X	X		$D = \sqrt{\dfrac{4\varepsilon(1-\varepsilon)}{M-1}} + \left(\dfrac{M-2}{M-1}\right)\varepsilon; \quad \varepsilon = \sum\limits_{l=1}^{M-1}\binom{M-1}{l}\dfrac{(-1)^{l+1}}{l+1}\exp\left[-\dfrac{l}{l+1}\left(\dfrac{E_c}{N_J}\right)\right]$
Partial-band	X			X	$D = \min\limits_{0\leq\lambda\leq1}\left\{\dfrac{\rho}{1-\lambda^2}\exp\left[-\dfrac{\lambda}{1+\lambda}\rho\left(\dfrac{E_c}{N_J}\right)\right]\right\}$
	X		X		$D = \rho\left[\sqrt{\dfrac{4\varepsilon(1-\varepsilon)}{M-1}} + \left(\dfrac{M-2}{M-1}\right)\varepsilon\right]; \quad \varepsilon = \sum\limits_{l=1}^{M-1}\binom{M-1}{l}\dfrac{(-1)^{l+1}}{l+1}\exp\left[-\dfrac{l}{l+1}\rho\left(\dfrac{E_c}{N_J}\right)\right]$
		X		X	$D = \min\limits_{0\leq\lambda\leq1}\left\{\dfrac{\rho}{1-\lambda^2}\exp\left[-\dfrac{\lambda}{1+\lambda}\rho\left(\dfrac{E_c}{N_J}\right)\right] + (1-\rho)\exp\left[-\lambda\rho\left(\dfrac{E_c}{N_J}\right)\right]\right\}$
		X	X		$D = \sqrt{\dfrac{4\varepsilon(1-\varepsilon)}{M-1}} + \left(\dfrac{M-2}{M-1}\right)\varepsilon; \quad \varepsilon = \rho\sum\limits_{l=1}^{M-1}\binom{M-1}{l}\dfrac{(-1)^{l+1}}{l+1}\exp\left[-\dfrac{l}{l+1}\rho\left(\dfrac{E_c}{N_J}\right)\right]$

Table 4.3.

Code functions.

		M	R	$P_b \leq G(D^m); \ E_c = RE_b/m$
Uncoded		2	1	$G(X) = \frac{1}{2}X$
		M	$\log_2 M$	$G(X) = \frac{1}{4}MX$
$K = 7$ Convolutional Codes		2	$\frac{1}{2}$	$G(X) = \frac{1}{2}[36\,X^{10} + 211\,X^{12} + 1404\,X^{14} + 11633\,X^{16} + \cdots]$
		2	$\frac{1}{3}$	$G(X) = \frac{1}{2}[7X^{15} + 8X^{16} + 22\,X^{17} + 44\,X^{18} + \cdots]$
		4	1	$G(X) = \frac{1}{2}[7X^7 + 39X^8 + 104\,X^9 + 352\,X^{10} + 1187X^{11} + \cdots]$
		8	1	$G(X) = \frac{1}{2}[X^7 + 4X^8 + 8X^9 + 49X^{10} + 92\,X^{11} + \cdots]$

knowledge and (4.134) when there is no jammer state knowledge. Table 4.2 also shows the diversity parameter m.

4.11.3 Coding Functions

Table 4.3 shows some typical coding functions which depend only on the alphabet size M. The $M = 2$ cases apply to both DS/BPSK and FH/BFSK. Diversity $m > 1$ is used only for the FH/MFSK systems. Examples of other codes will be given in Volume II, Chapter 2.

4.12 REFERENCES

[1] R. G. Gallager, *Information Theory and Reliable Communication*, New York: John Wiley, 1968.

[2] A. J. Viterbi and J. K. Omura, *Principles of Digital Communication and Coding*, New York: McGraw-Hill, 1979

[3] G. C. Clark, Jr., and J. B. Cain, *Error-Correction Coding for Digital Communications*, New York: Plenum Press, 1981.

[4] J. M. Wozencraft and R. S. Kennedy, "Modulation and demodulation for probabilistic coding," *IEEE Trans. Inform. Theory*, IT-12, July 1966, pp. 291–297.

[5] J. L. Massey, "Coding and modulation in digital communications," *Proceedings International Zurich Seminar on Digital Communications*, Switzerland, March 12–15, 1974.

[6] A. J. Viterbi, "Spread-spectrum communications—myths and realities," *IEEE Commun. Soc. Mag.*, vol. 17, no. 3, pp. 11–18, May 1979.

[7] L. Biederman, J. K. Omura, and P. C. Jain, "Decoding with approximate channel statistics for bandlimited nonlinear satellite channels," *IEEE Trans. Inform. Theory*, IT-27, pp. 697–708, November 1981.

[8] I. M. Jacobs, "Probability of error bounds for binary transmission on the slowly fading Rician channel," *IEEE Trans. Inform. Theory*, IT-12, pp. 431–441, October 1966.

[9] M. E. Hellman and J. Raviv, "Probability of error, equivocation, and the Chernoff bound," *IEEE Trans. Inform. Theory*, IT-16, pp. 368–372, July 1970.

[10] J. P. Odenwalder, *Optimum Decoding of Convolutional Codes*, Doctoral Disser-

tation, School of Engineering and Applied Science, University of California, Los Angeles, 1970, p. 64.

[11] W. C. Lindsey and M. K. Simon, *Telecommunication Systems Engineering*, Englewood Cliffs, NJ: Prentice-Hall, 1973.

[12] B. Trumpis, *Convolutional Codes for M-ary Channels*, Doctoral Dissertation, School of Engineering and Applied Science, University of California, Los Angeles, 1975.

[13] G. D. Forney, Jr., *Concatenated Codes*, Cambridge, MA: M.I.T. Press, 1966.

[14] L. J. Deutsch, and R. L. Miller, "Burst statistics of Viterbi decoding," TDA Progress Report 42-64, Jet Propulsion Laboratory, Pasadena, California, pp. 187–193, May-June 1981.

[15] R. Blahut, *Theory and Practice of Error Control Codes*, Reading, MA: Addison-Wesley, 1983.

[16] E. R. Berlekamp, "The technology of error-correcting codes," *Proc. IEEE*, vol. 68, pp. 564–593, May 1980.

APPENDIX 4A. CHERNOFF BOUND

Figure 4A.1 shows two functions

$$u(t) = \begin{cases} 1, & t \geq 0 \\ 0, & t < 0 \end{cases} \tag{4A.1}$$

and

$$e_\lambda(t) = e^{\lambda t} \qquad \text{all } t. \tag{4A.2}$$

For any $\lambda \geq 0$ clearly

$$u(t) \leq e_\lambda(t). \tag{4A.3}$$

Next consider a random variable X with probability density function $p_X(x)$ for all x. Then

$$\Pr\{X \geq 0\} = \int_0^\infty p_X(t)\, dt$$

$$= \int_{-\infty}^\infty u(t) p_X(t)\, dt$$

$$\leq \int_{-\infty}^\infty e_\lambda(t) p_X(t)\, dt$$

$$= E\{e^{\lambda X}\}. \tag{4A.4}$$

Hence,

$$\Pr\{X \geq 0\} \leq E\{e^{\lambda X}\} \qquad \text{for any } \lambda \geq 0. \tag{4A.5}$$

This is the Chernoff bound [1].

In the application of the Chernoff bound in this chapter, the random variable X has the form

$$X = \sum_{n=1}^N \left[m(y_n, \hat{x}_n; z_n) - m(y_n, x_n; z_n) \right] \tag{4A.6}$$

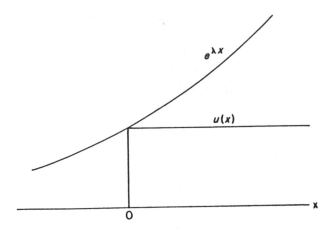

Figure 4A.1. Comparing $u(x)$ and $e^{\lambda x}$.

where the terms in this sum are independent of each other. Thus,

$$\Pr\left\{ \sum_{n=1}^{N} m(y_n, \hat{x}_n; z_n) \geq \sum_{n=1}^{N} m(y_n, x_n; z_n) \right\}$$

$$= \Pr\left\{ \sum_{n=1}^{N} \left[m(y_n, \hat{x}_n; z_n) - m(y_n, x_n; z_n) \right] \geq 0 \right\}$$

$$\leq E\left\{ \exp\left[\lambda \sum_{n=1}^{N} \left[m(y_n, \hat{x}_n; z_n) - m(y_n, x_n; z_n) \right] \right] \right\}$$

$$= \prod_{n=1}^{N} E\left\{ e^{\lambda[m(y_n, \hat{x}_n; z_n) - m(y_n, x_n; z_n)]} \right\}. \qquad (4A.7)$$

This is used as a bound for the pairwise error probability $P(\mathbf{x} \to \hat{\mathbf{x}})$ in Section 4.2.

APPENDIX 4B. FACTOR OF ONE-HALF IN ERROR BOUNDS

Based on the results of Jacobs [8] and Hellman and Raviv [9], general error bounds on the average error probability for cases where there are two hypotheses are derived. Special cases of these bounds result in a factor of one-half times the usual Bhattacharyya bound and the Chernoff bound.

Let Z be a continuous random variable that can have one of two probability densities:

$$H_1 : f_1(z), \quad -\infty < z < \infty$$

$$H_2 : f_2(z), \quad -\infty < z < \infty \qquad (4B.1)$$

where *a priori* probabilities for these two hypotheses are denoted

$$\pi_1 = \Pr\{H_1\} \text{ and } \pi_2 = \Pr\{H_2\}.$$

An arbitrary deterministic decision rule is characterized by the following decision function: given an observed value z of the random variable Z, let

$$u(z) = \begin{cases} 1, & \text{decide } H_1 \\ 0, & \text{decide } H_2. \end{cases} \qquad (4B.2)$$

In terms of this decision function define conditional error probabilities

$$P_{E_1} = \Pr\{\text{decide } H_2 | H_1\}$$

$$= \int_{-\infty}^{\infty} [1 - u(z)] f_1(z)\, dz \qquad (4B.3)$$

and

$$P_{E_2} = \Pr\{\text{decide } H_1 | H_2\}$$

$$= \int_{-\infty}^{\infty} u(z) f_2(z)\, dz. \qquad (4B.4)$$

The average error probability is

$$P_E = \pi_1 P_{E_1} + \pi_2 P_{E_2}$$

$$= \int_{-\infty}^{\infty} \{\pi_1 f_1(z)[1 - u(z)] + \pi_2 f_2(z) u(z)\}\, dz. \qquad (4B.5)$$

Often error probabilities are difficult to evaluate and so easily computed bounds are sometimes used. In the following Bhattacharyya and Chernoff bounds for various decision rules are examined.

Maximum *A Posteriori* (MAP)

The decision rule that minimizes P_E is the MAP rule,

$$u(z) = \begin{cases} 1, & \pi_1 f_1(z) \geq \pi_2 f_2(z) \\ 0, & \pi_1 f_1(z) < \pi_2 f_2(z) \end{cases} \qquad (4B.6)$$

which satisfies the inequalities

$$u(z) \leq \left[\frac{\pi_1 f_1(z)}{\pi_2 f_2(z)} \right]^{\alpha} \qquad (4B.7)$$

$$1 - u(z) \leq \left[\frac{\pi_2 f_2(z)}{\pi_1 f_1(z)} \right]^{\beta} \qquad (4B.8)$$

for any $\alpha \geq 0$, $\beta \geq 0$. These inequalities are typically used to derive the bounds

$$P_{E_1} \leq \int_{-\infty}^{\infty} \left[\frac{\pi_2 f_2(z)}{\pi_1 f_1(z)} \right]^{\beta} f_1(z) \, dz$$

$$= \left(\frac{\pi_2}{\pi_1} \right)^{\beta} \int_{-\infty}^{\infty} f_1(z)^{1-\beta} f_2(z)^{\beta} \, dz \qquad (4B.9)$$

and

$$P_{E_2} \leq \int_{-\infty}^{\infty} \left[\frac{\pi_1 f_1(z)}{\pi_2 f_2(z)} \right]^{\alpha} f_2(z) \, dz$$

$$= \left(\frac{\pi_1}{\pi_2} \right)^{\alpha} \int_{-\infty}^{\infty} f_1(z)^{\alpha} f_2(z)^{1-\alpha} \, dz. \qquad (4B.10)$$

Next, define the function

$$B(\alpha) = \int_{-\infty}^{\infty} f_1(z)^{\alpha} f_2(z)^{1-\alpha} \, dz. \qquad (4B.11)$$

Then an upper bound on the average error probability is

$$P_E \leq \pi_1^{1-\beta} \pi_2^{\beta} B(1 - \beta) + \pi_1^{\alpha} \pi_2^{1-\alpha} B(\alpha) \qquad (4B.12)$$

for any $\alpha \geq 0$, $\beta \geq 0$ where in general α and β are chosen to minimize these bounds. The special case where

$$\alpha = \beta = \tfrac{1}{2}$$

results in the Bhattacharyya bound [2]

$$P_E \leq 2\sqrt{\pi_1 \pi_2} \, B\left(\tfrac{1}{2}\right)$$

$$\leq B\left(\tfrac{1}{2}\right)$$

$$= \int_{-\infty}^{\infty} \sqrt{f_1(z) f_2(z)} \, dz \qquad (4B.13)$$

since

$$\sqrt{\pi_1 \pi_2} \leq \tfrac{1}{2}. \qquad (4B.14)$$

In most cases of interest such as when

$$f_1(z) = f_2(-z) \text{ for all } z \qquad (4B.15)$$

$\alpha = 1/2$ minimizes the function $B(\alpha)$. When $f_1(z)$ and $f_2(z)$ are conditional probabilities derived from a communication channel model, this is usually the case.

Let us now re-examine the general form for the average error probability using the MAP decision rule. Note that

$$P_E = \int_{-\infty}^{\infty} \{ \pi_1 f_1(z)[1 - u(z)] + \pi_2 f_2(z) u(z) \} \, dz$$

$$= \int_{-\infty}^{\infty} \min\{ \pi_1 f_1(z), \pi_2 f_2(z) \} \, dz. \tag{4B.16}$$

Following Hellman and Raviv [9], for any $a \geq 0$, $b \geq 0$ and $0 \leq \alpha \leq 1$

$$\min\{ a, b \} \leq a^\alpha b^{1-\alpha}. \tag{4B.17}$$

This yields the upper bound on the average error probability

$$P_E \leq \int_{-\infty}^{\infty} [\pi_1 f_1(z)]^\alpha [\pi_2 f_2(z)]^{1-\alpha} \, dz$$

$$= \pi_1^\alpha \pi_2^{1-\alpha} B(\alpha). \tag{4B.18}$$

If the minimizing choice of α is in the unit interval $[0, 1]$ then this bound is always a factor of one-half smaller than the bound given in (4B.12). In particular, for the Bhattacharyya bound where $\alpha = 1/2$, this bound (Hellman and Raviv) shows that there is always a factor of one-half,

$$P_E \leq \tfrac{1}{2} \int_{-\infty}^{\infty} \sqrt{f_1(z) f_2(z)} \, dz. \tag{4B.19}$$

Thus, the commonly used Bhattacharyya bound (often used in transfer function bit error bounds for convolutional codes) can be tightened by a factor of one-half.

Maximum Likelihood (ML)

The ML decision rule,

$$u(z) = \begin{cases} 1, & f_1(z) \geq f_2(z) \\ 0, & f_1(z) < f_2(z) \end{cases} \tag{4B.20}$$

tends to keep both conditional probabilities closer in value but only minimizes P_E when $\pi_1 = \pi_2 = 1/2$, the equal *a priori* probability case. In general, inequalities

$$u(z) \leq \left[\frac{f_1(z)}{f_2(z)} \right]^\alpha \tag{4B.21}$$

and

$$[1 - u(z)] \leq \left[\frac{f_2(z)}{f_1(z)} \right]^\beta \tag{4B.22}$$

result in conditional error bounds

$$P_{E_1} \leq B(1 - \beta) \tag{4B.23}$$

and

$$P_{E_2} \leq B(\alpha). \tag{4B.24}$$

The average error probability is simply

$$P_E \leq \pi_1 B(1 - \beta) + \pi_2 B(\alpha). \qquad (4B.25)$$

The choice $\alpha = \beta = 1/2$ which often minimizes this bound yields the usual Bhattacharyya bound

$$P_E \leq B(\tfrac{1}{2}). \qquad (4B.26)$$

Again using the inequality (4B.17) a tighter bound is as follows:

$$P_E = \int_{-\infty}^{\infty} \{\pi_1 f_1(z)[1 - u(z)] + \pi_2 f_2(z)u(z)\} \, dz$$

$$\leq \max\{\pi_1, \pi_2\} \int_{-\infty}^{\infty} \{f_1(z)[1 - u(z)] + f_2(z)u(z)\} \, dz$$

$$= \max\{\pi_1, \pi_2\} \int_{-\infty}^{\infty} \min\{f_1(z), f_2(z)\} \, dz$$

$$\leq \max\{\pi_1, \pi_2\} \int_{-\infty}^{\infty} f_1(z)^{\alpha} f_2(z)^{1-\alpha} \, dz$$

$$= \max\{\pi_1, \pi_2\} B(\alpha) \qquad (4B.27)$$

for $0 \leq \alpha \leq 1$. For the case

$$\pi_1 = \pi_2 = \tfrac{1}{2}$$

and $\alpha = 1/2$, we again have a reduction of the bound by a factor of one-half when compared with (4B.26). Most cases of interest have equal *a priori* probabilities where the MAP and ML decision rules are the same.

Maximum-Likelihood Metric and Chernoff Bounds

Assume that Z is some sort of metric used to make the decision such that when $Z = z$,

$$z \geq 0 \quad \text{choose } H_1$$
$$z < 0 \quad \text{choose } H_2. \qquad (4B.28)$$

The decision function is then

$$u(z) = \begin{cases} 1, & z \geq 0 \\ 0, & z < 0 \end{cases} \qquad (4B.29)$$

and the conditional errors are

$$P_{E_1} = \int_{-\infty}^{\infty} [1 - u(z)] f_1(z) \, dz \qquad (4B.30)$$

and

$$P_{E_2} = \int_{-\infty}^{\infty} u(z) f_2(z) \, dz. \qquad (4B.31)$$

For any $\alpha \geq 0$ and any $\beta \geq 0$

$$1 - u(z) \leq e^{-\alpha z} \qquad (4B.32)$$

and

$$u(z) \leq e^{\beta z} \qquad (4B.33)$$

resulting in the Chernoff bounds

$$P_{E_1} \leq \int_{-\infty}^{\infty} e^{-\alpha z} f_1(z)\, dz \qquad (4\text{B}.34)$$

and

$$P_{E_2} \leq \int_{-\infty}^{\infty} e^{\beta z} f_2(z)\, dz. \qquad (4\text{B}.35)$$

Thus, the averaged error probability is

$$P_E \leq \pi_1 C_1(\alpha) + \pi_2 C_2(\beta) \qquad (4\text{B}.36)$$

where

$$C_1(\alpha) = \int_{-\infty}^{\infty} e^{-\alpha z} f_1(z)\, dz \qquad (4\text{B}.37)$$

and

$$C_2(\beta) = \int_{-\infty}^{\infty} e^{\beta z} f_2(z)\, dz. \qquad (4\text{B}.38)$$

Jacobs [8] considered the conditions

$$f_1(-z) \geq f_1(z) \quad \text{all } z \leq 0 \qquad (4\text{B}.39\text{a})$$

and

$$f_2(-z) \geq f_2(z) \quad \text{all } z \geq 0. \qquad (4\text{B}.39\text{b})$$

Then using the inequality

$$\frac{e^{\omega} + e^{-\omega}}{2} = \cosh \omega$$

$$\geq 1 \quad \text{all } \omega \qquad (4\text{B}.40)$$

and a change of variables of integration, he derived the following inequalities:

$$
\begin{aligned}
C_1(\alpha) &= \int_{-\infty}^{\infty} e^{-\alpha z} f_1(z)\, dz \\
&= \int_{-\infty}^{0} e^{-\alpha z} f_1(z)\, dz + \int_{0}^{\infty} e^{-\alpha z} f_1(z)\, dz \\
&= \int_{-\infty}^{0} e^{-\alpha z} f_1(z)\, dz + \int_{-\infty}^{0} e^{\alpha z} f_1(-z)\, dz \\
&\geq \int_{-\infty}^{0} e^{-\alpha z} f_1(z)\, dz + \int_{-\infty}^{0} e^{\alpha z} f_1(z)\, dz \\
&= 2 \int_{-\infty}^{0} \cosh \alpha z \cdot f_1(z)\, dz \\
&\geq 2 \int_{-\infty}^{0} f_1(z)\, dz \qquad (4\text{B}.41)
\end{aligned}
$$

or

$$P_{E_1} \leq \tfrac{1}{2} C_1(\alpha) \qquad (4\text{B}.42)$$

and similarly

$$P_{E_2} \leq \tfrac{1}{2} C_2(\beta). \qquad (4\text{B}.43)$$

Thus, the often satisfied conditions given by Jacobs in (4B.39) result in a factor of one-half in the usual Chernoff bounds.

Less restrictive but more difficult to prove conditions are that

$$\int_0^\infty e^{-\alpha^* z} f_1(z) \, dz \geq \int_{-\infty}^0 e^{\alpha^* z} f_1(z) \, dz \qquad (4B.44a)$$

and

$$\int_{-\infty}^0 e^{\beta^* z} f_2(z) \, dz \geq \int_0^\infty e^{-\beta^* z} f_2(z) \, dz \qquad (4B.44b)$$

where α^* minimizes $C_1(\alpha)$ and β^* minimizes $C_2(\beta)$. Note that for the special case of $\alpha^* = 0$

$$\int_0^\infty f_1(z) \, dz \geq \int_{-\infty}^0 f_1(z) \, dz$$

$$= P_{E_1} \qquad (4B.45)$$

which is always satisfied when

$$P_{E_1} < \tfrac{1}{2}. \qquad (4B.46)$$

Indeed, conditions (4B.44) are also true for some non-negative range of α values. We assume it is true for the minimizing choices of the Chernoff bound parameters. Note that conditions (4B.44) are less restrictive than those of (4B.39) since (4B.39) implies (4B.44). Now consider the inequalities,

$$C_1(\alpha) \geq C_1(\alpha^*)$$

$$= \int_{-\infty}^\infty e^{-\alpha^* z} f_1(z) \, dz$$

$$= \int_{-\infty}^0 e^{-\alpha^* z} f_1(z) \, dz + \int_0^\infty e^{-\alpha^* z} f_1(z) \, dz$$

$$\geq \int_{-\infty}^0 e^{-\alpha^* z} f_1(z) \, dz + \int_{-\infty}^0 e^{\alpha^* z} f_1(z) \, dz$$

$$= 2 \int_{-\infty}^0 \cosh \alpha^* z \cdot f_1(z) \, dz$$

$$\geq 2 \int_{-\infty}^0 f_1(z) \, dz$$

$$= 2 P_{E_1} \qquad (4B.47)$$

or

$$P_{E_1} \leq \tfrac{1}{2} C_1(\alpha) \qquad (4B.48)$$

and similarly

$$P_{E_2} \leq \tfrac{1}{2} C_2(\beta). \qquad (4B.49)$$

Next, for the special case where

$$\pi_1 = \pi_2 = \tfrac{1}{2} \qquad (4B.50)$$

and

$$\alpha^* = \beta^* \tag{4B.51}$$

sufficient conditions can be given by

$$\int_0^\infty e^{-\alpha^* z} f_1(z)\, dz \geq \int_0^\infty e^{-\alpha^* z} f_2(z)\, dz \tag{4B.52a}$$

and

$$\int_{-\infty}^0 e^{\alpha^* z} f_2(z)\, dz \geq \int_{-\infty}^0 e^{\alpha^* z} f_1(z)\, dz. \tag{4B.52b}$$

Note that these conditions are always satisfied if our decision rule is a maximum-likelihood decision rule where

$$f_2(z) \leq f_1(z) \quad \text{for all } z \geq 0 \tag{4B.53a}$$

and

$$f_2(z) > f_1(z) \quad \text{for all } z < 0. \tag{4B.53b}$$

Assuming conditions (4B.52)

$$
\begin{aligned}
C_1(\alpha) + C_2(\beta) &\geq C_1(\alpha^*) + C_2(\alpha^*) \\
&= \int_{-\infty}^\infty e^{-\alpha^* z} f_1(z)\, dz + \int_{-\infty}^\infty e^{\alpha^* z} f_2(z)\, dz \\
&= \int_{-\infty}^0 e^{-\alpha^* z} f_1(z)\, dz + \int_0^\infty e^{-\alpha^* z} f_1(z)\, dz \\
&\quad + \int_{-\infty}^0 e^{\alpha^* z} f_2(z)\, dz + \int_0^\infty e^{\alpha^* z} f_2(z)\, dz \\
&\geq \int_{-\infty}^0 e^{-\alpha^* z} f_1(z)\, dz + \int_0^\infty e^{-\alpha^* z} f_2(z)\, dz \\
&\quad + \int_{-\infty}^0 e^{\alpha^* z} f_1(z)\, dz + \int_0^\infty e^{\alpha^* z} f_2(z)\, dz \\
&= 2\int_{-\infty}^0 \cosh\alpha^* z \cdot f_1(z)\, dz \\
&\quad + 2\int_0^\infty \cosh\alpha^* z \cdot f_2(z)\, dz \\
&\geq 2P_{E_1} + 2P_{E_2}
\end{aligned}
\tag{4B.54}
$$

or

$$
\begin{aligned}
P_E &= \tfrac{1}{2}P_{E_1} + \tfrac{1}{2}P_{E_2} \\
&\leq \tfrac{1}{4}C_1(\alpha) + \tfrac{1}{4}C_2(\beta),
\end{aligned}
\tag{4B.55}
$$

which is again a factor of one-half less than the original Chernoff bound on the average error probability (4B.36) for $\pi_1 = \pi_2 = 1/2$.

For the special case where Z happens to be a maximum-likelihood metric

$$Z = \ln\left[\frac{f_1(Z)}{f_2(Z)}\right], \tag{4B.56}$$

then the conditions (4B.52) hold and

$$\begin{aligned}
C_1(\alpha) &= \int_{-\infty}^{\infty} e^{-\alpha z} f_1(z)\, dz \\
&= \int_{-\infty}^{\infty} \left[\frac{f_2(z)}{f_1(z)}\right]^{\alpha} f_1(z)\, dz \\
&= \int_{-\infty}^{\infty} f_1(z)^{1-\alpha} f_2(z)^{\alpha}\, dz \\
&= B(1-\alpha) \tag{4B.57}
\end{aligned}$$

and

$$\begin{aligned}
C_2(\beta) &= \int_{-\infty}^{\infty} e^{\beta z} f_2(z)\, dz \\
&= \int_{-\infty}^{\infty} \left[\frac{f_1(z)}{f_2(z)}\right]^{\beta} f_2(z)\, dz \\
&= \int_{-\infty}^{\infty} f_1(z)^{\beta} f_2(z)^{1-\beta}\, dz \\
&= B(\beta) \tag{4B.58}
\end{aligned}$$

where $B(\cdot)$ is given by (4B.11). Recall that $B(1/2)$ is the Bhattacharyya bound.

Applications

In most applications of interest, we consider two sequences of length N,

$$\underline{x}_1, \underline{x}_2 \in X^N \tag{4B.59}$$

that can be transmitted over a memoryless channel with input alphabet X and output alphabet Y and conditional probability

$$p(y|x), \quad x \in X, \quad y \in Y. \tag{4B.60}$$

This is shown in the following figure.

$$p_N(\mathbf{y}|\mathbf{x}) = \prod_{n=1}^{N} p(y_n|x_n).$$

The receiver obtains a sequence

$$\mathbf{y} \in Y^N \tag{4B.61}$$

from the channel and must decide between hypotheses

$$H_1 : \mathbf{x}_1 \text{ is sent}$$
$$H_2 : \mathbf{x}_2 \text{ is sent} \tag{4B.62}$$

which have *a priori* probabilities

$$\pi_1 = \Pr\{H_1\} \quad \text{and} \quad \pi_2 = \Pr\{H_2\}.$$

The receiver will typically use a metric

$$m(y, x) \qquad x \in X, \quad y \in Y \tag{4B.63}$$

and the decision rule of choosing H_1 if and only if

$$\sum_{n=1}^{N} m(y_n, x_{1n}) \geq \sum_{n=1}^{N} m(y_n, x_{2n}). \tag{4B.64}$$

By defining

$$Z = \sum_{n=1}^{N} \left[m(y_n, x_{1n}) - m(y_n, x_{2n}) \right] \tag{4B.65}$$

we have the case considered in this chapter.

APPENDIX 4C. REED-SOLOMON CODE PERFORMANCE

A Reed-Solomon (RS) code [15] is a block code with an alphabet size of $Q = 2^m$. Its block length is $N = Q - 1$ symbols which can be extended to $N = Q$ and $N = Q + 1$. Here, the performance of only the $N = Q$ block length RS codes is presented since adjacent block lengths are only slightly different.

If an RS code has r redundant symbols, then the minimum Hamming distance between codewords is

$$d = r + 1 \tag{4C.1}$$

and the code is able to correct any pattern of t symbol errors and s symbol erasures for which

$$2t + s < d. \tag{4C.2}$$

Thus, this code can correct up to $t_0 = r/2$ errors when there are no erasures and up to $s_0 = r$ erasures when there are no errors.

Suppose we have a symmetric Q-ary memoryless channel with conditional probabilities

$$p(y|x) = \begin{cases} 1 - P_s, & y = x \\ \dfrac{P_s}{N-1}, & y \neq x \end{cases} \tag{4C.3}$$

where P_s is the uncoded symbol error probability. The decoded bit error probability is given approximately by

$$P_b = \sum_{i=t_0+1}^{N} \frac{i}{2(N-1)} \binom{N}{i} P_s^i (1 - P_s)^{N-i} \qquad (4C.4)$$

where t_0 is the number of symbol errors that can be corrected. Table 4C.1 shows values of the uncoded symbol error probabilities needed for each RS code of block length $N = Q = 2^m$ and parameter $t_0 = r/2$ to achieve decoded bit error rates of 10^{-3}, 10^{-5}, 10^{-7}, 10^{-9}, and 10^{-11}. (The last row for each N, however, does not apply here.) Consider, for example, the RS code whose symbols are eight-bit bytes ($m = 8$), whose block length is $N = 256$ symbols and whose redundancy is $r = 32$. This code can correct all symbol error patterns of up to $t_0 = 16$ errors. Suppose this code must attain a decoded bit error rate of $P_b = 10^{-5}$, then from Table 4C.1 the maximum uncoded symbol error probability is $P_s = 0.025$.

Next consider a memoryless erasure channel with conditional probabilities

$$p(y|x) = \begin{cases} 1 - P_E, & y = x \\ P_E, & y = \text{erasure} \end{cases} \qquad (4C.5)$$

where P_E is the probability a symbol is erased. The decoded bit error probability is

$$P_b = \sum_{i=s_0+1}^{N} \frac{i}{2(N-1)} \binom{N}{i} P_E^i (1 - P_E)^{N-i} \qquad (4C.6)$$

where s_0 is the number of erasures that can be corrected. Table 4C.1 also shows values of the erasure probabilities needed for each RS code of blocklength $N = Q = 2^m$ and parameter $s_0 = r$ to achieve decoded bit error rates of 10^{-3}, 10^{-5}, 10^{-7}, 10^{-9}, and 10^{-11}. Again, for the example with $N = 256$ and $r = 32$, up to $s_0 = 32$ erasures can be corrected. For a required decoded bit error rate of $P_b = 10^{-5}$, from Table 4C.1 the maximum erasure probability is $P_E = 0.0651$.

Table 4C.1 is provided courtesy of Dr. Elwyn Berlekamp [16].

Table 4C.1.
Required Probabilities [16]

	$P_b = 10^{-3}$	$P_b = 10^{-5}$	$P_b = 10^{-7}$	$P_b = 10^{-9}$	$P_b = 10^{-11}$
			$N = 8$		
$t_0, s_0 = 2$	4.60E-2	9.51E-3	2.03E-3	4.37E-4	9.41E-5
$= 4$	1.49E-1	5.65E-2	2.21E-2	8.74E-3	3.47E-3
			$N = 16$		
$t_0, s_0 = 2$	2.84E-2	5.73E-3	1.22E-3	2.62E-4	5.63E-5
$= 4$	7.72E-2	2.81E-2	1.09E-2	4.27E-3	1.69E-3
$= 8$	2.23E-1	1.23E-1	7.10E-2	4.17E-2	2.47E-2

Table 4C.1.
Continued

	$p_b = 10^{-3}$	$p_b = 10^{-5}$	$p_b = 10^{-7}$	$p_b = 10^{-9}$	$p_b = 10^{-11}$
			$N = 32$		
$t_0, s_0 = 2$	1.80E-2	3.55E-3	7.51E-4	1.61E-4	3.47E-5
$= 4$	4.36E-2	1.54E-2	5.87E-3	2.30E-3	9.11E-4
$= 8$	1.11E-1	5.88E-2	3.32E-2	1.93E-2	1.13E-2
$= 16$	2.90E-1	2.01E-1	1.45E-1	1.07E-1	7.96E-2
			$N = 64$		
$t_0, s_0 = 2$	1.17E-2	2.23E-3	4.68E-4	1.00E-4	2.16E-5
$= 4$	2.55E-2	8.70E-3	3.29E-3	1.29E-3	5.08E-4
$= 8$	5.96E-2	3.05E-2	1.70E-2	9.78E-3	5.74E-3
$= 16$	1.43E-1	9.50E-2	6.71E-2	4.87E-2	3.59E-2
$= 32$	3.45E-1	2.74E-1	2.25E-1	1.88E-1	1.58E-1
			$N = 128$		
$t_0, s_0 = 2$	7.72E-3	1.41E-3	2.94E-4	6.29E-5	1.35E-5
$= 4$	1.52E-2	5.00E-3	1.87E-3	7.29E-4	2.88E-4
$= 8$	3.30E-2	1.63E-2	8.99E-3	5.15E-3	3.01E-3
$= 16$	7.44E-2	4.81E-2	3.35E-2	2.41E-2	1.76E-2
$= 32$	1.69E-1	1.30E-1	1.05E-1	8.59E-2	7.16E-2
$= 64$	3.88E-1	3.34E-1	2.95E-1	2.64E-1	2.38E-1
			$N = 256$		
$t_0, s_0 = 2$	5.20E-3	8.94E-4	1.85E-4	3.96E-5	8.51E-6
$= 4$	9.29E-3	2.90E-3	1.08E-3	4.17E-4	1.64E-4
$= 8$	1.86E-2	8.89E-3	4.83E-3	2.76E-3	1.61E-3
$= 16$	3.98E-2	2.50E-2	1.72E-2	1.23E-2	8.97E-3
$= 32$	8.68E-2	6.51E-2	5.17E-2	4.21E-2	3.49E-2
$= 64$	1.91E-1	1.60E-1	1.39E-1	1.23E-1	1.10E-1
			$N = 512$		
$t_0, s_0 = 2$	3.62E-3	5.70E-4	1.17E-4	2.49E-5	5.35E-6
$= 4$	5.80E-3	1.70E-3	6.20E-4	2.39E-4	9.41E-5
$= 8$	1.07E-2	4.88E-3	2.62E-3	1.49E-3	8.64E-4
$= 16$	2.16E-2	1.32E-2	8.97E-3	6.37E-3	4.64E-3
$= 32$	4.54E-2	3.33E-2	2.62E-2	2.12E-2	1.75E-2
$= 64$	9.68E-2	7.97E-2	6.85E-2	6.01E-2	5.34E-2
			$N = 1024$		
$t_0, s_0 = 2$	2.65E-3	3.66E-4	7.38E-5	1.57E-5	3.37E-6
$= 4$	3.75E-3	9.95E-4	3.58E-4	1.38E-4	5.40E-5
$= 8$	6.29E-3	2.70E-3	1.43E-3	8.05E-4	4.67E-4
$= 16$	1.19E-2	6.99E-3	4.71E-3	3.33E-3	2.41E-3
$= 32$	2.40E-2	1.72E-2	1.34E-2	1.08E-2	8.90E-3
$= 64$	4.99E-2	4.04E-2	3.45E-2	3.01E-2	2.67E-2

Table 4C.1.
Continued

	$p_b = 10^{-3}$	$p_b = 10^{-5}$	$p_b = 10^{-7}$	$p_b = 10^{-9}$	$p_b = 10^{-11}$
		$N = 2048$			
$t_0, s_0 = 2$	2.14E-3	2.36E-4	4.67E-5	9.89E-6	2.12E-6
$= 4$	2.58E-3	5.88E-4	2.07E-4	7.92E-5	3.10E-5
$= 8$	3.82E-3	1.50E-3	7.81E-4	4.37E-4	2.53E-4
$= 16$	6.68E-3	3.73E-3	2.48E-3	1.74E-3	1.26E-3
$= 32$	1.29E-2	8.95E-3	6.91E-3	5.55E-3	4.55E-3
$= 64$	2.59E-2	2.06E-2	1.75E-2	1.52E-2	1.34E-2
		$N = 4096$			
$t_0, s_0 = 2$	2.00E-3	1.53E-4	2.96E-5	6.24E-6	1.34E-6
$= 4$	2.06E-3	3.49E-4	1.20E-4	4.57E-5	1.78E-5
$= 8$	2.50E-3	8.36E-4	4.28E-4	2.38E-4	1.37E-4
$= 16$	3.87E-3	2.00E-3	1.31E-3	9.17E-4	6.60E-4
$= 32$	6.99E-3	4.67E-3	3.57E-3	2.86E-3	2.34E-3
$= 64$	1.36E-2	1.06E-2	8.92E-3	7.73E-3	6.81E-3

Chapter 5

PSEUDONOISE GENERATORS

5.1 THE STORAGE / GENERATION PROBLEM

Pseudonoise (PN) sequences are used as spectrum-spreading modulations for direct sequence SS designs, as hopping pattern sources in frequency and/or time hopping systems, and as filter section controllers in matched filter SS systems. Ideally, most of the PN sequence design problems would be solved if it were possible to produce a sample of a sequence of independent random variables, uniformly distributed on the available alphabet, for use at the SS transmitter, and an identical sample sequence at the receiver for use in the detection process. This is the SS equivalent of the one-time pad used in cryptographic systems requiring the highest level of security. Unfortunately, the generation, recording, and distribution of these sample sequences at a rate (per information bit to be communicated) equal to the processing gain of the SS system, is generally not feasible.

Once the one-time pad is discarded as a viable approach, the designer must come to grips with the problem of storing/generating a signal which looks random (e.g., like a one-time pad), despite the fact that it must be done with a real system possessing finite storage capacity/generating capability. Let's illustrate the possibilities by considering methods for the generation of a specified binary (0 or 1) sequence $\{b_n, \ n = 1, 2, \ldots, N\}$. If N bits of the random sequence are not sufficient to complete the communication process, the sequence must be reused, i.e.,

$$b_{n+N} = b_n \quad \text{for all } n, \tag{5.1}$$

where N is the period of the extended sequence. Any other method of reusing the stored sequence other than periodic extension as indicated above, will require additional memory to supervise the reuse algorithm.

Two approaches, in which the complete sequence is stored simultaneously within the generator, are shown in Figure 5.1. One is ROM-based, the other a simple cyclic shift register. Both have the property that they work equally well with any desired sequence $\{b_n\}$. Obviously, the price for this versatility is memory: N binary storage elements and supporting hardware are needed.

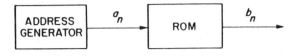

(a) ROM-Based Generator

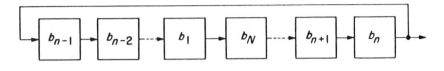

(b) Cyclic-Shift-Register Generator

Figure 5.1.   Sequence generators based on complete sequence storage.

The storage requirements indicated by these designs can be reduced by the insertion of Boolean functions at the proper places. The ROM in Figure 5.1(a) can be replaced by a Boolean function which calculates b_n directly from the address a_n. Similarly, all but M of the binary memory elements of the cyclic shift register in Figure 5.1(b) can be eliminated if a Boolean function can be mechanized which, for all n, computes b_{n+M} from the stored values b_{n+j}, $j = 0, 1, \ldots, M - 1$. The following two examples illustrate these procedures.

Example 5.1. A ROM-based generator for the sequence

$$1010110010110010$$

having period 16 can be specified as follows: For simplicity assume the address generator is a 4-bit binary counter producing consecutive addresses in the range 0 to 15, that is,

$$n = a_n = a_n^{(0)} + a_n^{(1)} \cdot 2 + a_n^{(2)} \cdot 2^2 + a_n^{(3)} \cdot 2^3$$
$$= \left(a_n^{(0)} a_n^{(1)} a_n^{(2)} a_n^{(3)} \right)_2 \tag{5.2}$$

where $(\ldots)_2$ is the binary representation of n. Table 5.1 identifies the mapping which has to be accomplished by the ROM or its replacement function. One possible implementation of this Boolean function is

$$b_n = a_n^{(0)} + a_n^{(1)} \cdot a_n^{(2)} + a_n^{(2)} \cdot a_n^{(3)} \tag{5.3}$$

where $+$ and $\cdot$ represent arithmetic modulo 2, $+$ being the EXCLUSIVE OR and $\cdot$ being the AND logic functions. The resulting PN generator is shown in Figure 5.2.

Table 5.1
Address-to-bit mapping

n	$a_n^{(0)}$	$a_n^{(1)}$	$a_n^{(2)}$	$a_n^{(3)}$	b_n
1	1	0	0	0	1
2	0	1	0	0	0
3	1	1	0	0	1
4	0	0	1	0	0
5	1	0	1	0	1
6	0	1	1	0	1
7	1	1	1	0	0
8	0	0	0	1	0
9	1	0	0	1	1
10	0	1	0	1	0
11	1	1	0	1	1
12	0	0	1	1	1
13	1	0	1	1	0
14	0	1	1	1	0
15	1	1	1	1	1
16	0	0	0	0	0

Notice that the function (5.3) in Example 5.1 is described in terms of two operations, + and · modulo 2, on the pair of elements 0 and 1. This structure has all the mathematical properties of a field, and is usually referred to as the Galois field GF(2) of two elements. Complete mastery of the techniques used to design PN generators will require knowledge of the finite field GF(2) and of the larger finite fields containing GF(2), namely GF(q), where the number q of elements in the field is a power of 2. The

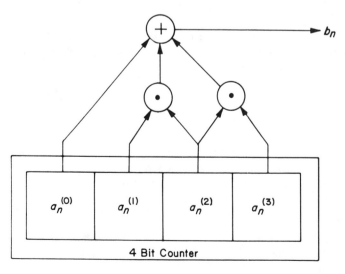

Figure 5.2. Counter-based PN generator.

reader is referred to Appendix 5A for a tutorial description of finite fields and their properties.

A completely specified Boolean function can be determined from tabulated values in at least two ways:

(a) The Quine-McCluskey algorithm [1]–[3], applied to the completely specified Boolean function of Table 5.1, will yield a minimum second-order realization of the function in terms of the logical operations AND, OR, and NEGATION.

(b) The sequences $a_n^{(i)}$, $i = 0, 1, 2, \ldots, k - 1$, each of length 2^k, along with the all-ones sequence of length 2^k, are the basis for a first-order Reed-Muller code [4]–[6]. Determination of the function described in Example 5.1 can be viewed as the result of decoding the 16-tuple, viewed as a codeword from a 4-th order Reed-Muller code of word length 16.

On the average, either of these techniques will produce rather complicated logical structures when applied to a random sequence of binary symbols. The expected number of terms in the resultant function determined by the Reed-Muller approach is half the sequence length. Hence, for a k-bit clock used to produce a randomly selected PN sequence of length 2^k, one would expect on the average the necessity of 2^{k-1} multiple input AND gates and an EXCLUSIVE OR capable of handling 2^{k-1} inputs. Being optimal, one would expect the Quine-McCluskey approach to require somewhat fewer operations, but nevertheless, the expected number of operations probably will still be a linear function of the period 2^k.

Although, on the average, the complexity of PN generators using counters is high, could there be a few exceptionally good designs of this type? There are strong reasons to believe that counters are always poor sources for inputs to functions generating PN sequences. First, the individual sequences $\{a_n^{(i)}\}$ have periods which are divisors of the desired period, and do not individually have a "random appearance." Furthermore, very few of the input sequences change value at any one time, i.e., $\{a_n^{(i)}\}$ changes every 2^i-th bit, thereby forcing the use of rather complicated logic to emulate a PN sequence which, on the average, changes every other bit.

Example 5.2. The sequence

$$001100111010110111110$$

with period 21 has a particularly simple mechanization based on a shift register with feedback logic. Since all five-tuples $(b_n, b_{n+1}, b_{n+2}, b_{n+3}, b_{n+4})$, $n = 1, 2, \ldots, 21$, are distinct, it follows that b_{n+5} can be determined uniquely from these previous five symbols. Hence, the designer must find a simple Boolean function which mechanizes the mapping shown in Table 5.2. One possible result, employing GF(2) arithmetic, is

$$b_{n+5} = b_n + b_{n+1} \cdot b_{n+2} \tag{5.4}$$

with the corresponding mechanization shown in Figure 5.3.

Table 5.2
Feedback logic function

n	b_n	b_{n+1}	b_{n+2}	b_{n+3}	b_{n+4}	b_{n+5}
1	0	0	1	1	0	0
2	0	1	1	0	0	1
3	1	1	0	0	1	1
4	1	0	0	1	1	1
5	0	0	1	1	1	0
6	0	1	1	1	0	1
7	1	1	1	0	1	0
8	1	1	0	1	0	1
9	1	0	1	0	1	1
10	0	1	0	1	1	0
11	1	0	1	1	0	1
12	0	1	1	0	1	1
13	1	1	0	1	1	1
14	1	0	1	1	1	1
15	0	1	1	1	1	1
16	1	1	1	1	1	0
17	1	1	1	1	0	0
18	1	1	1	0	0	0
19	1	1	0	0	0	1
20	1	0	0	0	1	1
21	0	0	0	1	1	0

The approach to determining the feedback function in Example 5.2 can be identical to that used in Example 5.1. The fact that Table 5.2 is incomplete, i.e., the Boolean function's values are specified for only 21 of the 32 possible input binary 5-tuples, may yield some flexibility in the design process. Obviously, then, knowledge of the desired function in incomplete tabular form is not sufficient for determining the behavior of the PN generator when its register is loaded with a 5-tuple not in the table.

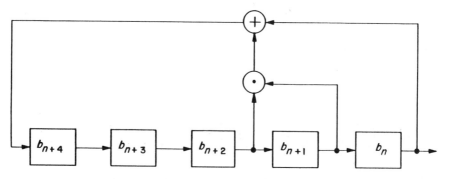

Figure 5.3. A shift-register generator using feedback logic.

Using the implemented function (5.4), it can be seen that the device diagrammed in Figure 5.3 supports three other periodic sequences, namely 0 (period 1), 10000 (period 5), and 01010 (period 5). On the other hand, if the implemented function had the value 1 for each of the ten input 5-tuples missing from Table 5.2, the resulting device would support no other periodic sequences, and regardless of the initial register contents, the output sequence would eventually become the specified period 21 sequence.

Pseudonoise generators are similar to oscillators in the sense that they provide an output signal but are not driven by an input signal, their next memory state being a prescribed function of their present memory state. The period N of the output sequence, therefore, is upper bounded by the number of distinct states, that number being 2^m when the memory is composed of m binary storage elements. Hence, when implemented in binary, the minimum number m of memory elements necessary to support period N operation is given by

$$m = \lceil \log_2(N) \rceil \tag{5.5}$$

where $\lceil x \rceil$ denotes the smallest integer greater than or equal to x. If designing a PN generator for a specified, randomly selected sequence, using a shift register design with feedback logic as in Example 5.2, one can typically expect that the number of binary memory elements needed will be approximately twice the minimum number given by (5.5).

It is worth noting that the period of a sequence should be long enough to preclude the possibility that natural delays caused by multipath or artificial delays created by repeater jamming result in a signal delay of an integral number of periods. Such integral period delays could result in situations in which an SS receiver could not discriminate between the desired SS signal and the multipath or jammer interference. It is possible to size the minimum memory requirements needed to preclude this situation by evaluating (5.5). As indicated by Table 5.3, it clearly is within the realm of possibility to achieve a practically non-repetitive PN generator with a modest investment in memory.

Table 5.3

Memory requirements (measured in binary storage elements) for memory-efficient PN generators operating at 10^6 or 10^9 bits/sec.

Period	Memory Requirement (10^6 bits/sec)	Memory Requirement (10^9 bits/sec)
1 second	20	30
1 minute	26	36
1 hour	32	42
1 day	37	47
1 year	45	55
1 century	52	62

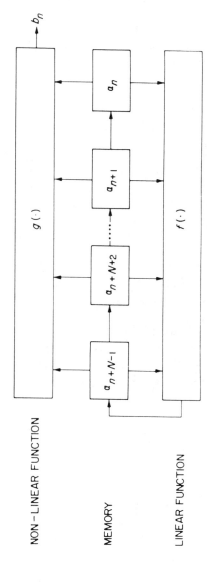

Figure 5.4. A linear feedback shift register with feedforward logic.

There do indeed exist shift registers which, along with a minimal amount of feedback logic, produce binary sequences possessing very large period and excellent pseudo-randomness properties. However, an extreme measure of luck would be necessary to randomly pick a sequence producible by a generator possessing relatively simple structural properties; analytical techniques must be used to determine structures which will serve adequately as PN generators. The PN generators to be explored in this chapter all are variations of a linear-feedback shift register (LFSR) with non-linear feedforward output logic (see Figure 5.4). LFSRs are amenable to analysis, and provide a register through which flows a steady stream of pseudorandom bits. Non-linear feedforward logic is added for several reasons; it suffices to say that the effect of this addition is to make the structure of the shift register difficult to determine rapidly from observations of the output sequence.

5.2 LINEAR RECURSIONS

5.2.1 Fibonacci Generators

A sequence $\{b_n\}$ of elements from a field $\mathscr{F}$ is said to satisfy a *linear recursion* if there exists a relation of the form

$$b_n = -\sum_{i=1}^{M} a_i b_{n-i} \quad \text{for all } n, \tag{5.6}$$

which allows each sequence element to be calculated from the M immediately preceding elements. The coefficients a_i in (5.6) are from the same field $\mathscr{F}$, and, assuming $a_M \neq 0$, M is called the *degree* of the linear recursion. A sequence generator based on (5.6) can be constructed as shown in Figure 5.5. When properly initialized with the first M elements of the desired sequence, this shift register will, without further inputs, produce the remainder of the periodic infinite-length sequence $\{b_n\}$. A generator, structured as shown in Figure 5.5, is often said to be in the *Fibonacci configuration* (after the mathematician who studied linear recursions) to distinguish it from other LFSR forms.

Every periodic sequence $\{b_n\}$ satisfies an infinite number of linear recursions, including the obvious recursions

$$b_n = b_{n+mN} \quad \text{for all integers } m \text{ and } n, \tag{5.7}$$

where N is the period of the sequence. For the case $m = 1$ in (5.7) the Fibonacci-configured LFSR is the cyclic register of Figure 5.1(b). This obvious recursion often is not the minimum-degree recursion satisfied by the sequence.

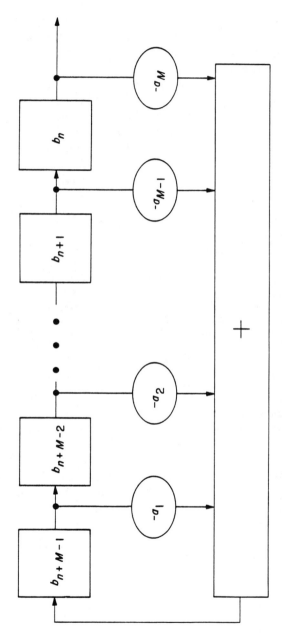

Figure 5.5. An M-stage linear-feedback shift register in Fibonacci form.

Table 5.4
Coefficients of linear recursions (5.6) of degree less than 7,
satisfied by the period-7 sequence of Example 5.3.

M	$a_1 a_2 \cdots a_M$
3	011
4	1101
5	00111
5	10001
6	010011
6	000101
6	111111
6	101001

Example 5.3. The binary sequence of period 7, beginning with

0010111,

satisfies several recursions of degree less than 7 over GF(2) (see Table 5.4).

The minimum number L of memory elements required to build a Fibonacci generator for $\{b_n\}$ is the degree of the minimum-degree recursion which generates $\{b_n\}$. This key number L is called the *linear span* of the sequence $\{b_n\}$, and is often used as an abstract measure of complexity of the sequence structure. In Example 5.3, the linear span L of the period-7 sequence is 3.

5.2.2 Formal Power Series and Characteristic Polynomials

The linear span and corresponding linear recursion for a given periodic sequence can be found theoretically by formal power series manipulations. Let $B(z)$ denote the formal power series having sequence elements as coefficients, i.e.,

$$B(z) = \sum_{n=1}^{\infty} b_n z^{-n}. \tag{5.8}$$

The connection between $B(z)$ and a linear recursion (5.6) satisfied by $\{b_n\}$ can be established in the following manner. Define

$$a_0 = 1, \tag{5.9}$$

multiply (5.6) by z^{-n}, sum over n, and manipulate to arrive at

$$0 = \sum_{i=0}^{M} a_i z^{-i} \sum_{n=M+1}^{\infty} b_{n-i} z^{-(n-i)}$$
$$= A(z)B(z) - D(z), \tag{5.10}$$

and, hence,

$$B(z) = \frac{D(z)}{A(z)}, \tag{5.11}$$

where

$$A(z) = \sum_{i=0}^{M} a_i z^{M-i}, \tag{5.12}$$

$$D(z) = \sum_{n=1}^{M} d_n z^{M-n}, \tag{5.13a}$$

$$d_n = \sum_{i=0}^{n-1} a_i b_{n-i}, \qquad n = 1, 2, \dots, M. \tag{5.13b}$$

Thus, the formal power series of every sequence which satisfies a linear recursion can be written as the ratio of two polynomials in z (no negative powers of z are contained in either $A(z)$ or $D(z)$). The denominator polynomial's coefficients are the weights used in the linear recursion, and the numerator polynomial's coefficients are related to the initial elements $b_1, b_2, \dots, b_M$, by the M independent equations (5.13b).

The following technique can be used to find the linear span and corresponding linear recursion for a sequence $\{b_n\}$, beginning with knowledge of $b_1, b_2, \dots, b_N$, and the fact that N is the sequence period. By substituting the periodic sequence structure into (5.8), $B(z)$ is identified as a ratio of polynomials.

$$B(z) = \sum_{m=0}^{\infty} \sum_{n=1}^{N} b_n z^{-(mN+n)}$$

$$= \frac{1}{1 - z^{-N}} \cdot \sum_{n=1}^{N} b_n z^{-n}$$

$$= \frac{\displaystyle\sum_{n=1}^{N} b_n z^{N-n}}{z^N - 1}. \tag{5.14}$$

The recursion and initial conditions identified with the denominator and numerator polynomials of (5.14) correspond to the cyclic generator structure of Figure 5.1(b). Cancellation of common factors in the numerator and denominator polynomials in (5.14) will yield lower-degree recursions and, hence, mechanizations with lower memory requirements.

Carrying this idea to its logical conclusions, Euclid's algorithm (see Appendix 5A, sections 5A.2 and 5A.3) can be applied to determine the greatest common divisor $G(z)$,

$$G(z) = \gcd\left(\sum_{n=1}^{N} b_n z^{N-n}, z^N - 1 \right), \tag{5.15}$$

of the numerator and denominator polynomials in (5.14). After cancellation of this common factor of highest possible degree, $B(z)$ is of the form

$$B(z) = \frac{P(z)}{Q(z)}, \qquad (5.16)$$

where

$$\gcd(P(z), Q(z)) = 1. \qquad (5.17)$$

It can be assumed without loss of generality that multiplicative constants have been transferred to the numerator in (5.16), so that the denominator polynomial $Q(z)$ is *monic*, i.e., the coefficient of the highest power of z in $Q(z)$ is unity. Clearly, there is no rational representation of $B(z)$ with a denominator polynomial of lesser degree, and, hence, the linear span L of $\{b_n\}$ is given by

$$L = \deg Q(z), \qquad (5.18)$$

and $\{b_n\}$ satisfies the linear recursion

$$b_n = - \sum_{i=1}^{L} q_i b_{n-i}, \qquad (5.19)$$

where q_i is the i-th coefficient of $Q(z)$, i.e.,

$$Q(z) = \sum_{i=0}^{L} q_i z^{L-i}. \qquad (5.20)$$

The monic denominator polynomial $Q(z)$ in (5.16) achieving this linear span is called the *characteristic polynomial* of the sequence $\{b_n\}$, and the corresponding numerator polynomial $P(z)$ is called the *initial condition polynomial* of $\{b_n\}$. The following theorem is based on the above derivation.

THEOREM 5.1. *Let $\{b_n\}$ be a sequence possessing characteristic polynomial $Q(z)$. Then the period N of $\{b_n\}$ is the smallest integer N such that $Q(z)$ divides $z^N - 1$.*

Clearly then, sequences possessing the same characteristic polynomial have the same period.

5.2.3 Galois Generators

One consequence of the above development is that the sequence $\{b_n\}$ can be generated alternatively by a logic circuit that performs the division of $P(z)$ by $Q(z)$. The first step in this formal long division is shown below:

$$
\begin{array}{r}
p_1 z^{-1} \\
\hline
z^L + q_1 z^{L-1} + \cdots + q_L \overline{)\; p_1 z^{L-1} + p_2 z^{L-2} + \cdots + p_L } \\
p_1 z^{L-1} + q_1 p_1 z^{L-2} + \cdots + q_{L-1} p_1 + q_L p_1 z^{-1} \\
\hline
r_1^{(1)} z^{L-2} + \cdots + r_{L-1}^{(1)} + r_L^{(1)} z^{-1}
\end{array}
$$

Note that the above formal computation is valid, even if $p_1 = 0$. The first cycle (shown above) in this never-ending long-division process produces a remainder polynomial via the equation

$$P(z) - p_1 z^{-1} Q(z) = z^{-1} R^{(1)}(z),$$ (5.21)

with the quotient coefficient p_1 being the first bit b_1 of the formal power series $B(z)$. Each succeeding cycle in the long-division process divides the previous cycle's remainder by $Q(z)$ to produce a quotient corresponding to the next element in $\{b_n\}$, and a new remainder. Specifically, let the n-th formal power series remainder in the division process be represented by

$$z^{-n} R^{(n)}(z) = \sum_{i=1}^{L} r_i^{(n)} z^{L-i-n}, \qquad n = 1, 2, \ldots .$$ (5.22)

The long division process imposes the following recursion on these remainders:

$$z^{-n} \left[R^{(n)}(z) - r_1^{(n)} z^{-1} Q(z) \right] = z^{-n-1} R^{(n+1)}(z).$$ (5.23)

In summary, the remainder polynomial (no negative powers of z) recursion

$$z R^{(n)}(z) - r_1^{(n)} Q(z) = R^{(n+1)}(z), \qquad n = 1, 2, \ldots,$$ (5.24)

initialized by

$$R^{(0)}(z) = P(z),$$ (5.25)

produces the desired output sequence through the series of quotient coefficients

$$b_n = r_1^{(n-1)}, \qquad n = 1, 2, \ldots .$$ (5.26)

The division of $P(z)$ by $Q(z)$ is embodied in the device, shown in Figure 5.6, which mechanizes the remainder recursion (5.22). This LFSR form is called the *Galois configuration* because it is related to Galois field multiplication (more on this in Section 5.5).

Example 5.4. A binary sequence of period $N = 30$ begins with the following thirty bits, indexed from the left:

000001100010010101110011101101

The formal power series representation (5.7) for this sequence begins

$$B(z) = z^{-6} + z^{-7} + z^{-11} + z^{-14} + z^{-16} + z^{-17} + \ldots,$$ (5.27)

and reduces by (5.14)–(5.17) to

$$B(z) = \frac{1}{z^6 + z^5 + z^4 + z^3 + z^2 + 1}.$$ (5.28)

The Galois and Fibonacci generators for the above sequence are shown in Figure 5.7.

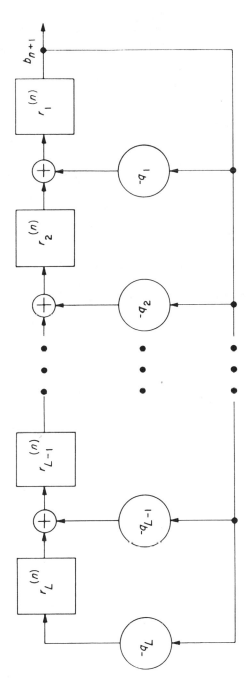

Figure 5.6. A minimum-memory LFSR in the Galois configuration for generating $\{b_n\}$.

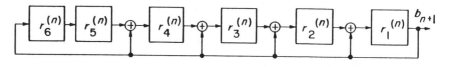

(a) Galois configuration

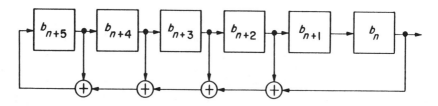

(b) Fibonacci configuration

Figure 5.7. Minimum-memory, linear feedback generators for the sequence represented by $1/(z^6 + z^5 + z^4 + z^3 + z^2 + 1)$.

5.2.4 State Space Viewpoint

The PN generators above can be viewed as autonomous linear sequential circuits, whose state vector s_n corresponds to the register contents, and whose operation is described formally by a state equation

$$s_{n+1} = As_n, \tag{5.29}$$

and an output equation

$$b_n = Bs_n, \tag{5.30}$$

which describe circuit operation for $n = 1, 2, 3, \ldots$. Then it follows that

$$b_n = BA^{n-1}s_1 \tag{5.31}$$

For minimum-memory linear generators, the matrix A is $L \times L$ and the vector B is $1 \times L$. The minimum-memory Fibonacci generator is described in these terms by

$$s_n = \begin{bmatrix} b_n \\ b_{n+1} \\ \vdots \\ b_{n+L-1} \end{bmatrix}, \tag{5.32}$$

$$A = \begin{bmatrix} 0 & 1 & 0 & \cdots & 0 \\ 0 & 0 & 1 & & 0 \\ \vdots & & & & \vdots \\ 0 & 0 & 0 & & 1 \\ -q_L & -q_{L-1} & -q_{L-2} & \cdots & -q_1 \end{bmatrix}, \qquad (5.33)$$

$$B = \begin{bmatrix} 1 & 0 & 0 & \cdots & 0 \end{bmatrix}. \qquad (5.34)$$

Similarly, the minimum-memory Galois generator is represented by

$$s_n = \begin{bmatrix} r_L^{(n-1)} \\ r_{L-1}^{(n-1)} \\ \vdots \\ r_1^{(n-1)} \end{bmatrix}, \qquad (5.35)$$

$$A = \begin{bmatrix} 0 & 0 & \cdots & 0 & -q_L \\ 1 & 0 & & 0 & -q_{L-1} \\ 0 & 1 & & 0 & -q_{L-2} \\ \vdots & & & & \\ 0 & 0 & \cdots & 1 & -q_1 \end{bmatrix}, \qquad (5.36a)$$

$$B = \begin{bmatrix} 0 & 0 & \cdots & 0 & 1 \end{bmatrix}. \qquad (5.36b)$$

The A matrices in (5.33) and (5.36a), which are transposes of each other, are referred to as *companion matrices* of the monic characteristic polynomial $Q(z)$, where

$$Q(z) = \sum_{i=0}^{L} q_i z^{L-i}. \qquad (5.37)$$

Other linear sequential circuits, which generate the same output sequence $\{b_n\}$, can be constructed from transformations of the state space.

Let T be a nonsingular matrix over the sequence element field $\mathcal{F}$, and define a new state vector s_n',

$$s_n' = T s_n. \qquad (5.38)$$

Then the linear sequential circuit of the new state sequence $\{s_n'\}$ has a state equation governed by the transition matrix

$$A' = T A T^{-1}, \qquad (5.39)$$

and an output equation

$$b_n = B T^{-1} s_n'. \qquad (5.40)$$

There is no significant difference between these generators when the output sequence is a linear function of the state sequence, as shown in (5.30), and (5.31). Certainly, the memory requirements are identical in each case, although some realizations may use fewer adders than others [50]. On the other hand, when the output b_n is a non-linear function of the circuit state s_n, i.e., non-linear feedforward logic is employed, then attempts to deduce the structure of the generator from a short segment of the output sequence may be complicated by the insertion of T.

5.2.5 Determination of Linear Recursions from Sequence Segments

When given a segment $b_1, b_2, \ldots, b_K$, of K symbols from a sequence $\{b_n\}$, it is impossible to determine with absolute certainty the remainder of the sequence. However, if one proceeds on the assumption that K is at least twice the linear span of the sequence, and if that assumption is in fact correct, then a linear generator of the complete sequence can be determined. For the moment, consider the possibility of knowing the linear span L of the sequence *a priori*, in addition to $2L$ consecutive sequence elements. In this case, the fact that (5.6) also is linear in the L unknown coefficients a_i, $i = 1, 2, \ldots, L$, can be used to determine the recursion.

Example 5.5. Suppose that a binary sequence of elements from GF(2) is known to satisfy a linear recursion of degree 5, and, furthermore, a portion of the sequence contains

$$\ldots 1011101100011 \ldots \tag{5.41}$$

where the sequence index is increasing from left to right. Applying (5.6) separately to overlapping blocks of six consecutive symbols from the sequence gives the following set of equations:

$$
\begin{aligned}
0 &= a_1 + a_2 + a_3 + + a_5 \\
1 &= a_2 + a_3 + a_4 \\
1 &= a_1 + a_3 + a_4 + a_5 \\
0 &= a_1 + a_2 + a_4 + a_5 \\
0 &= a_2 + a_3 + a_5.
\end{aligned} \tag{5.42}
$$

In constructing these L (in this case five) linearly independent equations, we used $L + 1$ specified sequence elements for the first equation plus one additional (previously unused) element for each additional equation. Hence, knowledge of $2L$ bits from the given sequence was used in the construction of (5.42). In the special case where the element field is GF(2) as it is here, the fact that the degree of the recursion is L and a_L is non-zero implies that a_L must be 1, the only non-zero field element. Therefore, only $L - 1$ equations are needed in the GF(2) case, in this example the first four in (5.42), which can be rewritten as

$$
\begin{aligned}
1 &= a_1 + a_2 + a_3 \\
1 &= a_2 + a_3 + a_4 \\
0 &= a_1 + a_3 + a_4 \\
1 &= a_1 + a_2 + a_4.
\end{aligned} \tag{5.43}
$$

Since addition in GF(2) is modulo 2, adding the first three equations directly gives $a_3 = 0$, and adding the last three gives $a_4 = 0$. Substitution of these values gives $a_2 = 1$ and $a_1 = 0$, and, hence, the resulting linear recursion is

$$b_n = b_{n-2} + b_{n-5}. \qquad (5.44)$$

When K symbols from the sequence are known but the linear span L is not, one can always guess L and follow the above procedure. If the guess is less than the true linear span, additional data will not be consistent with the deduced (incorrectly) linear generator. On the other hand, if the guess is greater than the true linear span, then multiple solutions will appear in the above procedure, all leading to correct, but usually non-minimal, generators of the complete sequence. Lacking knowledge of L, a more organized approach uses techniques for constructing rational approximations to the sequence's formal power series $B(z)$, when only the first K terms are known. These techniques, suggested by Massey for use in the Berlekamp-Massey decoding algorithm [7], [8], are closely related to continued fraction computations [9], [10], and will produce the characteristic and initial condition polynomials, $Q(z)$ and $P(z)$ of (5.16), whenever the number of available symbols K is at least twice the linear span L. Furthermore, this analysis can be performed in real time, in a manner such that more sequence symbols can be inserted into the computation as they become available.

If the above procedure is perceived as a threat to the objectives of the SS system design, then the designer must take steps to insure that $2L$ sequence elements cannot be observed directly from the transmitted signal. Three possible methods for accomplishing this are:

(a) Ensure that the linear span of the sequence $\{b_n\}$ (presumably used as a SS carrier of some sort) is much larger than the processing gain of the system, so that data modulation effects will, with high probability, preclude direct observation of $2L$ sequence elements.
(b) Insert erroneous symbols in the transmission of $\{b_n\}$ so that error-free observation of $2L$ consecutive sequence elements is virtually impossible. This can be accomplished with very little degradation to system performance when high processing gain is employed.
(c) Transmit in-band noise along with the SS signal based on $\{b_n\}$, to insure that no observer will get a "clean look" at the sequence. The cost of this can be made a small fraction of the processing gain.

Other methods for insuring transmission security against this threat are left to the ingenuity of the reader.

5.3 MEMORY-EFFICIENT LINEAR GENERATORS

5.3.1 Partial Fraction Decompositions

Initial conditions notwithstanding, the basic structural information about a linear generator is contained in its characteristic polynomial, $Q(z)$. Assuming that the generated sequence $\{b_n\}$ consists of elements from GF(q), the polynomial $Q(z)$ will then be over GF(q), i.e., have coefficients in GF(q). However, its roots may not be in GF(q). Let's assume that $Q(z)$ can be factored so that

$$Q(z) = \prod_{j=1}^{J} Q_j^{p_j}(z), \tag{5.45}$$

where each polynomial $Q_j(z)$ is irreducible over GF(q), and the linear span L of $\{b_n\}$ is given by

$$L = \sum_{j=1}^{J} p_j \deg Q_j(z). \tag{5.46}$$

The term *irreducible* over GF(q), when applied to a polynomial $A(z)$ over GF(q), means that $A(z)$ cannot be factored into a product of polynomials over GF(q). Hence, the factorization described in (5.45) is similar to the decomposition of an integer into prime factors.

The factorization (5.45) indicates that the rational representation (5.16) of the formal power series $B(z)$ can be expanded by partial fractions.

$$B(z) = \frac{P(z)}{\prod\limits_{j=1}^{J} Q_j^{p_j}(z)} = \sum_{j=1}^{J} \frac{P_j(z)}{Q_j^{p_j}(z)}. \tag{5.47}$$

Therefore, the sequence represented by $B(z)$ also can be generated by summing the component sequences represented by

$$B_j(z) = \frac{P_j(z)}{Q_j^{p_j}(z)}, \qquad j = 1, 2, \ldots, J, \tag{5.48}$$

the j-th sequence having characteristic polynomial $Q_j^{p_j}(z)$ and initial condition polynomial $P_j(z)$. Let's define the period of the j-th sequence to be N_j. Then the period of the composite sequence $\{b_n\}$ is given by

$$N = \text{lcm}(N_1, N_2, \ldots, N_J), \tag{5.49}$$

where lcm( ) represents the least common multiple of the listed integers.

Example 5.6. The sequence of Example 5.4 was represented by

$$B(z) = \frac{1}{z^6 + z^5 + z^4 + z^3 + z^2 + 1} = \frac{P_1(z)}{(z+1)^2} + \frac{P_2(z)}{z^4 + z^3 + 1}, \tag{5.50}$$

where $P_1(z)$ and $P_2(z)$ are polynomials of degrees at most 1 and 3,

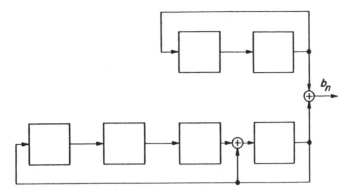

Figure 5.8. An LFSR generator for the sequence of Examples 5.4 and 5.6.

respectively. Solving for these polynomials gives the component generators

$$B_1(z) = \frac{z}{(z+1)^2} \tag{5.51a}$$

$$B_2(z) = \frac{z^3 + z^2 + z + 1}{z^4 + z^3 + 1}. \tag{5.51b}$$

The first generator produces the period 2 sequence

$$10$$

while the second generator outputs the period 15 sequence

$$101011001000111.$$

Extending these sequences periodically and summing gives the period 30 sequence of Example 5.4. The Galois-configured, component-structured, LFSR shown in Figure 5.8 can be initialized to have an output identical to the generators of Figure 5.7.

5.3.2 Maximization of Period for a Fixed Memory Size

The fundamental memory constraint (5.5) indicates that the maximum period achievable by any q-ary generator with M memory elements is q^M. This maximum period is reduced by 1 if the generator uses linear feedback, because the zero state is self-perpetuating, i.e., an LFSR initialized with zeros produces the uninteresting all-zeros sequence. Thus, the most efficient LFSR generators must cycle through all possible non-zero states before repeating, and the period N of an LFSR's state sequence is bounded by

$$N \le q^M - 1. \tag{5.52}$$

In Example 5.6, the four-stage component LFSR generator achieved its maximum possible period of 15, while the two-stage generator did not; and the composite generator of six stages produced a period 30 sequence, nowhere near the upper bound of 63 dictated by (5.52).

Assuming a $Q(z)$ of the form (5.45), applying the bound (5.52) to the component sequences, and using (5.49) gives

$$N \le \prod_{j=1}^{J} N_j$$

$$\le \prod_{j=1}^{J} q^{p_j \deg Q_j(z)} - 1$$

$$\le q^L - 1. \tag{5.53}$$

It is clear from (5.53) that the fundamental bound (5.52) on N can be achieved if and only if all the above inequalities hold with equality. Equivalently, considering each inequality in turn, equality holds in (5.53) if and only if

(a) the periods of the component sequences are relatively prime,
(b) each component sequence achieves the bound (5.52) on its period for its memory size, and
(c) the number J of component sequences must be 1.

When the degrees of the component generator polynomials are large, then item (c) is not critical to the efficient use of memory in creating large periods.

5.3.3 Repeated Factors in the Characteristic Polynomial

Consider now the special case in which the characteristic polynomial of a sequence over $GF(q)$ is the q-th power of a polynomial over $GF(q)$. Theorem 5A.10 of Appendix 5A indicates that

$$Q^q(z) = \left(\sum_{i=0}^{d} q_{d-i} z^i \right)^q = \sum_{i=0}^{d} q_{d-i} z^{qi}. \tag{5.54}$$

Hence, the coefficient list of $Q^q(z)$ is simply

$$q_0 0 \ldots 0 q_1 0 \ldots 0 q_2 0 \ldots 0 q_{d-1} 0 \ldots 0 q_d,$$

where $0 \ldots 0$ represents a string of $q - 1$ zeros. A shift register governed by this form of generator polynomial will produce q interleaved sequences. That is, a symbol b_n in the composite output sequence $\{b_n\}$ is linked recursively only to the prior symbols b_{n-mq}, $m = 1, 2, 3, \ldots$.
    This line of reasoning can be iterated to yield the following result.

THEOREM 5.2. *Let N be the period of a sequence having characteristic polynomial $Q(z)$ over $GF(q)$. Then the period N' of the sequence having characteristic polynomial $Q^p(z)$ is given by*

$$N' = q^I N, \tag{5.55}$$

Table 5.5

Maximum periods N for twenty-stage binary LFSRs with characteristic polynomials of the form $Q^p(z)$ over GF(2), where $Q(z)$ has degree d.

degree	power		period
d	p	I	N
20	1	0	1048575
10	2	1	2046
5	4	2	124
4	5	3	120
2	10	4	48

where the interleaving exponent I is the unique integer satisfying the relation

$$q^{I-1} < p \le q^I. \tag{5.56}$$

Assuming that the period of the sequence generated by $Q(z)$ is maximal, i.e., is $q^d - 1$ where d is the degree of $Q(z)$, it is apparent that interleaving is not an efficient way to achieve sequences with long periods. As an example, the achievable periods for a register of twenty binary storage elements are tabulated in Table 5.5.

5.3.4 *M*-Sequences

Linear-feedback shift registers with L stages which produce the maximum possible period $q^L - 1$ do in fact exist, and the sequences which they produce are called *maximal LFSR sequences* or *m-sequences*. When the connection between shift register sequences over GF(q) and larger finite fields is established in Section 5.5, it will be apparent that irreducible polynomials of degree L over GF(q) generate sequences whose periods must be divisors of $q^L - 1$. Those special irreducible polynomials, which are the characteristic polynomials of *m*-sequences, are called *primitive polynomials*, and they exist for every degree over every finite field. Appendix 5B contains factorizations of $2^L - 1$ and a table of primitive polynomials over GF(2). Techniques described in Appendix 5A.8 can be applied to find other primitive/irreducible polynomials from those listed in Appendix 5B. It suffices at this point to note:

THEOREM 5.3. *A linear generator of a given memory size produces a sequence of elements from* GF(q), *with the largest possible period if and only if its characteristic polynomial is primitive over* GF(q).

Two sequences $\{b_n\}$ and $\{b'_n\}$, each having period N, are called *cyclically equivalent* if there exists an integer τ such that $b_{n+\tau} = b'_n$ for all n. A pair of

Table 5.6
The number $N_p(L)$ of cyclically distinct m-sequences with linear span L.

L	$N_p(L)$	L	$N_p(L)$
2	1	16	2048
3	2	17	7710
4	2	18	7776
5	6	19	27594
6	6	20	24000
7	18	21	84672
8	16	22	120032
9	48	23	356962
10	60	24	276480
11	176	25	1296000
12	144	26	1719900
13	630	27	4202496
14	756	28	4741632
15	1800	29	18407808

sequences, which are not cyclically equivalent, are termed *cyclically distinct*. Since m-sequences are cyclically distinct if and only if their linear recursions differ, the number of cyclically distinct m-sequences over GF(q) with linear span L is equal to the number $N_p(L)$ of primitive polynomials of degree L over GF(q), and is given by

$$N_p(L) = \frac{q^L - 1}{L} \prod_{i=1}^{J} \frac{p_i - 1}{p_i} \qquad (5.57)$$

where the J prime numbers p_i, $i = 1, 2, \ldots, J$ are determined from the prime decomposition of $q^L - 1$, i.e.,

$$q^L - 1 = \prod_{i=1}^{J} p_i^{e_i} \qquad (5.58)$$

where e_i, $i = 1, \ldots, J$ are positive integers. Table 5.6 illustrates the exponential nature of $N_p(L)$ for binary m-sequences.

5.4 STATISTICAL PROPERTIES OF M-SEQUENCES

Primitive polynomials provide LFSR connections for m-sequence generators which are virtually as efficient as counters in providing state sequences with large periods, the LFSR's state sequence being shorter by one than the counter's period for the same memory size. However, the binary counter did not provide a good basis for a PN generator because the individual memory cells of the counter produced simple "square wave" sequences, thereby requiring significant amounts of logic to produce sequences with the appearance of randomness. As will now be demonstrated, each memory cell of a Fibonacci m-sequence generator produces a sequence with statistical properties far superior, for PN applications, to a counter's cell output.

5.4.1 Event Counts

Since m-sequences over GF(q) with linear span L have period $q^L - 1$ and since the all-zeros sequence is self-perpetuating in a linear generator, it immediately follows that an m-sequence generator must cycle through the $q^L - 1$ non-zero state L — tuples. In particular, a Fibonacci generator cycles through the state L-tuples

$$s_n = \begin{bmatrix} b_n \\ b_{n+1} \\ \vdots \\ b_{n+L-1} \end{bmatrix}, \qquad n = 1, 2, \ldots, q^L - 1, \qquad (5.59)$$

which correspond to L-symbol segments of the m-sequence $\{b_n\}$ (see (5.30)–(5.32)). This viewpoint leads to verification of the following random-appearance properties:

Property R-1.

The number N_b of times that a non-zero symbol b occurs in one period of an m-sequence $\{b_n\}$ exceeds the number N_0 of times 0 occurs in one period, by 1.

$$N_b = N_0 + 1, \quad \text{for all } b \neq 0. \qquad (5.60)$$

This property, which states that symbols occur as equally often as possible within one period, is referred to as the *balance property*. Its proof follows directly from counting the number of times that a state L-tuple begins with a specific symbol in one cycle of the state sequence.

Property R-2.

Let $\{b_n\}$ be an m-sequence of elements from GF(q) with linear span L. The number N_a of positions within one period of $\{b_n\}$ at which a J-tuple $a_1 a_2 \ldots a_J$ occurs, is given by

$$N_a = \begin{cases} q^{L-J}, & \text{for } a \neq 0, \quad 1 \leq J \leq L \\ q^{L-J} - 1, & \text{for } a = 0, \quad 1 \leq J \leq L \\ 0 \text{ or } 1, & L < J. \end{cases} \qquad (5.61)$$

This generalization of the balance property also follows from state counting arguments.

The probability of obtaining a particular J-tuple a in J independent trials, each trial having q possible equally likely outcomes, is given by q^{-J}.

Table 5.7

Comparison of binary m-sequence J-tuple statistics with Bernoulli sequence
probabilities for $L > J$.

J	Bernoulli J-tuple Probability	m-sequence statistics N_0/N, N_a/N		
		$L = 4$	$L = 6$	$L = 8$
1	.500	.467, .533	.492, .508	.498, .502
2	.250	.200, .267	.238, .254	.247, .251
3	.125	.067, .133	.111, .127	.122, .125

If an m-sequence of $GF(q)$ elements is to have the appearance of such a random process's sample function, it is necessary that the time-average statistic N_a/N of the J-tuple event a be as close to q^{-J} as possible. For $J \le L$, Property R-2 gives

$$\frac{N_0}{N} = \frac{q^{L-J} - 1}{q^L - 1} \qquad (5.62)$$

for the all-zeros J-tuple (the worst case). Clearly N_0/N rapidly approaches q^{-J} as L increases (see Table 5.7).

5.4.2 The Shift-and-Add Property

The next property is the foundation for several important characteristics of m-sequences.

Property R-3.

Let $\{b_n\}$ be an m-sequence over $GF(q)$ with linear span L. Then for any τ, $\tau \ne 0 \bmod q^L - 1$, the difference of the m-sequence $\{b_n\}$ and its τ-shift $\{b_{n+\tau}\}$ is another shift $\{b_{n+\tau'(\tau)}\}$ of the same m-sequence. That is,

$$b_{n+\tau'(\tau)} = b_{n+\tau} - b_n \quad \text{for all } n, \qquad (5.63)$$

where $\tau'(\tau)$ is defined for all $\tau \ne 0 \bmod q^L - 1$.

Proof. The m-sequence and its τ-shift both satisfy the linear recursion of degree L determined by its characteristic polynomial $Q(z)$.

$$b_{n+\tau} = -\sum_{i=1}^{L} q_i b_{n+\tau-i} \quad \text{for all } n, \qquad (5.64)$$

and the difference of $\{b_{n+\tau}\}$ and $\{b_n\}$ therefore satisfies the same recursion.

$$b_{n+\tau} - b_n = -\sum_{i=1}^{L} q_i (b_{n+\tau-i} - b_{n-i}) \quad \text{for all } n. \qquad (5.65)$$

Since $\{s_n\}$ represents the state sequence for a Fibonacci generator producing $\{b_n\}$, then the initial state s'_n for $\{b_{n+\tau} - b_n\}$ is

$$s'_1 = s_{1+\tau} - s_1. \tag{5.66}$$

This initial state s'_1 is non-zero if, and only if, τ is not a multiple of the period $q^L - 1$ of the state sequence $\{s_n\}$. When s'_1 is non-zero, it is a state $s_{1+\tau'(\tau)}$ of the Fibonacci generator of $\{b_n\}$, and, hence, $\{b_{n+\tau'(\tau)}\}$ is the result of the recursion (5.64) operating on this initial state. ∎

When the symbol field has characteristic 2 as it does in the case of binary (0 or 1) sequences, subtraction and addition are identical operations, and Property R-3 is usually called the *shift-and-add* (or *cycle-and-add*) property. It also is true that the sum of an m-sequence and its proper cyclic shift is another shift of the m-sequence, but the form (5.63) is more useful in the general results to follow.

5.4.3 Hamming Distance Properties of Derived Real-Integer Sequences

Sometimes an m-sequence generator over $GF(q)$ is used to produce a pseudo-random sequence of integers between 0 and $q^J - 1$ by viewing the n-th J-tuple $(b_n, b_{n+1}, \ldots, b_{n+J-1})$ in the m-sequence $\{b_n\}$ as the base q expansion of the n-th integer i_n in the output integer sequence.

$$i_n = \sum_{j=0}^{J-1} b_{n+j} q^j \tag{5.67}$$

Some applications using integer sequences like $\{i_n\}$ require that the sequence be as distinct from its cyclic shifts as possible. One measure of this quality is the *Hamming metric* which counts the number of positions in which two N-tuples x and y differ. This metric takes the form

$$H(x, y) = \sum_{i=1}^{N} h(x_i, y_i) \tag{5.68}$$

where $h(x, y)$ is zero if x and y are identical, and one otherwise. The following attribute indicates that $\{i_n\}$ behaves much like a random sequence, with symbol matches between it and its cyclic shift occurring at a rate very close to one in every q^J tests.

Property R-4.

Let $\{b_n\}$ be an m-sequence over $GF(q)$ with linear span L, and let $\{i_n\}$ be formed according to (5.67). Let i_k denote one period of the sequence $\{i_n\}$, beginning with the k-th integer. Then for $\tau \neq 0 \bmod q^L - 1$, the Hamming distance between i_1 and $i_{1+\tau}$ is given by

$$H(i_1, i_{1+\tau}) = q^L(1 - q^{-J}). \tag{5.69}$$

Proof. The proof of this result follows from Property R-3 which establishes that

$$i_n = i_{n+\tau}$$
$$\text{iff } (b_n, \ldots, b_{n+J-1}) = (b_{n+\tau}, \ldots, b_{n+\tau+J-1})$$
$$\text{iff } (b_{n+\tau'(\tau)}, \ldots, b_{n+\tau'(\tau)+J-1}) = \mathbf{0} \tag{5.70}$$

with a count of the latter event evaluated in Property R-2. ∎

Using a variation of Plotkin's bound, developed by Lempel and Greenberger [11], it can be proven that the integer sequence $\{i_n\}$ of (5.67) possesses optimal Hamming distance properties.

THEOREM 5.4. *Let $\{a_n\}$ be a sequence of period N, composed of symbols from an alphabet $\mathscr{A}$, and let $\boldsymbol{a}_j$ denote the N-tuple of consecutive symbols from $\{a_n\}$, beginning at the j-th symbol. Let*

$$H_{\min} = \min_{0 < \tau < N} H(\boldsymbol{a}_1, \boldsymbol{a}_{1+\tau}), \tag{5.71}$$

and determine the integers α and β from

$$N = \alpha|\mathscr{A}| + \beta, \qquad 0 \le \beta < |\mathscr{A}|. \tag{5.72}$$

Then

$$H_{\min} \le N - \frac{\alpha}{N-1}(N + \beta - |\mathscr{A}|). \tag{5.73}$$

Proof. Certainly the minimum value $H_{\min}$ of Hamming distance between $\{a_n\}$ and its cyclic shifts must be less than the average value H_{avg} of the same quantities. Now

$$H_{\text{avg}} = \frac{1}{N-1} \sum_{\tau=1}^{N-1} H(\boldsymbol{a}_1, \boldsymbol{a}_{1+\tau})$$
$$= \frac{1}{N-1} \sum_{n=1}^{N} \sum_{\tau=1}^{N-1} h(a_n, a_{n+\tau}). \tag{5.74}$$

Let N_a denote the number of times that the symbol "a" occurs in one period of $\{a_n\}$. The number of times that $h(a_n, a_{n+\tau})$ is zero as τ varies over its range is simply $N_{a_n} - 1$. Hence,

$$H_{\text{avg}} = \frac{1}{N-1} \sum_{n=1}^{N} (N - N_{a_n})$$
$$= \frac{1}{N-1} \left(N^2 - \sum_{a \in \mathscr{A}} N_a^2 \right). \tag{5.75}$$

Removing the dependence of (5.75) on the set of integers $\{N_a\}$ yields the bound

$$H_{\min} \le H_{\text{avg}} \le \max_{\{N_a\}} \frac{1}{N-1} \left(N^2 - \sum_{a \in \mathscr{A}} N_a^2 \right), \tag{5.76}$$

where $\{N_a\}$ is a set of non-negative integers summing to N. It can be verified that the right side of (5.76) is maximized by making the integers N_a as nearly equal as possible. That is, the maximizing choice for $\{N_a\}$ contains the integer $\alpha + 1$ a total of β times and the integer α a total of $|\mathscr{A}| - \beta$ times. Substitution of these values into (5.76) and simplification using (5.72) yields the final result. ∎

Evaluation of the bound (5.73) for the parameters

$$N = q^L - 1, \qquad |\mathscr{A}| = q^J, \qquad (5.77)$$

whence

$$\beta = q^J - 1, \qquad \alpha = q^{L-J} - 1, \qquad (5.78)$$

gives the bound

$$H_{\min} \le q^L(1 - q^{-J}). \qquad (5.79)$$

Property R-4 indicates that $\{i_n\}$ satisfies this bound with equality and, therefore, possesses the desirable attribute of being uniformly maximally distant from its proper cyclic shifts.

5.4.4 Correlation Properties of Derived Complex Roots-of-Unity Sequences

The case in which the field size q is a prime p, usually 2 in practice, deserves added attention. In this case, GF(p) arithmetic is modulo p arithmetic, and the field GF(p) is composed solely of integers $0, 1, \ldots, p - 1$. By viewing these integers in GF(p) as being identical to their counterparts in the real numbers, the following mapping from b_n in GF(p) to a complex p-th root of unity can be constructed:

$$a_n = \rho^{b_n}, \qquad (5.80)$$

where ρ is a primitive p-th root of unity in the field of complex numbers, e.g.,

$$\rho = \exp(i2\pi/p). \qquad (5.81)$$

(In the special case when $p = 2$, then $\rho = -1$ and $b_n \in$ GF(2) $= \{0, 1\}$ is mapped into $a_n \in \{1, -1\}$.) Consequently, addition modulo p of two elements b_m and b_n from GF(p) in the domain of the mapping (5.80) is isomorphic to complex multiplication of two elements in the range of the mapping, i.e.,

$$a_m a_n = \rho^{b_m + b_n}, \qquad (5.82)$$

with addition of integers in the exponent being both real and modulo p since ρ is a p-th root of unity.

An m-sequence $\{b_n\}$ of elements from GF(p), satisfying the linear recursion

$$b_n = -\sum_{i=1}^{L} q_i b_{n-i}, \qquad (5.83)$$

can be mapped by (5.80) into a sequence $\{a_n\}$ of complex p-th roots of unity satisfying the multiplicative recursion

$$a_n = \prod_{i=1}^{L} a_{n-i}^{-q_i}. \tag{5.84}$$

Note that $\{a_n\}$ generally does not satisfy a linear recursion of degree L over the complex numbers. However, the resulting complex sequence $\{a_n\}$ often retains the name m-sequence.

Since the m-sequence $\{a_n\}$ consists of complex numbers, it is possible to evaluate its *periodic autocorrelation function* $P_{aa}(\tau)$, defined by

$$P_{aa}(\tau) = \sum_{n=1}^{N} a_{n+\tau} a_n^*, \tag{5.85}$$

where $(\ )^*$ denotes conjugation. The nearly ideal periodic correlation function of m-sequences is described in the following result.

Property R-5.

Let $\{a_n\}$ be a complex m-sequence of period N, composed of p-th roots of unity. The periodic correlation $P_{aa}(\tau)$ of $\{a_n\}$ has the form

$$P_{aa}(\tau) = \begin{cases} N, & \tau = 0 \bmod N \\ -1, & \tau \neq 0 \bmod N. \end{cases} \tag{5.86}$$

Proof. The case $\tau = 0 \bmod N$ is obvious. Note that $N = p^L - 1$, L being the linear span of the corresponding m-sequence $\{b_n\}$ over GF(p), and by R-2 the number of occurrences of any specified non-zero symbol in one period of $\{b_n\}$ is p^{L-1}. Hence, for $\tau \neq 0 \bmod N$,

$$P_{aa}(\tau) = \sum_{n=1}^{N} \rho^{b_{n+\tau} - b_n} \qquad [\text{by } (5.80)]$$

$$= \sum_{n=1}^{N} \rho^{b_{n+\tau'(\tau)}} \qquad [\text{by R-3}]$$

$$= -1 + \rho^{L-1} \sum_{j=0}^{p-1} \rho^j \qquad [\text{by R-2}]$$

$$= -1. \tag{5.87}$$

∎

When the sequence period N is large compared to the processing gain, as it is in many systems, the full period correlation $P_{aa}(\tau)$ loses some of its value as a design parameter. Correlation calculations in this case typically are carried out over blocks of K symbols, where K may be larger than the linear span L and much smaller than the period N. A more appropriate statistic for study in this case is the *partial-period correlation* defined as

$$P_{aa}(K, n, \tau) = \sum_{j=0}^{K-1} a_{n+j+\tau} a_{n+j}^* \tag{5.88}$$

which computes the cross-correlation between two blocks of K symbols from $\{a_n\}$, one block located τ symbols from the other.

Unlike full-period correlation calculations, partial-period correlation values depend on the initial location n in the sequence where the correlation computation begins. Hence, an explicit description of the function $P_{aa}(K, n, \tau)$ must include values over the range $0 \le n < N$ of n. Often the size of N precludes direct calculation of all these values, and computable time averages therefore are substituted to give a statistical description of the partial-period correlation function. Denoting the time-average operation by $\langle \cdot \rangle$, the first and second time-average moments of $P_{aa}(K, n, \tau)$ are given by

$$\langle P_{aa}(K, n, \tau) \rangle = \frac{1}{N} \sum_{n=1}^{N} P_{aa}(K, n, \tau) \tag{5.89}$$

$$\langle |P_{aa}(K, n, \tau)|^2 \rangle = \frac{1}{N} \sum_{n=1}^{N} |P_{aa}(K, n, \tau)|^2, \tag{5.90}$$

where the parameter over which the average is being computed is n.

Property R-6.

Let $\{a_n\}$ be a complex *m*-sequence of period N, composed of p-th roots of unity. The time-averaged first and second moments of the partial-period correlation function of $\{a_n\}$ are given by

$$\langle P_{aa}(K, n, \tau) \rangle - \begin{cases} -K/N, & \tau \ne 0 \bmod N \\ K, & \tau = 0 \bmod N \end{cases} \tag{5.91}$$

and

$$\langle |P_{aa}(K, n, \tau)|^2 \rangle = \begin{cases} K\left(1 - \dfrac{K-1}{N}\right), & \tau \ne 0 \bmod N \\ K^2, & \tau = 0 \bmod N \end{cases} \tag{5.92}$$

respectively, for $K \le N$.

Proof. The first moment calculation is straightforward:

$$\langle P_{aa}(K, n, \tau) \rangle = \frac{1}{N} \sum_{n=1}^{N} \sum_{k=0}^{K-1} a_{n+k+\tau} a_{n+k}^*$$

$$= \frac{1}{N} \sum_{k=0}^{K-1} P_{aa}(\tau), \tag{5.93}$$

and (5.91) follows directly from Property R-5.

The derivation of the second moment uses the additional fact that Property R-3 for an *m*-sequence over GF(p) translates via (5.80) into the property

$$a_{n+\tau} a_n^* = a_{n+\tau'(\tau)}, \quad \text{for all } n, \tag{5.94}$$

when $\tau \neq 0 \bmod N$. Hence,

$$
\begin{aligned}
\langle |P_{aa}(K,n,\tau)|^2 \rangle &= \frac{1}{N} \sum_{n=1}^{N} \sum_{j=0}^{K-1} a_{n+j+\tau} a_{n+j}^* \left(\sum_{k=0}^{K-1} a_{n+k+\tau} a_{n+k}^* \right)^* \\
&= \frac{1}{N} \sum_{n=1}^{N} \sum_{j=0}^{K-1} \sum_{k=0}^{K-1} a_{n+j+\tau'(\tau)} a_{n+k+\tau'(\tau)}^* \\
&= \frac{1}{N} \sum_{j=0}^{K-1} \sum_{k=0}^{K-1} P_{aa}(j-k).
\end{aligned}
\tag{5.95}
$$

Noting that there are K terms for which $j = k$, and $K(K - 1)$ terms for which $j \neq k$, the final result follows immediately by applying Property R-5 to (5.95). $\blacksquare$

As a check, note that the results of Property R-6 reduce to those of the full period case R-5 when $K = N$.

For comparison, consider a periodic sequence $\{x_n\}$ composed of N independent, identically distributed (i.i.d.) random variables, uniformly distributed over the elements of GF(p), p prime. Furthermore, let $\{z_n\}$ be the corresponding complex sequence determined by the usual mapping (5.80) to p-th roots of unity. Clearly the elements of $\{z_n\}$ are i.i.d. random variables, uniformly distributed on the p-th roots of unity. Both the full-period and partial-period time-average autocorrelation functions of $\{z_n\}$ are random variables whose ensemble-average moments can be evaluated, using the independence assumption and the fact that

$$
\mathbf{E}\{z_n\} = 0,
\tag{5.96}
$$

where $\mathbf{E}$ denotes the ensemble average operator. The first moment of the partial period correlation of $\{z_n\}$ is easily shown to be

$$
\mathbf{E}\{P_{zz}(K,n,\tau)\} = \begin{cases} K, & \tau = 0 \bmod N \\ 0, & \tau \neq 0 \bmod N \end{cases}
\tag{5.97}
$$

Evaluation of the second moment of $P_{zz}(K,n,\tau)$ uses the fourth moment

$$
\mathbf{E}\{z_{n+j+\tau} z_{n+j}^* z_{n+k+\tau}^* z_{n+k}\} = \begin{cases} 1, & \tau = 0 \bmod N \\ 1, & k = j \text{ and } \tau \neq 0 \bmod N \\ 0, & k \neq j \text{ and } \tau \neq 0 \bmod N \end{cases}
\tag{5.98}
$$

to yield

$$
\mathbf{E}\{|P_{zz}(K,n,\tau)|^2\} = \sum_{j=0}^{K-1} \sum_{k=0}^{K-1} \mathbf{E}\{z_{n+j+\tau} z_{n+j}^* z_{n+k+\tau}^* z_{n+k}\}
$$

$$
= \begin{cases} K^2, & \tau = 0 \bmod N \\ K, & \tau \neq 0 \bmod N. \end{cases}
\tag{5.99}
$$

Comparisons of (5.97) and (5.99) for a random sequence with (5.91) and

(5.92) for an *m*-sequence both indicate that when $K \ll N$, then the time-averaged mean and correlation values of an *m*-sequence are very close to the corresponding ensemble averages for a randomly chosen sequence. This fact and the balance properties of *m*-sequences are used to justify their approximation by a random sequence of bits in later analyses of SS system performance.

Both the full-period and partial-period correlation computations have a particularly simple characterization when $\{b_n\}$ is a sequence of elements from GF(2) and $\{a_n\}$ is a sequence of $+1$'s and -1's. In this case when $\tau \neq 0 \bmod N$,

$$P_{aa}(K, n, \tau) = \sum_{j=0}^{K-1} (-1)^{b_{n+j+\tau} - b_{n+j}}$$

$$= \sum_{j=0}^{K-1} (-1)^{b_{n+j+\tau'(\tau)}}$$

$$= K - wt\big((b_{n+\tau'(\tau)}, \ldots, b_{n+\tau'(\tau)+K-1})\big) \qquad (5.100)$$

where the weight $wt(x)$ of a vector x denotes the number of non-zero elements in x. Hence, $P_{aa}(K, n, \tau)$ is a simple affine transformation of the weight of a K-tuple from $\{b_n\}$, beginning at element index $n + \tau'(\tau)$. This relation, a direct result of the shift-and-add property, simplifies the tabulation and analysis of partial-period correlation statistics.

While the results of the last three sections appear to present a relatively complete theory and are to a great extent available in [12], the study of partial-period correlation has remained a topic of research interest for many years. Bartee and Wood [13] used an exhaustive search to find the binary *m*-sequence, of each possible period up to $2^{14} - 1$, which possessed the largest value for the minimum K-tuple weight, D.

$$D = \min_{1 \leq n \leq N} wt\big((b_n, \ldots, b_{n+K-1})\big) \qquad (5.101)$$

Equivalently by (5.101), Bartee and Wood found the *m*-sequences $\{a_n\}$ which possess the maximum value of $\max_{0 \leq n \leq N} \max_{0 < \tau < N} P_{aa}(K, n, \tau)$, for fixed linear span L and correlation length K.

Table 5.8 displays characteristic polynomials of *m*-sequences which possess the largest value of D among all *m*-sequences of linear span L. These polynomials are specified in octal, i.e., the coefficients of the polynomial are the bits in the binary representation of the octal number. The binary representation of the octal number is determined simply by converting each octal symbol to its equivalent three-bit binary representation. For example,

$$27405 \text{ (octal)} = \underline{010}\,\underline{111}\,\underline{100}\,\underline{000}\,\underline{101} \text{ (binary)} \qquad (5.102)$$

represents $z^{13} + z^{11} + z^{10} + z^9 + z^8 + z^2 + 1$. Leading zeros resulting from the octal to binary conversion may be ignored.

Table 5.8

Characteristic polynomials (in octal) of m-sequences with linear span L, having largest value D (in parentheses) of minimum K-tuple weight. The optimum polynomial is not unique. (Abstracted from [13].)

L \ K	10	20	30	40	50	60	70	80	90	100	200
2	(6)	(13)	(20)	7 (26)	7 (33)	7 (40)	7 (46)	7 (53)	7 (60)	7 (66)	7 (133)
3	(5)	(11)	(16)	13 (22)	13 (28)	13 (33)	13 (40)	13 (45)	13 (51)	13 (56)	13 (113)
4	(4)	(9)	(16)	23 (20)	23 (25)	23 (32)	23 (36)	23 (41)	23 (48)	23 (52)	23 (105)
5	(3)	(8)	(15)	75 (19)	45 (24)	45 (30)	75 (34)	45 (39)	45 (45)	75 (50)	75 (101)
6	(2)	(7)	(12)	155 (17)	103 (22)	103 (29)	147 (33)	147 (38)	147 (42)	155 (47)	(99)
7	(2)	(7)	(12)	367 (17)	367 (21)	313 (26)	211 (31)	345 (36)	345 (41)	211 (46)	(96)
8	(2)	(6)	(10)	703 (15)	703 (20)	747 (25)	747 (30)	703 (34)	453 (39)	703 (44)	(96)
9	(1)	(5)	(10)	1131 (14)	1715 (19)	1715 (24)	1773 (29)	1773 (34)	1423 (38)	1773 (43)	(91)
10		(5)	(9)	2033 (14)	3507 (18)	3507 (22)	3507 (27)	2461 (32)	3525 (37)	3525 (42)	(89)
11		(4)	(8)	4173 (13)	7461 (17)	7655 (22)	7655 (27)	7107 (31)	4055 (36)	5607 (40)	(89)
12		(4)	(8)	17147 (12)	14271 (17)	10605 (21)	17147 (25)	17025 (30)	12727 (35)	17025 (39)	(86)
13		(4)	(7)	34035 (11)	21103 (16)	27405 (20)	31231 (24)	23231 (29)	37335 (33)	32467 (38)	(85)
14				41657 (11)	73071 (15)	65575 (19)	64457 (23)	64167 (28)	61333 (32)	65277 (37)	(83)

Lindholm [14] derived expressions for the first five time-average moments of the partial period correlation values of m-sequences (including those of Property R-6). He noted that only the first two moments were independent of the characteristic polynomial of an m-sequence with specified period. Cooper and Lord [15], following a study of Mattson and Turyn [16], noted that for n in the vicinity of a run of $L - 1$ zeros in an m-sequence, $P_{aa}(K, n, \tau)$ had lower correlation values, as τ varied around zero, for m-sequences whose characteristic polynomials contained more non-zero coefficients. Wainberg and Wolf [17] compared the moments of K-tuple weight distributions for m-sequences over GF(2) to the corresponding moments for a purely random sequence, and, extending Lindholm's work, carried this out through the sixth moment with $K \leq 100$ for several sequences.

Fredricsson [18] carried this idea closer to the realm of coding theory by noting that the set of N-tuples consisting of one period of an m-sequence and all its cyclic shifts, and the all-zeros N-tuple, together form a linear code which is the dual of a single-error-correcting Hamming code. Since the set $\{(b_n, \ldots, b_{n+K-1}): n = 1, 2, \ldots, N\}$ of K-tuples along with the all-zeros K-tuple form a punctured version of this m-sequence code, the set's dual is a shortened Hamming code. The weight distribution of the shortened Hamming code can be related in turn to the moments of the weight distribution of K-tuples from $\{b_n\}$ by the MacWilliams-Pless identities [19]. Bekir [20] extended this approach by applying moment techniques [21], [22] to generate upper and lower bounds on the distribution function of partial-period correlation values.

5.5 GALOIS FIELD CONNECTIONS

5.5.1 Extension Field Construction

The reader with no prior knowledge of finite field structures is urged at this point to review Appendix 5A, for the connection between certain LFSRs and finite fields will now be clarified. Assume that one has mechanized arithmetic units to carry out addition and multiplication in a finite field GF(q) which we will refer to as the *ground field*. In many practical designs the ground field is GF(2) since modulo 2 arithmetic units are available. Let $m_\alpha(z)$ be an irreducible polynomial of degree d over GF(q), $d > 1$, with a root called α; $m_\alpha(z)$ is termed the *minimum polynomial of α over* GF(q). Certainly, α is not a member of GF(q), or else $m_\alpha(z)$ would have a factor $z - \alpha$ over GF(q). Hence, α is a member of a larger field, called an *extension field* of GF(q), which will now be constructed.

Let $\mathscr{S}_d$ be the set of q^d distinct polynomials over GF(q) of degree less than d in an indeterminate z. That is,

$$\mathscr{S}_d = \left\{ X(z): X(z) = \sum_{i=0}^{d-1} x_i z^i, x_i \in \text{GF}(q) \right\}. \qquad (5.103)$$

It is well known that the q^d distinct elements of $GF(q^d)$ correspond to the polynomials of $\mathscr{S}_d$ evaluated at a root of an irreducible polynomial of degree d over $GF(q)$. Therefore, let the field element x in $GF(q^d)$ corresponding to the polynomial $X(z)$ in $\mathscr{S}_d$ be given by

$$x = X(\alpha) = \sum_{i=0}^{d-1} x_i \alpha^i. \tag{5.104}$$

The representation (5.104) clearly indicates that $GF(q^d)$ can be viewed as a vector space of dimension d over a scalar field $GF(q)$ with $1, \alpha, \alpha^2, \ldots, \alpha^{d-1}$ serving as a basis. While there is also an obvious correspondence between field elements x and polynomials $X(z)$, the field element itself is not a polynomial in an indeterminate z, but rather should be viewed as a "pseudopolynomial" in the basis element α.

Addition of the elements x and y in the extension field, corresponding to the polynomials $X(z)$ and $Y(z)$ in $\mathscr{S}_d$, corresponds to vector addition.

$$x + y = X(\alpha) + Y(\alpha) = \sum_{i=0}^{d-1} (x_i + y_i) \alpha^i. \tag{5.105}$$

More effort is required to express the result of a multiplication operation in the same basis; specifically a polynomial multiplication modulo $m_\alpha(z)$ must first be carried out to determine $R(z)$, i.e.,

$$X(z)Y(z) = W(z)m_\alpha(z) + R(z), \tag{5.106}$$

where $R(z)$ is a member of $\mathscr{S}_d$. Then, since α is a root of $m_\alpha(z)$, it follows that

$$xy = X(\alpha)Y(\alpha) = R(\alpha). \tag{5.107}$$

Hence, computation in $GF(q^d)$ is often referred to as arithmetic modulo $m_\alpha(z)$.

5.5.2 The LFSR as a Galois Field Multiplier

In Section 5.2.3 the contents of a shift register after n shifts were considered to be coefficients of a polynomial $R^{(n)}(z)$, and successive register contents were related by the polynomial recursion (see (5.24)–(5.26))

$$zR^{(n)}(z) = b_n Q(z) + R^{(n+1)}(z), \tag{5.108}$$

with

$$R^{(0)}(z) = P(z), \tag{5.109}$$

where $P(z)/Q(z)$ is the formal power series representation of $\{b_n\}$. Suppose that

$$Q(z) = m_\alpha(z), \tag{5.110}$$

i.e., the characteristic polynomial of the sequence $\{b_n\}$ produced by a Galois-configured LFSR is irreducible over $GF(q)$ and has root α. Evaluating (5.108) at $z = \alpha$ gives

$$\alpha R^{(n)}(\alpha) = R^{(n+1)}(\alpha). \qquad (5.111)$$

Hence, one shift of a d-element LFSR with connections specified by $m_\alpha(z)$, corresponds to multiplication of a field element $R^{(n)}(\alpha)$ of $GF(q^d)$ by the element α, the register contents before the shift being the coefficients of the representation of $R^{(n)}(\alpha)$ in the basis $1, \alpha, \ldots, \alpha^{d-1}$, and after the shift being the coefficients for $R^{(n+1)}(\alpha)$. Solving the recursion in the finite field indicates that the register contents after n shifts represent

$$R^{(n)}(\alpha) = \alpha^n P(\alpha). \qquad (5.112)$$

Certainly, the period of the state sequence (s_n) for this LFSR (see (5.35)) is the same as the smallest value of N such that

$$\alpha^n = 1. \qquad (5.113)$$

This smallest value of N satisfying (5.113) is called the *exponent* (or *order*) *of* α, and can be shown to be a divisor of the number of non-zero elements of $GF(q^d)$, namely $q^d - 1$ (see Appendix 5A.6). Furthermore, it has been shown that the multiplicative group of a finite field is cyclic; hence, each Galois field $GF(q^d)$ contains elements whose exponents are exactly $q^d - 1$, these elements being called *primitive elements* of the Galois field's multiplicative group. The minimum polynomials of these primitive elements are the *primitive polynomials* appearing as characteristic polynomials of *m*-sequences.

5.5.3 Determining the Period of Memory Cell Outputs

While the multiplier viewpoint relates the period of certain LFSR state sequences to field element properties, it is not yet clear that the period of the state sequence $\{s_n\}$ is identical to the period of the output sequence $\{b_n\}$. Referring to Figure 5.6, let $R_i(z)$ denote the formal power series representation of $\{r_i^{(n-1)}\}$, the sequence of contents of the i-th memory element in an arbitrary Galois-configured LFSR.

$$R_i(z) \triangleq \sum_{n=1}^{\infty} r_i^{(n-1)} z^{-n}. \qquad (5.114)$$

The output memory element has index 1 and, therefore,

$$R_1(z) = \frac{P(z)}{Q(z)}, \qquad (5.115)$$

where $Q(z)$ is the characteristic polynomial of the output sequence $\{b_n\}$. The series representation of the sequence produced by the second memory element can be related to $R_1(z)$ by noting that the input to the first memory element obeys the relation

$$r_1^{(n)} = r_2^{(n-1)} - q_1 r_1^{(n-1)}, \qquad n = 1, 2, \ldots, \qquad (5.116)$$

where q_1 is a coefficient of $Q(z)$ (see (5.20)). Hence, carefully checking initial conditions,

$$R_1(z) = z^{-1}\big[r_1^{(0)} + R_2(z) - q_1 R_1(z)\big], \qquad (5.117)$$

whence

$$R_2(z) = (z + q_1)R_1(z) - r_1^{(0)}. \qquad (5.118)$$

By iterating this computation and substituting (5.115), it can be shown that the formal power series representation for the sequence generated by the k-th memory cell is

$$R_k(z) = \left(\sum_{i=0}^{k-1} q_i z^{k-1-i}\right)\frac{P(z)}{Q(z)} - \left(\sum_{i=1}^{k-1} r_i^{(0)} z^{k-1-i}\right), \qquad k = 2, 3, \ldots, L.$$

$$(5.119)$$

where L is the number of memory elements (the degree of $Q(z)$) and also the linear span of $\{b_n\}$. Since $P(z)$ and $Q(z)$ contain no common factors, the following result is now evident.

THEOREM 5.5. *Let $Q(z)$ be the characteristic polynomial for the output of a Galois-configured LFSR. Then (with indexing as shown in Figure 5.6) the sequence generated in the k-th memory element, $k > 1$, has characteristic polynomial given by $Q(z)/\gcd(Q(z), \sum_{i=0}^{k-1} q_i z^{k-1-i})$.*

COROLLARY 5.1. *Let the minimum polynomial $m_\alpha(z)$ over GF(q) be the characteristic polynomial of the output sequence from a Galois-configured LFSR. Then all memory cells in the register produce sequences having characteristic polynomial $m_\alpha(z)$ and possess identical periods equal to the exponent of α.*

Proof. Since $m_\alpha(z)$ is irreducible, all cells have the same characteristic polynomial by Theorem 5.5, and all possess identical periods by Theorem 5.1. ∎

Example 5.7. As an aside to illustrate what can happen when $Q(z)$ is composite, consider a Galois LFSR with output sequence possessing the following characteristic polynomial over GF(2).

$$Q(z) = z^8 + z^7 + z^6 + z^4 + z^3 + z + 1$$

$$= (z^3 + z + 1)^2(z^2 + z + 1). \qquad (5.120)$$

Since both $(z^3 + z + 1)$ and $(z^2 + z + 1)$ are primitive with periods 7 and 3, respectively, Theorem 5.2 indicates that the period of the output sequence for a properly initialized register should be $2 \times 3 \times 7$, i.e., 42. Theorem 5.5 predicts that the characteristic polynomial for the sequences of memory cells 3 and 4 is $(z^3 + z + 1)^2$, while that for the sequence of cell 5 is $(z^3 + z + 1)(z^2 + z + 1)$. Hence, cells 3 and 4 will generate period 14 sequences (see Theorem 5.2), while cell 5 will produce a period 21 sequence. The register

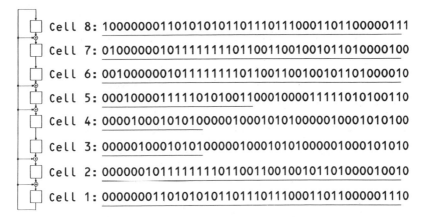

Cell 8: 100000001101010101101110111000110110000111

Cell 7: 01000000101111111101100110010010110100000100

Cell 6: 0010000001011111111011001100100101101000010

Cell 5: 000100001111101010011000100001111010100110

Cell 4: 0000100010101000001000101010000010001010100

Cell 3: 0000010001010100000100010101000001000101010

Cell 2: 0000001011111111011001100100101101000010010

Cell 1: 0000000110101010110111011100011011000001110

Figure 5.9. A Galois LFSR producing sequences with differing periods. Single periods are underlined.

configuration and memory cell sequences for this example are shown in Figure 5.9.

5.5.4 The Trace Representation of M-Sequences

A fundamental mathematical tool used in the further investigation of PN generators is a particular linear mapping from a finite field onto a subfield. This mapping, called the *trace function*, will now be reviewed, and an explicit expression for the elements of an m-sequence will be constructed.

Let $\mathrm{GF}(q)$ be any finite field contained within a larger field $\mathrm{GF}(q^d)$. Then the *trace polynomial* $\mathrm{Tr}_q^{q^d}(z)$ from $\mathrm{GF}(q^d)$ to $\mathrm{GF}(q)$ is defined as

$$\mathrm{Tr}_q^{q^d}(z) \triangleq \sum_{i=0}^{d-1} z^{q^i}. \tag{5.121}$$

The trace (function) in $\mathrm{GF}(q)$ of an element α in $\mathrm{GF}(q^d)$ is defined as $\mathrm{Tr}_q^{q^d}(\alpha)$, i.e., the trace polynomial evaluated at α. The values of q and d are often obvious in the context of a particular application, in which case the cumbersome superscript and subscript are dropped from the trace notation.

The trace function has the following useful properties which are proved in Appendix 5A.9.

Property T-1.

When α is in $\mathrm{GF}(q^d)$, then $\mathrm{Tr}_q^{q^d}(\alpha)$ is in $\mathrm{GF}(q)$.

Property T-2.

All roots of an irreducible polynomial $m_\alpha(z)$ over $\mathrm{GF}(q)$, with root α in $\mathrm{GF}(q^d)$, have the same trace, i.e.,

$$\mathrm{Tr}_q^{q^d}(\alpha^{q^i}) = \mathrm{Tr}_q^{q^d}(\alpha) \quad \text{for all } i. \tag{5.122}$$

Property T-3.

The trace function is linear. That is, for a and b in $GF(q)$, and α and β in $GF(q^d)$,

$$\mathrm{Tr}_q^{q^d}(a\alpha + b\beta) = a\,\mathrm{Tr}_q^{q^d}(\alpha) + b\,\mathrm{Tr}_q^{q^d}(\beta). \qquad (5.123)$$

Property T-4.

For each choice of b in $GF(q)$ there are q^{d-1} elements α in $GF(q^d)$ for which

$$\mathrm{Tr}_q^{q^d}(\alpha) = b. \qquad (5.124)$$

Property T-5.

If $GF(q) \subset GF(q^k) \subset GF(q^d)$, then

$$\mathrm{Tr}_q^{q^d}(\alpha) = \mathrm{Tr}_q^{q^k}\!\left(\mathrm{Tr}_{q^k}^{q^d}(\alpha)\right), \qquad (5.125)$$

for all α in $GF(q^d)$.

Those familiar with finite fields will recognize that when the minimum polynomial $m_\alpha(z)$ of α over $GF(q)$ has degree d, then $\alpha, \alpha^q, \alpha^{q^2}, \ldots, \alpha^{q^{d-1}}$ are its distinct roots, $-\mathrm{Tr}_q^{q^d}(\alpha)$ is the coefficient of z^{d-1} in $m_\alpha(z)$, and the trace is therefore in $GF(q)$. However, when $m_\alpha(z)$ has degree k, $k < d$ (k must divide d), then the roots of $m_\alpha(z)$ appear d/k times in (5.121) and $\mathrm{Tr}_q^{q^d}(\alpha)$ is $(d/k)\mathrm{Tr}_q^{q^k}(\alpha)$.

Once a primitive element α has been specified by choosing its minimum polynomial $m_\alpha(z)$ of degree d over $GF(q)$, then the trace of any non-zero element α^k in $GF(q)$ can be evaluated directly.

$$\mathrm{Tr}_q^{q^d}(\alpha^j) = \mathrm{Tr}_q^{q^d}(z^j)\,\mathrm{mod}\,m_\alpha(z). \qquad (5.126)$$

The substitution of z^j for z in the trace polynomial allows the evaluation of the trace of α^j by substitution of α, and the reduction of the resulting polynomial mod $m_\alpha(z)$ eliminates terms which will be zero upon evaluation at α. The result on the right side of (5.126) is an element of $\mathscr{S}_d$ (see (5.103)), while Property T-2 implies that the quantity on the left must be in $GF(q)$. Since the elements of $\mathscr{S}_d$ corresponding to $GF(q)$ elements are in fact those same elements, i.e., only x_0 in (5.104) is non-zero in this case, the substitution of $z = \alpha$ on the right side of (5.126) is not necessary.

Example 5.8. Let α be a root of the primitive polynomial $z^5 + z^2 + 1$. The root α and its conjugates (other roots of the same primitive polynomial), namely α^2, α^4, α^8, and α^{16}, are elements of $GF(32)$. Other elements of $GF(32)$ can also be grouped into root sets. The powers on α corresponding to a root set must be relatively prime to the order of α to insure that the corresponding polynomial is primitive. In this case primitivity is guaranteed

Table 5.9
Cyclotomic cosets and associated trace function values
when α is a root of $z^5 + z^2 + 1$ over GF(2).

Cyclotomic coset elements x	$\mathrm{Tr}_2^{32}(\alpha^x)$
1, 2, 4, 8, 16	0
3, 6, 12, 24, 17	1
5, 10, 20, 9, 18	1
7, 14, 28, 25, 19	0
11, 22, 13, 26, 21	1
15, 30, 29, 27, 23	0
0	1

for all degree 5 irreducible polynomials because the order of α, namely 31, is prime. Each set of root powers is called a *cyclotomic coset* (see Appendix 5A.8). Property T-2 indicates that all the roots of the same irreducible polynomial have the same trace value; hence, trace values can be computed by (5.126) and associated with cyclotomic cosets. This is illustrated in Table 5.9 for this example.

The following theorem provides a convenient representation for certain LFSR sequences having irreducible characteristic polynomials. Here, and in discussions to follow, the superscripts and subscripts on the trace function will be omitted when their values are evident.

THEOREM 5.6. *Let* GF(q^d) *be the smallest field containing the element* α, *and let* $m_\alpha(z)$ *be the degree d minimum polynomial of* α *over* GF(q). *Then the sequence* $\{\mathrm{Tr}_q^{q^d}(\alpha^n)\}$, *n being the sequence index, has characteristic polynomial* $m_\alpha(z)$.

Proof. Consider a d-cell Galois LFSR producing an output sequence having characteristic polynomial $m_\alpha(z)$. Combining the representation (5.22) for the remainder polynomial in terms of memory contents, with the field element interpretation of (5.112) gives

$$\alpha^n = \beta R^{(n)}(\alpha) = \beta \sum_{i=1}^{L} r_i^{(n)} \alpha^{L-i} \tag{5.127}$$

where

$$\beta = [P(\alpha)]^{-1} \tag{5.128}$$

with α and β in GF(q^d). Applying the trace function with Property T-3 to both sides of (5.127) and noting that the cell sequences are elements of GF(q), gives

$$\mathrm{Tr}(\alpha^n) = \sum_{i=1}^{L} \mathrm{Tr}(\beta \alpha^{L-i}) r_i^{(n)}. \tag{5.129}$$

Corollary 5.1 indicates that the L memory cell sequences $\{r_i^{(n)}\}$, $i = 1, \ldots, L$, all have the same characteristic polynomial $m_\alpha(z)$; therefore, the linear combination of those sequences specified by (5.129), namely the sequence $\{\text{Tr}(\alpha^n)\}$, has the same characteristic polynomial.   ∎

The following corollary provides an explicit and compact mathematical representation of the elements of an m-sequence, which will be used in several forthcoming analyses.

COROLLARY 5.2. *If $\{b_n\}$ is an m-sequence over* $\text{GF}(q)$ *with the minimum polynomial $m_\alpha(z)$ as its characteristic polynomial, then there exists a non-zero element γ in* $\text{GF}(q^d)$ *such that*

$$\{b_n\} = \left\{ \text{Tr}_q^{q^d}(\gamma\alpha^n) \right\}. \qquad (5.130)$$

Proof. By Theorem 5.6, the same recursion which generates $\{b_n\}$ must also produce $\{\text{Tr}(\alpha^n)\}$, as well as any shift thereof. Since $\{b_n\}$ is an m-sequence, its characteristic polynomial $m_\alpha(z)$ is primitive, and the corresponding LFSR supports only two cyclically distinct state sequences, namely the perpetual zero-state sequence and the state sequence of $\{b_n\}$. Property T-4 indicates that $\text{Tr}(\alpha^n)$ must be non-zero for some n; hence, the state sequence corresponding to the trace sequence cannot be the zero-state sequence. Therefore, the trace sequence or some shift of it must be the m-sequence. That is,

$$\{\text{Tr}(\alpha^{m+n})\} = \{b_n\} \qquad (5.131)$$

for some m, and setting $\gamma = \alpha^m$ completes the proof.   ∎

5.5.5 A Correlation Computation

To illustrate the use of trace representations in the study of periodic correlation properties, consider the calculation of the periodic correlation of $\{a_n\}$, a roots-of-unity m-sequence derived from an m-sequence $\{b_n\}$ over $\text{GF}(q)$ via (5.80). Let $m_\alpha(z)$ denote the degree d primitive characteristic polynomial of $\{b_n\}$. Then the periodic correlation function of $\{a_n\}$, defined in (5.85), can be rewritten in terms of the trace function by using Corollary 5.2 and simplifying with Property T-3.

$$P_{aa}(\tau) = \sum_{n=1}^{N} \rho^{\text{Tr}(\gamma\alpha^{n+\tau}) - \text{Tr}(\gamma\alpha^n)}$$

$$= \sum_{n=1}^{N} \rho^{\text{Tr}(\delta\alpha^n)}, \qquad (5.132)$$

where

$$\delta = \gamma(\alpha^\tau - 1). \qquad (5.133)$$

Since α is primitive, α^n takes on the values of all the non-zero elements in $GF(q^d)$ as n goes from 1 to N, N being the sequence period. Adding the zero element to this list and noting that its trace is zero, yields

$$P_{aa}(\tau) = -1 + \sum_{\beta \in GF(q^d)} \rho^{\delta\beta}. \tag{5.134}$$

As β varies over $GF(q^d)$, so does $\delta\beta$ unless δ is zero, and applying Property T-4, gives

$$P_{aa}(\tau) = \begin{cases} q^d - 1, & \delta = 0 \\ -1, & \delta \neq 0. \end{cases} \tag{5.135}$$

The condition $\delta = 0$ occurs if, and only if, $\alpha^\tau = 1$, and since α is primitive,

$$\delta = 0 \text{ iff } \tau = 0 \bmod q^d - 1. \tag{5.136}$$

While this has been a somewhat more involved proof of randomness Property R-5 than that given in Section 5.4.4, the trace function methodology employed here readily generalizes to several more complicated full-period correlation calculations.

5.5.6 Decimations of Sequences

Let $\{b_n\}$ be an arbitrary sequence of period N, and consider the sequence $\{c_n\}$ defined by

$$c_n = b_{Jn}, \quad \text{for all } n. \tag{5.137}$$

The sequence $\{c_n\}$ is said to be the *decimation by J* of the sequence $\{b_n\}$. If J divides N, then $\{c_n\}$ has period N/J. On the other hand, if J and N are relatively prime, then the smallest multiple of J that is a multiple of N is NJ, and the period of $\{c_n\}$ is N. In general, the decimation by J of a sequence with period N produces a sequence with period $N/\gcd(J, N)$.

It should be clear to the reader that when $\gcd(J, N)$ is 1, the resulting rearrangement of sequence elements does not change the first order statistics of the sequence. A more interesting result indicates that the set of correlation values of real number sequences is preserved under decimation by a number relatively prime to the sequence period.

THEOREM 5.7. *Let $\{a_n\}$ be an arbitrary sequence of complex numbers with period N, and let $\{c_n\}$ be the decimation by J of $\{a_n\}$, where $\gcd(J, N) = 1$. Then*

$$P_{cc}(\tau) = P_{aa}(J\tau). \tag{5.138}$$

Proof. Evaluation of the periodic correlation of $\{c_n\}$ gives

$$P_{cc}(\tau) = \sum_{n=1}^{N} a_{J(n+\tau)} a_{Jn}^*$$

$$= \sum_{k \in \mathcal{X}} a_{k+J\tau} a_k^*, \tag{5.139}$$

where

$$\mathscr{K} = \{ k \colon Jn = k \bmod N, 1 \leq n \leq N \}. \tag{5.140}$$

If two distinct choices of n, say n_1 and n_2, in the range 1 to N, yield the same value k in (5.140), then

$$Jn_1 = Jn_2 \bmod N, \tag{5.141}$$

or, assuming $n_1 > n_2$,

$$J(n_1 - n_2) = mN \tag{5.142}$$

where m is integer. However, this is impossible since N has no factors in common with J, and $n_1 - n_2$ is less than N. Therefore, the elements of $\mathscr{K}$ are the N distinct integers from 1 to N. ∎

The trace representation gives considerable insight into the effects of decimation on m-sequences. Suppose that by Corollary 5.2 we have constructed the m-sequence

$$\{ b_n \} = \mathrm{Tr}_q^{q^d}(\alpha^n), \tag{5.143}$$

and decimation by J of $\{ b_n \}$ results in the sequence

$$\{ f_n \} = \mathrm{Tr}_q^{q^d}(\alpha^{Jn}). \tag{5.144}$$

When J is relatively prime to the period $q^d - 1$ of $\{ b_n \}$, then α^J is another primitive element of $\mathrm{GF}(q^d)$ and, thus, $\{ f_n \}$ is an m-sequence with the minimum polynomial $m_{\alpha^J}(z)$ for its characteristic polynomial. The results on decimations of m-sequences can now be summarized.

THEOREM 5.8. Let $\{ b_n \}$ denote an m-sequence over $\mathrm{GF}(q)$ with linear span L. Then:

    (a) A decimation $\{ b_{Jn} \}$ of $\{ b_n \}$ is an m-sequence if and only if J is relatively prime to $q^L - 1$.

    (b) Two sequences $\{ b_{Jn} \}$ and $\{ b_{Kn} \}$ produced by decimations, with J and K both relatively prime to $q^L - 1$, are cyclically distinct if and only if

$$J \neq q^k K \bmod q^L - 1, \tag{5.145}$$

    for all integers k.

    (c) All m-sequences of period $q^L - 1$ can be constructed by decimations of $\{ b_n \}$.

Example 5.9. Consider the set of m-sequences over $\mathrm{GF}(2)$ with period 31. Let α be a root of the primitive polynomial $z^5 + z^2 + 1$. One can construct the m-sequence $\mathrm{Tr}(\alpha^n)$ and its decimations by various values of J simply by

Table 5.10

M-sequences $\mathrm{Tr}(\alpha^{Jn})$, $n = 0, 1, \ldots$, of period 31 over GF(2), constructed by decimations by J, along with their characteristic polynomials.

J	Sequence	Characteristic Polynomial (in octal)
1	1001011001111100011011101010000	45
3	1111101110001010110100001100100	75
5	1110100010010101100001110011011	67
7	1001001100001011010100011101111	57
11	1110110011100001101010010001011	73
15	1000010101110110001111100110100	51

reading the trace values from Table 5.9. The results of this procedure are shown in Table 5.10. The characteristic polynomials of sequences constructed by the decimation process can be found directly by the techniques of Section 5.2.5, or by calculation of $m_{\alpha^J}(z)$ from $m_{\alpha}(z)$ as described in Appendix 5A.8.

5.6 NON-LINEAR FEED-FORWARD LOGIC

Two related disadvantages of m-sequences and other LFSR sequences have already been discussed: (1) Simple linear analysis will allow an observer to predict the output of an L-stage LFSR from an observation of $2L$ consecutive output symbols. (2) The shift-and-add property of m-sequences allows an observer to produce different "time-advanced" versions of an m-sequence while observing the sequence, although the amount of shift cannot be predicted without further analysis. In either case, the observer, possessing no *a priori* information about the nature of the SS code generator, has the potential to predict the generator output and use this information to read the message, or to jam or spoof the intended receiver. As will be demonstrated in this section, vulnerability to this type of countermeasure can be significantly reduced by the inclusion of non-linear feed-forward logic (NLFFL) operating on the contents of the m-sequence generator's shift register, to produce an SS code sequence having high linear span.

5.6.1 A Powers-of-α Representation Theorem

The mathematical analysis of NLFFL effects requires use of the representation of a sequence $\{b_n\}$ of GF(q) elements with period $q^M - 1$, in terms of the $q^M - 1$ sequences $\{1\}$ (the all ones sequence), $\{\alpha^n\}, \{\alpha^{2n}\}, \ldots,$ $\{\alpha^{(q^M-2)n}\}$, where α is a primitive element of GF(q^M). Such a representation of $\{b_n\}$ must be unique since the basis sequences in the representation are linearly independent. This can be proven by noting that if there exist coefficients $a_0, \ldots, a_{q^M-2}$, not all zero, such that

$$P(\alpha^n) \triangleq \sum_{i=0}^{q^M-2} a_i \alpha^{in} = 0, \qquad 0 \le n < q^M - 1, \qquad (5.146)$$

then the polynomial $P(z)$ has $q^M - 1$ roots and degree at most $q^M - 2$, a contradiction.

The linear span L of a sequence (i.e., the degree of its characteristic polynomial) is related to the sequence's powers-of-α representation in the following result.

THEOREM 5.9. *Let $\{b_n\}$ be a sequence over GF(q), and let α be a primitive element of GF(q^M). Let $\{b_n\}$ be represented as*

$$b_n = \sum_{\delta \in \Delta} a_\delta \alpha^{\delta n} \qquad (5.147)$$

for all n, where Δ is the set of indices of non-zero coefficients in the expansion. Then the linear span L of $\{b_n\}$ is equal to the number of terms in the representation (5.147), i.e.,

$$L = |\Delta|. \qquad (5.148)$$

Proof. Since b_n is in GF(q) and all elements of that field have multiplicative order dividing $q - 1$, it follows that

$$b_n = b_n^q$$

$$= \sum_{\delta \in \Delta} a_\delta^q \alpha^{q \delta n}, \qquad (5.149)$$

for all n, by Theorem 5A.7.1 of Appendix 5A. Since the representation (5.147) is unique, it must be identical to (5.149). Therefore,

$$a_{q\delta} = a_\delta^q \quad \text{for all} \quad \delta \in \Delta. \qquad (5.150)$$

This implies that Δ can be decomposed into cyclotomic cosets (see Appendix 5A.8), the terms of (5.147) with indices in the same coset summing to a trace function.

Let Δ' be composed of the coset leaders (single representative elements) of the cyclotomic cosets present in the decomposition of Δ. Then

$$b_n = \sum_{\delta \in \Delta'} \text{Tr}_q^{q^{d(\delta)}} (a_\delta \alpha^{\delta n}). \qquad (5.160)$$

Note that the trace function for the δ-th term is calculated from the smallest field GF($q^{d(\delta)}$) containing α^δ down to GF(q). In the form (5.160), $\{b_n\}$ can be viewed as the sum of $|\Delta'|$ sequences, each over GF(q). Furthermore, the number of terms $d(\delta)$ of (5.147) which are combined to produce the sequence $\{\text{Tr}(a_\delta \alpha^{\delta n})\}$, is identical to the degree of the minimum polynomial

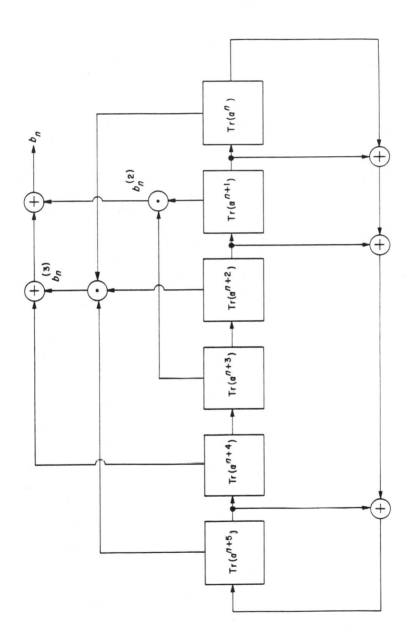

Figure 5.10(a) The LFSR with NLFFL used in Example 5.10. Arithmetic/logic operations are GF(2)/EXCLUSIVE OR and AND.

$m_{\alpha^\delta}(z)$, which by Theorem 5.6 is the characteristic polynomial of that sequence. Therefore, the characteristic polynomial $Q(z)$ of $\{b_n\}$ is given by

$$Q(z) = \prod_{\delta \in \Delta'} m_{\alpha^\delta}(z), \tag{5.161}$$

and has degree L given by

$$L = \sum_{\delta \in \Delta'} d(\delta) = |\Delta|. \tag{5.162}$$

Thus, the linear span of a sequence can be evaluated simply by counting terms in its powers-of-α representation. ∎

Since the index n in $\{b_n\}$ usually denotes a discrete time value and since δ multiplies n in (5.147), we will refer to the elements δ of the set Δ in (5.147) as *time coefficients*. Theorem 5.9 can be rephrased to state that the linear span of $\{b_n\}$ is the number of distinct time coefficients in the powers-of-α representation of $\{b_n\}$.

Example 5.10. Consider the binary 6-stage LFSR with primitive characteristic polynomial $z^6 + z^5 + z^2 + z + 1$, to which has been added NLFFL as shown in Figure 5.10(a). By loading the register with $\mathrm{Tr}_2^{64}(\alpha^i)$, $i = 0, \ldots, 5$, i.e., 011111, where α is a root of the LFSR's characteristic polynomial, the sequences in Figure 5.10(b) are produced at different points in the generator. One full period is shown, with the first bit corresponding to index $n = 0$.

The contents of the individual storage elements have been expressed as the trace functions indicated in the above diagram, in accordance with Theorem 5.6. These functions, having domain GF(64) and range GF(2), are defined as

$$\mathrm{Tr}_2^{64}(\alpha^j) = \sum_{i=0}^{5} \alpha^{j 2^i}. \tag{5.163}$$

The NLFFL output sequence $\{b_n\}$ can be represented in terms of the

---

$\mathsf{Tr}(\alpha^n)$:
011111101011100011001110110000011110010010101001101000010001011
$\mathsf{Tr}(\alpha^{n+4})$:
111010111000110011011000001111001001010100110100001000101011000101110111
$b_n^{(2)}$:
111101010100000000010100000000110000000101000001000000000010010
$b_n^{(3)}$:
010100100000000000001000000000001000000001000000000000000001001
b_n:
0100110011001100111100000001110111001011111110110001000100101100

Figure 5.10(b). Sequences produced at key locations within the generator of Example 5.10.

storage element sequences by

$$b_n = \text{Tr}(\alpha^{n+4}) + \text{Tr}(\alpha^{n+1})\text{Tr}(\alpha^{n+3})$$
$$+ \text{Tr}(\alpha^n)\text{Tr}(\alpha^{n+2})\text{Tr}(\alpha^{n+5})$$
$$= \text{Tr}_2^{64}(\alpha^{n+17}) + \text{Tr}_2^{64}(\alpha^{3n+28})$$
$$+ \text{Tr}_2^{64}(\alpha^{5n+47}) + \text{Tr}_2^{64}(\alpha^{7n+6})$$
$$+ \text{Tr}_2^{64}(\alpha^{11n+28}) + \text{Tr}_2^{64}(\alpha^{13n+29})$$
$$+ \text{Tr}_2^4(\alpha^{21n+21}). \tag{5.164}$$

Formed by substitution of (5.163) and simplification, the right side of (5.164) contains trace expressions with different domains; hence, domain and range are specified for each term. The above calculation is tedious and, even for this example, is more easily done by computer. By applying Theorem 5.9 and counting the terms in the trace functions of (5.164), namely six functions with six terms and one function with two terms, the linear span L of $\{b_n\}$ is determined to be 38.

The expansion (5.164) indicates that the characteristic polynomial $Q(z)$ of the NLFFL output sequence $\{b_n\}$ is

$$Q(z) = m_\alpha(z)m_{\alpha^3}(z)m_{\alpha^5}(z)m_{\alpha^7}(z)m_{\alpha^{11}}(z)m_{\alpha^{13}}(z)m_{\alpha^{21}}(z). \tag{5.165}$$

The factors of $Q(z)$ can be determined from knowledge that $m_\alpha(z)$ is $z^6 + z^5 + z^2 + z + 1$, and the techniques of Appendix 5A.8, to give

$$Q(z) = z^{38} + z^{37} + z^{31} + z^{27} + z^{26} + z^{25}$$
$$+ z^{23} + z^{22} + z^{20} + z^{19} + z^{18} + z^{17}$$
$$+ z^{16} + z^{15} + z^8 + z^7 + z^4 + z^3$$
$$+ z^2 + z + 1. \tag{5.166}$$

As a check, it can be verified that the output sequence shown in Figure 5.10(b) satisfies a recursion based on the characteristic polynomial $Q(z)$.

5.6.2 Key's Bound on Linear Span

Suppose that a binary NLFFL function is put in Reed-Muller canonic form, i.e., described as a sum of products of its inputs. Let's investigate the potential number of terms contributed to the powers-of-α expansion of the logic's output by one product term containing J factors, each factor being the contents of some memory element in an m-sequence generator. Assuming the characteristic polynomial of the m-sequence generator is $m_\alpha(z)$, the trace function $\text{Tr}(\gamma_j\alpha^n)$ can be used to represent the j-th factor (sequence) in the product p_n, giving

$$p_n = \prod_{j=1}^{J} \text{Tr}(\gamma_j\alpha^n) = \prod_{j=1}^{J} \sum_{i=0}^{M-1} (\gamma_j\alpha^n)^{2^i}, \tag{5.167}$$

where γ_j, $j = 1, \ldots, J$, and α are elements of $GF(2^M)$, M being the number of memory elements in the m-sequence generator and also the degree of $m_\alpha(z)$. (This representation is valid for a linear m-sequence generator in any configuration.)

Converting the product of sums in (5.167) to a sum of products gives

$$p_n = \sum_{i_1=0}^{M-1} \cdots \sum_{i_J=0}^{M-1} \gamma_i \alpha^{c(i)n} \tag{5.168}$$

where

$$\gamma_i = \prod_{j=1}^{J} \gamma_j^{2^{i_j}}, \tag{5.169}$$

$$c(i) = \sum_{j=1}^{J} 2^{i_j} \mod 2^M - 1, \tag{5.170}$$

the latter sum being modulo the multiplicative order of the element α. The base 2 representation of $c(i)$, thus, is limited to a binary M-tuple, and (5.170) indicates that at most J of these M symbols can be ones. Since each of the integers i_j in (5.170) ranges independently between zero and $M - 1$, it follows that the number $_M N_J$ of distinct values that $c(i)$ can assume is given by the number of binary M-tuples with at most J ones, excluding the all-zeros n-tuple.

$$_M N_J = \sum_{j=1}^{J} \binom{M}{j}. \tag{5.171}$$

Two different products (AND gates), with distinct sets of J inputs from the memory cells of an m-sequence generator, potentially produce the same set of powers-of-α sequences for their output representations. Furthermore, when one product has J inputs and another has K inputs, $J < K$, the representation of the K-fold product can potentially contain all powers-of-α sequences which occur in the J-fold product, and more. The word "potentially" is used because multiple terms with the same time coefficient $c(i)$ (5.170) are produced, and there is a small likelihood that the coefficients of these terms will add to zero.

The *order D* of an NLFFL function is defined as the largest number of factors in any product from its sum-of-products representation. The following theorem is a consequence of the above arguments.

THEOREM 5.10 (Key's Bound [23]). *The linear span L of a sequence $\{b_n\}$ produced by NLFFL of order D operating on the contents of an M-stage m-sequence generator is bounded by*

$$L \leq _M N_D = \sum_{j=1}^{D} \binom{M}{j}. \tag{5.172}$$

Table 5.11
Cyclic equivalence class representatives of non-zero coefficients
in the powers-of-α representation in Example 5.11.

	coefficients
base-2	base-10
000001	1
000011	3
000101	5
000111	7
001011	11
001101	13
010101	21

If the logic function includes the possibility of complementation by the use of an additional constant input term "1" in its sum-of-products representation, then the lower limit on the sum in Key's bound (5.172) must be reduced to zero.

Example 5.11. Key's bound on the linear span L for the output of third-order logic operating on a six-stage m-sequence generator is

$$L \le \binom{6}{1} + \binom{6}{2} + \binom{6}{3} = 41. \qquad (5.173)$$

The linear span achieved under these conditions in Example 5.10 is 38. Binary 6-tuple representations of the time coefficients (5.170) occurring in $\{b_n\}$'s powers-of-α expansion in the example include the ones shown in Table 5.11 and their cyclic shifts. This list includes cyclically equivalent representatives of all 6-tuples having weight 3 or less, with the exception of 001001 which in decimal is 9. It can be verified that α^{9n} is produced eight different ways in the product-of-sums to sum-of-products conversion (5.164) in the example, and that the eight corresponding coefficients from GF(64) sum to zero, thereby eliminating $\{\alpha^{9n}\}$ as a component in the representation of $\{b_n\}$. Since α^9 has order 7, it is an element of GF(8) and, therefore, has a three term trace and a degree-three minimum polynomial. Hence, this missing term accounts for the difference of three between Key's bound and the achieved linear span.

It has been pointed out [24] that for cryptographic applications, linear span is only one of several measures that must be considered in selecting a key sequence. This note of caution applies equally well to the selection of spectrum-spreading sequences. For example, the logic imposed on the m-sequence generator of Figure 5.11 produces an output whose linear span is 31, the period of the m-sequence generator. On the other hand, the AND gate output will be a "1" only when the all-ones 5-tuple appears in the register, an event which occurs only once in the 31 bit period of the generator. The output of such a generator can hardly be called pseudo-

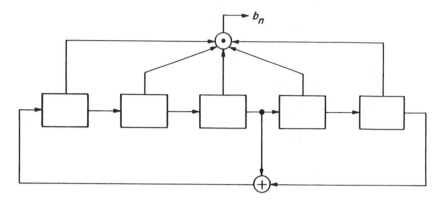

Figure 5.11. A high-linear-span generator.

random, when a prediction that an output bit will be zero is correct 96.7% of the time.

The following result suggests a simple way to guarantee nearly equal numbers of zeros and ones in the output of a generator.

THEOREM 5.11. *Let* $s = (r_1, r_2, \ldots, r_M)$ *denote the memory state of an m-sequence generator. If the output of NLFFL on the memory state can be represented in the form*

$$b = r_i + f(r_1, \ldots, r_{i-1}, r_{i+1}, \ldots, r_M) \qquad (5.174)$$

for some i, where $f(\cdot)$ *is any Boolean function independent of* r_i, *then the NLFFL output sequence will be balanced.*

Proof. Since the state M-tuple takes on all possible values except all-zeros during one period of operation, then the states $(r_1, \ldots, r_{i-1}, 0, r_{i+1}, \ldots, r_M)$ and $(r_1, \ldots, r_{i-1}, 1, r_{i+1}, \ldots, r_M)$ each occur once for each choice of the remaining values of the r_j's, $j \neq i$, excluding all-zeros. Each such state pair will contribute one 0 and one 1 in one period of the NLFFL output when (5.174) holds. Therefore, one full period of the output will have either (a) one more 0 than 1, or (b) one more 1 than 0, depending on the output value when the register memory is in the state having $r_j = 0$ for all $j \neq i$, and $r_i = 1$. ∎

Figure 5.12, which illustrates a modification of the generator of Figure 5.11, provides a balanced pseudorandom sequence. However, balance and large linear span do not together guarantee the generation of unpredictable sequences.

A preliminary design for the NLFFL to be attached to an M-stage m-sequence generator can be mapped out with the aid of Key's bound, the result being an estimate of the number of multipliers required and the

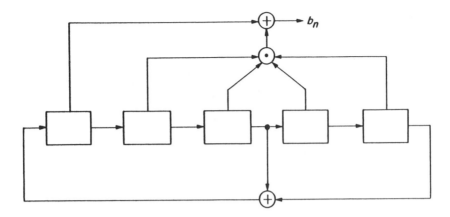

Figure 5.12. A high-linear-span (30) generator with balance.

number of inputs for each multiplier in a sum-of-products form. To complete the design, one must choose the connections between multiplier inputs and shift register memory elements, and then determine, on a case by case basis, whether or not the other statistics (e.g., correlation properties, run-length statistics, etc.) constitute a satisfactory design. Very little has appeared in the open literature on this problem, although it was stated in [24], in the context of quadratic NLFFL and on the basis of experimental results, that each memory element should connect to at most one multiplier, and that the spacing between memory elements connected to one multiplier should differ from the spacing between connections to any other multiplier. These suggestions may serve as initial guidelines in selecting an NLFFL design which will generate a sequence with good statistical properties.

5.6.3 Difference Set Designs

A v, k, λ *cyclic difference set* [26] is a collection $\mathscr{D} = \{d_1, d_2, \ldots, d_k\}$ of k integers modulo v ($0 \le d_i < v$ for all i) such that for every choice of the integer $\beta, 0 < \beta < v$, the equation

$$d_i - d_j - \beta \bmod v \tag{5.175}$$

has exactly λ solution pairs (d_i, d_j) with d_i and d_j being distinct elements of $\mathscr{D}$. Since there are exactly $k(k-1)$ non-zero differences between the k elements of $\mathscr{D}$, and since these must assume each of $v - 1$ values exactly λ times, the parameters v, k, λ are constrained by the relation

$$k(k - 1) = \lambda(v - 1). \tag{5.176}$$

It is easily verified that modulo v addition of a constant to each element of a difference set yields a new difference set with the same parameters. The complement $\{0, 1, \ldots, v - 1\} - \mathscr{D}$ of a v, k, λ difference set $\mathscr{D}$ is a v^*, k^*, λ^*

difference set $\mathcal{D}^*$ with

$$v^* = v, \quad k^* = v - k, \quad \lambda^* = v - 2k + \lambda. \tag{5.177}$$

The incidence vector $\boldsymbol{b}$ of a difference set has elements defined by

$$b_n = \begin{cases} 0, & n \notin \mathcal{D} \\ 1, & n \in \mathcal{D}, \end{cases} \tag{5.178}$$

for $0 \le n < v$. The sequence $\{b_n\}$ constructed by extending the incidence vector periodically with period v has a two-level periodic correlation function $P_{bb}(\tau)$ given by

$$P_{bb}(\tau) = \sum_{n=0}^{v-1} b_{n+\tau} b_n$$
$$= \begin{cases} k, & \tau = 0 \bmod v \\ \lambda, & \tau \neq 0 \bmod v. \end{cases} \tag{5.179}$$

Conversely, if a periodic binary (0 or 1) sequence $\{b_n\}$ has two-level periodic correlation function, then one period of $\{b_n\}$ is the index vector of a difference set.

A binary roots-of-unity sequence $\{a_n\}$, based on the difference set $\mathcal{D}$, can be constructed via the relation

$$a_n = (-1)^{b_n} = 1 - 2b_n, \tag{5.180}$$

b_n being 0 or 1. The periodic correlation function $P_{aa}(\tau)$ of the sequence $\{a_n\}$ also is two valued, being

$$P_{aa}(\tau) = \sum_{n=0}^{v-1} a_{n+\tau} a_n$$
$$= v - 4k + 4P_{bb}(\tau)$$
$$= \begin{cases} v, & \tau = 0 \bmod v \\ v - 4k + 4\lambda, & \tau \neq 0 \bmod v. \end{cases} \tag{5.181}$$

Any binary ($+1$ or -1) sequence with two-level periodic correlation can be related to the incidence matrix of a difference set by (5.180).

One period of an m-sequence, generated by an M-stage shift register, is the incidence vector for a difference set with parameters

$$v = 2^M - 1, \quad k = 2^{M-1}, \quad \lambda = 2^{M-2}, \tag{5.182}$$

and, complementing this set gives another difference set with parameters

$$v^* = 2^M - 1, \quad k^* = 2^{M-1} - 1, \quad \lambda^* = 2^{M-2} - 1. \tag{5.183}$$

These difference sets, corresponding to m-sequences, are called *Singer sets*, and were developed as a result of work in finite projective geometry [27].

Turyn [68] and Baumert [26] survey the known difference set designs. Decimation of any of the resulting sequences may produce a class of potentially interesting pseudorandom sequences. The sequences to be discussed next are one such class which has many nice attributes.

5.6.4 GMW Sequences

Gordon, Mills, and Welch (GMW) [28] have generalized the structure of Singer sets, thereby creating a larger collection of difference sets possessing the same parameters as displayed in (5.182). More precisely, when the integer M in (5.182) is composite, i.e.,

$$M = JK, \tag{5.184}$$

then some of the difference sets created by the GMW construction do not correspond to Singer sets. The binary sequences, called GMW sequences, derived from GMW difference sets, have correlation properties identical to those of m-sequences, but possess larger linear span than m-sequences.

Some of the difference sets in the GMW construction have incidence vectors whose elements, considered here as being in GF(2), can be specified as

$$b_n = \text{Tr}_2^{2^J}\!\left(\left[\text{Tr}_{2^J}^{2^M}(\alpha^n)\right]^r\right) \tag{5.185}$$

where α is a primitive element of GF(2^M), and r is any integer relatively prime to $2^J - 1$, r in the range $0 < r < 2^J - 1$. When $r = 1$, then (5.185) reduces by (5.125) to the trace representation of an m-sequence. Notice also that the interior trace function in (5.185) is an m-sequence over GF(2^J), and has the following property.

LEMMA 5.1. *Let an m-sequence over* GF(q), $q \geq 2$, *be defined by*

$$b_n' = \text{Tr}_q^{q^K}(\alpha^n), \tag{5.186}$$

where α is a primitive element of GF(q^K), *and let*

$$T = \frac{q^K - 1}{q - 1}. \tag{5.187}$$

Then every segment of T consecutive symbols from $\{b_n'\}$ contains exactly $(q^{K-1} - 1)/(q - 1)$ zeros.

Proof. The field element α^T has order $q - 1$ and, hence, belongs to the ground field GF(q). Therefore, by the linearity property (5.123) of the trace function, it follows that for any integer k,

$$\text{Tr}_q^{q^K}(\alpha^n) = \alpha^{-kT}\text{Tr}(\alpha^{n+kT}). \tag{5.188}$$

Since α^{-kT} is not zero, it follows that when one trace in (5.188) is zero, so is the other; thus, the zero locations in $\{b_n'\}$ are subject to a T periodicity. Consequently, every segment of T symbols from $\{b_n'\}$ contains the same number of zeros, that number being easily evaluated from the fact that $q^{K-1} - 1$ zeroes occur in one period, $q^K - 1$ symbols, of $\{b_n'\}$. ∎

It will now be verified directly that the GMW sequences of (5.185) have the same autocorrelation properties as m-sequences.

THEOREM 5.12. *Let* $\{a_n\}$ *be a GMW sequence whose elements are given by*

$$a_n = (-1)^{\mathrm{Tr}_2^{2^J}([\mathrm{Tr}_2^{2^M}(\alpha^n)]^r)}, \tag{5.189}$$

where α *is a primitive element of* $\mathrm{GF}(2^M)$, *and* $r, 0 < r < 2^J - 1$, *is relatively prime to* $2^J - 1$. *Then the periodic autocorrelation function* $P_{aa}(\tau)$ *of* $\{a_n\}$ *is given by*

$$P_{aa}(\tau) \triangleq \sum_{n=0}^{2^M-2} a_{n+\tau} a_n$$

$$= \begin{cases} 2^M - 1, & \tau = 0 \bmod 2^M - 1 \\ -1, & \tau \neq 0 \bmod 2^M - 1. \end{cases} \tag{5.190}$$

Proof. It follows immediately from trace function linearity that

$$P_{aa}(\tau) = \sum_{n=0}^{2^M-2} (-1)^{\mathrm{Tr}_2^{2^J}([\mathrm{Tr}_2^{2^M}(\alpha^{n+\tau})]^r + [\mathrm{Tr}_2^{2^M}(\alpha^n)]^r)} \tag{5.191}$$

Let T be the smallest power of α yielding an element of $\mathrm{GF}(2^J)$, i.e.,

$$T = \frac{2^M - 1}{2^J - 1}, \tag{5.192}$$

and express the index n in (5.191) as

$$n = n_0 + n_1 T, \tag{5.193}$$

where $0 \leq n_0 < T$ and $0 \leq n_1 \leq 2^J - 1$. Using the linearity of the inner traces in (5.191) gives

$$P_{aa}(\tau) = \sum_{n_0=0}^{T-1} \sum_{n_1=0}^{2^J-2} (-1)^{\mathrm{Tr}_2^{2^J}(\alpha^{rTn_1}\delta(\tau,n_0))}, \tag{5.194}$$

where

$$\delta(\tau, n_0) = \left[\mathrm{Tr}_{2^J}^{2^M}(\alpha^{\tau+n_0})\right]^r + \left[\mathrm{Tr}_{2^J}^{2^M}(\alpha^{n_0})\right]^r. \tag{5.195}$$

Since r is relatively prime to $2^J - 1$, it follows that α^{rT} is a primitive element of $\mathrm{GF}(2^J)$, and α^{rTn_1} takes on the values of all non-zero elements of that field as n_1 varies over its range. Hence, by including the zero element of $\mathrm{GF}(2^J)$ in the sum of (5.194), one arrives at

$$P_{aa}(\tau) = -T + \sum_{n_0=0}^{T-1} \sum_{\beta \in \mathrm{GF}(2^J)} (-1)^{\mathrm{Tr}_2^{2^J}(\beta\delta(\tau,n_0))}. \tag{5.196}$$

When $\delta(\tau, n_0)$ is not zero, the inner sum vanishes because half the exponents are zero and half one, by trace Property T-4. Let $N_0(\tau)$ denote the number of values of n_0 in the range $0 \leq n_0 < T$ for which $\delta(\tau, n_0)$ is zero. Then (5.196) reduces to

$$P_{aa}(\tau) = -T + 2^J N_0(\tau). \tag{5.197}$$

Since r is relatively prime to $2^J - 1$ and therefore has an inverse modulo $2^J - 1$, it follows that

$$\delta(\tau, n_0) = 0 \Leftrightarrow \mathrm{Tr}_{2^J}^{2^M}(\alpha^{\tau+n_0}) = \mathrm{Tr}_{2^J}^{2^M}(\alpha^{n_0})$$

$$\Leftrightarrow \mathrm{Tr}_{2^J}^{2^M}((\alpha^\tau - 1)\alpha^{n_0}) = 0. \tag{5.198}$$

When $\alpha^\tau - 1$ is zero, the right-hand equation in (5.198) is satisfied for all n_0. If $\alpha^\tau - 1$ is not zero, then Lemma 5.1 can be applied with (5.198) to determine $N_0(\tau)$. Thus,

$$N_0(\tau) = \begin{cases} T, & \tau = 0 \bmod 2^M - 1 \\ \dfrac{2^{M-J} - 1}{2^J - 1}, & \tau \neq 0 \bmod 2^M - 1. \end{cases} \tag{5.199}$$

and the result follows by substitution of (5.199) and (5.192) into (5.197). ∎

The linear span of a GMW sequence can be evaluated exactly by finding the number of terms in its powers-of-α representation and applying Theorem 5.9.

THEOREM 5.13. *Let $\{b_n\}$ be a GMW sequence whose elements are given by*

$$b_n = \mathrm{Tr}_2^{2^J}\left(\left[\mathrm{Tr}_{2^J}^{2^M}(\alpha^n)\right]^r\right), \tag{5.200}$$

where α is a primitive element of $\mathrm{GF}(2^M)$ and r, $0 < r < 2^J - 1$, is relatively prime to $2^J - 1$. Then the linear span L of $\{b_n\}$ is given by

$$L = J(M/J)^w, \tag{5.201}$$

where w is the number of ones in the base-2 representation of r.

Proof. The exponent r can be written as

$$r = \sum_{i=1}^{w} 2^{j_i}, \tag{5.202}$$

where the j_i's are distinct integers in the range $0 \leq j_i < J$ for all i. Hence,

$$b_n = \mathrm{Tr}_2^{2^J}\left(\prod_{i=1}^{w}\left[\mathrm{Tr}_{2^J}^{2^M}(\alpha^n)\right]^{2^{j_i}}\right). \tag{5.203}$$

Since the inner trace is a sum of elements from a field of characteristic 2, all cross-product terms disappear when the trace is squared and (5.203) reduces to

$$b_n = \mathrm{Tr}_2^{2^J}\left(\prod_{i=1}^{w}\sum_{k=0}^{K-1}\alpha^{n 2^{Jk+j_i}}\right)$$

$$= \mathrm{Tr}_2^{2^J}\left(\sum_{k_1=0}^{K-1}\cdots\sum_{k_w=0}^{K-1}\alpha^{nc(k,r)}\right), \tag{5.204}$$

where K is defined by (5.184) and

$$c(\mathbf{k},r) = \sum_{i=1}^{w} 2^{Jk_i + j_i}. \tag{5.205}$$

The time coefficients $c(\mathbf{k},r)$ for the representation of the nonlinear sequence over $GF(2^J)$, as expressed in (5.205), are all less than $2^M - 1$ (the order of α). Since the exponents of 2 in (5.205) are distinct modulo J, it is easily verified that no two distinct $\mathbf{k}$ vectors produce the same time coefficient, $c(\mathbf{k},r)$ and, therefore, no terms in the w-fold sum in (5.204) can be combined.

When expanding the outer trace function in (5.204), cross-product terms again must disappear, and

$$b_n = \sum_{m=0}^{J-1} \sum_{k_1=0}^{K-1} \cdots \sum_{k_w=0}^{K-1} \alpha^{nc(\mathbf{k},r)2^m}. \tag{5.206}$$

Therefore, the time coefficients in the powers-of-α representation of $\{b_n\}$ are simply $c(\mathbf{k},r)2^m$. To determine whether or not these coefficients are distinct modulo the order of α, solutions $\mathbf{k}_1, \mathbf{k}_2, m_1, m_2$ to the equation

$$c(\mathbf{k}_1,r)2^{m_1} = c(\mathbf{k}_2,r)2^{m_2} \bmod 2^M - 1, \tag{5.207}$$

must be sought. Since $2^J - 1$ divides $2^M - 1$, any solution to (5.207) must also satisfy the same equation modulo $2^J - 1$. But $c(\mathbf{k},r)$ modulo $2^J - 1$ is simply r, and (5.207) modulo $2^J - 1$ can be rewritten as

$$r(2^{m_1} - 2^{m_2}) = 0 \bmod 2^J - 1. \tag{5.208}$$

However, r is relatively prime to $2^J - 1$ and, hence, in the allowed range of m_1 and m_2, (5.208) exhibits only the solution $m_1 = m_2$. Therefore, the only solution to (5.207) is the identity $\mathbf{k}_1 = \mathbf{k}_2$ and $m_1 = m_2$, and all time coefficients in (5.206) are distinct. Application of Theorem 5.9 and counting the terms in (5.206) yields the final result. ∎

Example 5.12. A GMW sequence of period 63 is defined by

$$b_n = \mathrm{Tr}_2^8\left(\left[\mathrm{Tr}_8^{64}(\alpha^n)\right]^3\right), \tag{5.209}$$

where $z^6 + z^5 + z^2 + z + 1$ is the minimum polynomial of α over $GF(2)$. One simple plan for mechanizing a generator for $\{b_n\}$ is to construct a generator for the m-sequence $\mathrm{Tr}_8^{64}(\alpha^n)$ and use a ROM to complete the mapping to $GF(2)$. The elements of $GF(8)$ are 0 and α^{9i}, $i = 0, 1, \ldots, 6$, and the minimum polynomial of α over $GF(8)$ is easily determined (see Appendix 5A.8) to be $z^2 + \alpha^{54}z + \alpha^9$. A block diagram of the generator employing $GF(8)$ arithmetic is shown in Figure 5.13.

The actual mechanization takes advantage of the fact that $GF(8)$ is a three-dimensional vector space over $GF(2)$ with basis $1, \alpha^9, \alpha^{18}$. That is, an element from $GF(8)$ can be written as

$$\gamma = \gamma_0 \cdot 1 + \gamma_1 \alpha^9 + \gamma_2 \alpha^{18}, \tag{5.210}$$

Table 5.12

A representation for GF(8) elements, and the ROM mapping for Example 5.12.

γ	γ_0	γ_1	γ_2	$\mathrm{Tr}_2^8(\gamma^3)$
0	0	0	0	0
1	1	0	0	1
α^9	0	1	0	1
α^{18}	0	0	1	1
α^{27}	1	1	0	0
α^{36}	0	1	1	1
α^{45}	1	1	1	0
α^{54}	1	0	1	0

with $\gamma_0, \gamma_1, \gamma_2$ in GF(2). Table 5.12 lists this representation along with the ROM mapping, $\mathrm{Tr}_2^8(\gamma^3)$. The multiplications required in the generator of Figure 5.13 are easily mechanized in this representation.

$$\alpha^{54}\gamma = \alpha^{54}\gamma_0 + \gamma_1 + \alpha^9\gamma_2$$

$$= (\alpha^{18} + 1)\gamma_0 + \gamma_1 + \alpha^9\gamma_2$$

$$= (\gamma_0 + \gamma_1)1 + \gamma_2\alpha^9 + \gamma_0\alpha^{18}. \tag{5.211}$$

Similarly,

$$\alpha^9\gamma = \gamma_2 \cdot 1 + (\gamma_0 + \gamma_2)\alpha^9 + \gamma_1\alpha^{18}. \tag{5.212}$$

This results in the mechanization of Figure 5.14(a). One period of each of the sequences produced in this generator is shown in Figure 5.14(b). Notice the periodically recurring zeros in the GF(8) sequences (T is 9 in this example), this demonstrating the structure exploited in the proof of Theorem 5.12.

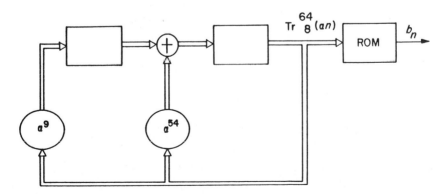

Figure 5.13. A GMW sequence generator in the Galois configuration with elements in GF(8).

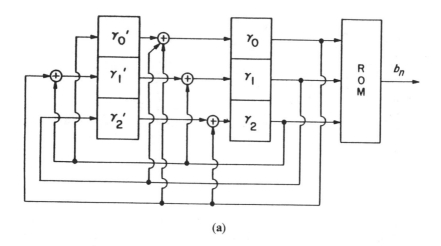

(a)

γ_0: 01110001100111011000001111001001010100110100001000101101111101
γ_1: 00001111001001010100110100000100010110111111010111000110011101110110
γ_2: 00111011000001111001001010100110100000100010110111111101011100011

γ_0': 10011101100000011110010010101001101000010001011011111010111100001
γ_1': 001001010100110100000100010110111111010111000110011101100001111
γ_2': 00000111100100101010011010000010001011011111101011100011001110011

b_n: 0100111010111010010111000110011111100100101110011101010000001010

(b)

Figure 5.14. (a) A GF(2) mechanization of the GMW sequence generator in Example 5.12. (b) Sequences produced in this generator.

Since r is 3, the base-2 expansion of r has weight 2, and Theorem 5.13 states that the linear span in this example is 12. The time coefficients $c(k, 3)$ in the powers-of-α expansion of the cubed GF(8) sequence are given in Table 5.13. When the final trace operation to GF(2) is carried out, the remaining cyclic shifts of the binary time coefficients listed in Table 5.12 are added to give twelve distinct coefficients in all. It is obvious from the cyclotomic coset representatives in this table that the characteristic polynomial of $Q(z)$ of $\{b_n\}$ is

$$Q(z) = m_{\alpha^3}(z)m_{\alpha^5}(z)$$

$$= z^{12} + z^8 + z^7 + z^6 + z^5 + z^3 + 1. \qquad (5.213)$$

This can be verified directly as specifying the recursion satisfied by the output sequence.

The k-tuple statistics of the output sequence in this example are shown in Table 5.14. As indicated by this data, the k-tuple statistics of a GMW

Table 5.13
Time coefficients $c(k, 3)$ of the sequence $[\text{Tr}_8^{64}(\alpha^n)]^3$.

Time Coefficients	
base-2	base-10
110000	3
100010	17
010100	10
000110	24

sequence are not uniformly distributed for all k less than the number of memory elements used, as they are for m-sequences by Property R-2 (5.61).

A uniform distribution on k-tuple occurrences is achieved in GMW sequences for a restricted range of k, as stated in the following theorem.

THEOREM 5.14. *Let* $\{b_n\}$ *be the GMW sequence defined by (5.200). Then the number* N_a *of positions within one period of* $\{b_n\}$ *at which the k-tuple* $a_1 a_2 \ldots a_k$ *occurs, is given by*

$$N_a = \begin{cases} 2^{M-k}, & \text{for } a \neq 0, \quad 1 \leq k \leq \dfrac{M}{J} \\ 2^{M-k} - 1, & \text{for } a = 0, \quad 1 \leq k \leq \dfrac{M}{J} \end{cases} \quad (5.214)$$

Proof. The intermediate field sequence $\text{Tr}_{2^J}^{2^M}(\alpha^n)$ is an m-sequence with linear span M/J, and hence k-tuples $a'_1 \ldots a'_k$ from this sequence occur $N'_{a'}$ times in one period according to the count specified in (5.61):

$$N'_{a'} = \begin{cases} 2^{J(M/J - k)}, & \text{for } a' \neq 0, \quad 1 \leq k \leq \dfrac{M}{J} \\ 2^{J(M/J - k)} - 1, & \text{for } a' = 0, \quad 1 \leq k \leq \dfrac{M}{J} \end{cases} \quad (5.215)$$

Since r is relatively prime to $2^J - 1$, the r-th power mapping of the intermediate field onto itself is one-to-one, and the above count $N'_{a'}$ also applies to the sequence $\{[\text{Tr}_{2^J}^{2^M}(\alpha^n)]^r\}$.

Table 5.14
K-tuple statistics for the GMW sequence of length 63.

k	(k-tuple: Occurrences per period)
1	(0:31), (1:32)
2	(00:15), (01:16), (10:16), (11:16)
3	(000:6), (100:9), (010:9), (001:9)
	(110:7), (101:7), (011:7), (111:9)
4	(0000:4), (1000:2), (0100:5), (0010:5)
	(0001:2), (1100:4), (0110:1), (0011:4)
	(1001:7), (1010:4), (0101:4), (1110:6)
	(0111:6), (1011:3), (1101:3), (1111:3)

Let $\mathscr{A}_a$ be the set of k-tuples a' over GF(2^J) which map into a under the trace mapping

$$a_i = \mathrm{Tr}_2^{2^J}(a_i'), \qquad \text{for } i = 1, \ldots, k. \tag{5.216}$$

Note that the all-zeros k-tuple maps into itself. Then the number of occurrences of a in one period of $\{b_n\}$ is given by

$$N_a = \sum_{a' \in \mathscr{A}_a} N_{a'}'. \tag{5.217}$$

Since the number of elements in GF(2^J) which map into zero by (5.216) is 2^{J-1} (trace property T-4), it is easily verified that

$$|\mathscr{A}_a| = (2^{J-1})^k, \quad \text{for all } k\text{-tuples } a. \tag{5.218}$$

Therefore, when $a \neq 0$ and hence $0 \notin \mathscr{A}_a$, then all the terms in (5.217) are identical by (5.215), and

$$N_a = |\mathscr{A}_a| \cdot 2^{J(M/J-k)} = 2^{M-k}, \quad \text{for } a \neq 0. \tag{5.219}$$

A similar computation for $a = 0$ completes the proof. ∎

The next result indicates that all proper decimations and allowable choices of the exponent r yield distinct GMW sequences.

THEOREM 5.15. *Let $\{b_n\}$ and $\{c_n\}$ be GMW sequences whose elements are given by*

$$b_n = \mathrm{Tr}_2^{2^J}\!\left(\left[\mathrm{Tr}_{2^J}^{2^M}(\alpha^n)\right]^r\right) \tag{5.220}$$

$$c_n = \mathrm{Tr}_2^{2^J}\!\left(\left[\mathrm{Tr}_{2^J}^{2^M}(\alpha^{an})\right]^s\right), \tag{5.221}$$

where r and s are integers relatively prime to $2^J - 1$ and less than $2^J - 1$, a is relatively prime to $2^M - 1$, and α is a primitive element of GF(2^M). Then $\{b_n\}$ and $\{c_n\}$ are cyclically equivalent if and only if

$$r = 2^j s \bmod 2^J - 1 \quad \text{for some} \quad 0 \le j < J \tag{5.222}$$

and

$$a = 2^m \quad \text{for some} \quad 0 \le m < M. \tag{5.223}$$

Proof. Define T, n_0, and n_1 as in (5.192) and (5.193). Let $\beta = \alpha^\tau$ and $\gamma = \alpha^T$ (a primitive element of GF(2^J)). Then the linearity property of the trace gives

$$\{b_n\} = \{c_{n+\tau}\}$$

$$\Leftrightarrow \mathrm{Tr}_2^{2^J}(\gamma^{rn_1}k_1) = \mathrm{Tr}_2^{2^J}(\gamma^{asn_1}k_2) \tag{5.224}$$

where

$$k_1 = \left[\mathrm{Tr}_{2^J}^{2^M}(\alpha^{n_0})\right]^r \tag{5.225}$$

$$k_2 = \left[\mathrm{Tr}_{2^J}^{2^M}(\beta\alpha^{an_0})\right]^s. \tag{5.226}$$

The trace functions in (5.224) represent two m-sequences over GF(2), each with element index n_1. These sequences are identical for some β if and only if γ^r and γ^{as} have the same minimum polynomial over GF(2); hence, equality requires

$$r = 2^j as \bmod 2^J - 1, \quad \text{for some} \quad 0 \le j < J \qquad (5.227)$$

(see Theorem 5A.14 in Appendix 5A).

Assuming (5.227) is satisfied, the fact that conjugate elements have the same trace (see 5.122) can be used to show that

$$\text{Tr}_2^{2^J}\!\left(\gamma^{rn_1}k_1\right) = \text{Tr}_2^{2^J}\!\left(\gamma^{2^j asn_1}k_1\right)$$

$$= \text{Tr}_2^{2^J}\!\left(\gamma^{asn_1}k_1^{2^{J-j}}\right), \qquad (5.228)$$

and, hence, equality holds in (5.224) when k_2 is identical to $k_1^{2^{J-j}}$, i.e., when

$$\left[\text{Tr}_{2^J}^{2^M}(\alpha^{n_0})\right]^{2^{J-j}r} = \left[\text{Tr}_{2^J}^{2^M}(\beta\alpha^{an_0})\right]^s. \qquad (5.229)$$

Substitution of (5.227) into (5.229) and raising each side of the result to a power equal to the multiplicative inverse of s modulo $2^J - 1$ yields the result

$$\left[\text{Tr}_{2^J}^{2^M}(\alpha^{n_0})\right]^a = \text{Tr}_{2^J}^{2^M}(\beta\alpha^{an_0}). \qquad (5.230)$$

If the two sequences over GF(2^J), whose elements with index n_0 are given by the right and left sides of (5.230), are cyclically equivalent, then their linear spans must be equal. It can be shown via the techniques of Theorem 5.13, that the sequence defined by the left side of (5.230) has linear span $(M/J)^w$, where w is the number of ones in the binary representation of $a \bmod 2^J - 1$. The linear span of the sequence represented by the right side of (5.230) is M/J, since α^a is also a primitive element of GF(2^M). Hence, equality requires that w be 1 and, therefore, (5.223) is necessary. Substitution of (5.223) into (5.227) results in (5.222) as a necessary requirement. It is easily verified that these necessary conditions are also sufficient. ∎

Hence, the number N_{GMW} of cyclically distinct GMW sequences, which can be constructed by this method for a fixed M and J, is given by

$$N_{\text{GMW}} = N_p(M)N_p(J), \qquad (5.231)$$

where $N_p(d)$ denotes the number of primitive polynomials of degree d over GF(2), or equivalently the number of cyclically distinct m-sequences of linear span d over GF(2) (see Section 5.3.4).

Example 5.13. Consider a design in which the ROM size constrains J to be 7. Since $2^7 - 1$ is the prime number 127, all binary 7-tuples, except 0000000 and 1111111, are radix-2 representations of numbers relatively prime to 127. One set of acceptable choices of r, with cyclically inequivalent radix-2

Table 5.15
Design parameter tradeoffs for GMW sequences with a constraint $J = 7$.

w	$\binom{J}{w}/J$	r values	$M = 14$		$M = 28$	
			$N_{\text{GMW}}(w)$	L	$N_{\text{GMW}}(w)$	L
1	1	1	756	14	4741632	28
2	3	3, 5, 9	2268	28	14224896	112
3	5	7, 11, 13, 19, 21	3780	56	23708160	448
4	5	15, 23, 27, 29, 43	3780	112	23708160	1792
5	3	31, 47, 55	2268	224	14224896	7168
6	1	63	756	448	4741632	28672

representations, is shown in Table 5.15, which contains a total of $N_p(7)$ entries, with $\binom{J}{w}/J$ entries for each value of w. This table also indicates the number $N_{\text{GMW}}(w)$ of GMW sequences constructed with weight-w values of r, along with their linear span L, for two possible values of memory size M.

5.7 DIRECT-SEQUENCE MULTIPLE-ACCESS DESIGNS

5.7.1 A Design Criterion

In many situations, more than one SS code generator is required to operate simultaneously within the same locale and bandwidth, i.e., in a multiple-access (MA) environment. The distinguishability of different SS signals by receivers participating in the SSMA scheme depends on many factors, including the SS sequences, modulation formats, and the receiver's detector structure. This section will explore one tractable criterion for DS-SSMA signal design, namely minimization of the absolute value of periodic correlation between signals.

Consider a set of J sequences, each with period N, denoted by $\{a_n^{(j)}\}$, $j = 1, \ldots, J$. The periodic cross-correlation $P_{jk}(\tau)$ at shift τ between sequences from this collection is defined as

$$P_{jk}(\tau) = \sum_{n=1}^{N} a_{n+\tau}^{(j)} \left(a_n^{(k)} \right)^*. \tag{5.232}$$

The maximum out-of-phase periodic autocorrelation magnitude P_A for this signal set is defined as

$$P_A = \max_j \max_{0 < \tau < N} |P_{jj}(\tau)|, \tag{5.233}$$

and the maximum cross-correlation magnitude P_C between signals in this set is given by

$$P_C = \max_{j \neq k} \max_{0 \leq \tau < N} |P_{jk}(\tau)|. \tag{5.234}$$

The signal sets to be discussed here all are designed to minimize

$$P_{\max} = \max(P_A, P_C). \tag{5.235}$$

This criterion for sequence selection may not correspond to any DS-SSMA network performance measure, but it can be argued that a signal set which optimizes network performance probably has a small value of $P_{\max}$; hence, one should find good signal sets among those with small $P_{\max}$. Furthermore, from an analytic viewpoint, no other design criteria have proven tractable in choosing long-period sequences.

In situations in which a capability against repeater jamming is not a requirement, the code period N may be relatively short, its value in most instances being underbounded by the multipath time-spread of the channel. Hence, in some multiple access designs which require no AJ provisions, correlations may be calculated over full periods. Conversely, when intelligent jamming is a possible threat, then the SS direct sequence must possess a large period and large linear span; shortening the DS period in this situation for the sake of somewhat improved multiple access capability would allow the possibility of catastrophic consequences.

5.7.2 Welch's Inner Product Bound

The correlation computation of (5.232) can be viewed as an inner product between the N-tuples $(a_{1+\tau}^{(j)}, \ldots, a_{N+\tau}^{(j)})$ and $(a_1^{(k)}, \ldots, a_N^{(k)})$. This connection will be exploited in applying the following theorem to determine a bound on $P_{\max}$.

THEOREM 5.16 (Welch's Bound [29]). *Let* $(c_1^{(\nu)}, \ldots, c_L^{(\nu)})$, $\nu = 1, \ldots, M$, *represent L-tuples of complex numbers, and define the inner product*

$$R_{\nu\lambda} = \sum_{n=1}^{L} c_n^{(\nu)} (c_n^{(\lambda)})^*. \tag{5.236}$$

Assume that each L-tuple has unit length, i.e., $R_{\nu\nu} = 1$ *for all* ν, *and define*

$$R_{\max} = \max_{\nu \neq \lambda} |R_{\nu\lambda}|. \tag{5.237}$$

Then

$$R_{\max}^2 \geq \frac{M - L}{(M - 1)L}. \tag{5.238}$$

Proof. Beginning with the definition of $R_{\max}$ and the fact that the vectors are unit length, it is easily verified that

$$M(M-1)R_{max}^2 + M \geq \sum_{\nu=1}^{M} \sum_{\lambda=1}^{M} |R_{\nu\lambda}|^2$$

$$= \sum_{\nu=1}^{M} \sum_{\lambda=1}^{M} \sum_{i=1}^{L} \sum_{j=1}^{L} c_i^{(\nu)} \left(c_i^{(\lambda)}\right)^* \left(c_j^{(\nu)}\right)^* c_j^{(\lambda)}$$

$$\overset{(1)}{=} \sum_{i=1}^{L} \sum_{j=1}^{L} \left| \sum_{\nu=1}^{M} c_i^{(\nu)} \left(c_j^{(\nu)}\right)^* \right|^2$$

$$\overset{(2)}{\geq} \sum_{i=1}^{L} \left| \sum_{\nu=1}^{M} |c_i^{(\nu)}|^2 \right|^2$$

$$\overset{(3)}{\geq} \frac{1}{L} \left| \sum_{i=1}^{L} \sum_{\nu=1}^{M} |c_i^{(\nu)}|^2 \right|^2 \overset{(4)}{=} \frac{M^2}{L}. \tag{5.239}$$

Here equality (1) follows from changing the order of summation, inequality (2) is achieved by neglecting terms with $i \neq j$, (3) is Cauchy's inequality for sums of squares, or equivalently is a result of the convexity of the quadratic function, and (4) simply uses the fact that the L-tuples are unit length. Solving the derived inequality (5.239) for R_{max}^2 gives the final result. ∎

Welch's inner-product bound can be applied to a variety of signal design situations. Here we apply it to the maximum absolute value of correlation for a set of roots-of-unity sequences. Let

$$\left(c_1^{(\nu)}, \ldots, c_L^{(\nu)}\right) = N^{-1/2} \left(a_{1+\tau}^{(j)}, \ldots, a_{N+\tau}^{(j)}\right) \tag{5.240}$$

for $0 \leq \tau < N$ and $1 \leq j \leq J$, the $N^{-1/2}$ being required to make the N-tuple unit length. Here one may think of ν as ranging over JN values corresponding to the distinct pairs (j, τ); hence, the required parameter substitution in Welch's bound is $L = N$ and $M = JN$. The quantity R_{max} can be shown to correspond to P_{max}/N, giving the following result.

COROLLARY 5.3. *The maximum absolute value P_{max} of the out-of-phase periodic autocorrelation and of the periodic cross-correlation for a set of J roots-of-unity sequences with period N is lower bounded by*

$$\left(\frac{P_{max}}{N}\right)^2 \geq \frac{J-1}{JN-1}. \tag{5.241}$$

When the number J of sequences is even moderately large, then (5.241) indicates that P_{max} must be at least on the order of $\sqrt{N}$.

When the set of sequences $\{a_n^{(j)}\}$, $1 \leq j \leq J$, of period N, are composed of ± 1 strings, a correspondence with a code $\mathscr{C}$ containing JN binary code words $\boldsymbol{b}_{j,m}$, $1 \leq j \leq J$, $0 \leq m < N$, of length N over GF(2), can be established through the relation

$$a_n^{(j)} = (-1)^{b_n^{(j)}}, \quad \boldsymbol{b}_{j,m} = \left(b_{m+1}^{(j)}, \dots, b_{m+N}^{(j)} \right). \tag{5.242}$$

Note that the n code words $\boldsymbol{b}_{j,m}$, $0 \leq m < N$, are cyclically equivalent for each value of j; therefore, the code $\mathscr{C}$ is called a *cyclic code*. A bound on periodic cross-correlation magnitude for the ± 1 sequence set can be rewritten in terms of Hamming distance for the corresponding cyclic code over GF(2), as

$$P_{\max} \geq |P_{jk}(\tau)| = \left| \sum_{n=1}^{N} (-1)^{b_{n+\tau}^{(j)} - b_n^{(k)}} \right|$$

$$= |N - 2H(\boldsymbol{b}_{j,\tau}, \boldsymbol{b}_{k,0})|, \tag{5.243}$$

whence,

$$(N - P_{\max})/2 \leq H(\boldsymbol{b}_{j,\tau}, \boldsymbol{b}_{k,0}) \leq (N + P_{\max})/2 \tag{5.244}$$

for all j, k, and τ. Thus, the design of a set of J, period N, binary, roots-of-unity sequences to have small correlation, is equivalent to the design of a cyclic error-control code $\mathscr{C}$ containing NJ words of length N with the Hamming distance between all pairs of code words close to $N/2$. This link with error-control coding makes many of the design techniques in the well-developed theory of coding available for DS-SSMA signal design.

Two well-known SSMA signal designs, namely Gold sequences [35], [36] and the small set of Kasami sequences [30], [31], are a direct result of design techniques for cyclic error-control codes. The Kasami sequences are among several known sets of sequences which nearly meet the Welch bound on $P_{\max}$, including bent sequences [32], and the prime-period group character sequences [33]. The Kasami and bent sequence designs are both binary and possess identical numbers, periods, and periodic correlation properties, but the bent sequences have considerably longer linear span. The group-character sequences are non-binary but each design contains one binary sequence, namely the quadratic residue (or Legendre) sequence (e.g., see [12], Chapter 3, Section 5.6). Gold sequence sets, although not meeting the Welch bound, are larger than the Kasami and bent sets, and achieve a more restricted bound of Sidelnikov [39], which states that for any N binary (± 1) sequences of period N,

$$P_{\max} > (2N - 2)^{1/2}. \tag{5.245}$$

The Gold, Kasami, and bent designs will be developed later in this section.

5.7.3 Cross-Correlation of Binary M-Sequences

It is clear from Table 5.6 that for reasonable register lengths, e.g., 20 to 30 stages, there exist large numbers of cyclically distinct m-sequences. Therefore, it is reasonable to determine whether or not several m-sequences can be used together as a DS-SSMA signal set. We approach this question by considering the cross-correlation between a roots-of-unity m-sequence $\{a_n\}$ and its r-th decimation $\{a_{rn}\}$, the latter being another m-sequence whenever r is relatively prime to the period of $\{a_n\}$.

Let $\{\mathrm{Tr}_2^{2^L}(\alpha^n)\}$ denote a binary m-sequence, α being a primitive element of $\mathrm{GF}(2^L)$, and let $\{a_n\}$ be the corresponding roots-of-unity m-sequence. The periodic cross-correlation between $\{a_n\}$ and its r-th decimation $\{a'_n\}$, $a'_n = a_{rn}$ for all n, is given by

$$P_{aa'}(\tau) = \sum_{n=1}^{2^L-1} (-1)^{\mathrm{Tr}(\alpha^{n+\tau})+\mathrm{Tr}(\alpha^{rn})}. \tag{5.246}$$

Since α is primitive, α^n scans through the non-zero elements of $\mathrm{GF}(2^L)$ as n varies in (5.246), and, after compensating for the insertion of the zero element into the calculation, (5.246) leads to

$$\Delta_r(y) \triangleq 1 + P_{aa'}(\tau) = \sum_{x \in \mathrm{GF}(2^L)} (-1)^{\mathrm{Tr}(xy+x^r)}, \tag{5.247}$$

where $y = \alpha^\tau$. Analytical simplifications of (5.247) have been achieved only for certain values of r. A comprehensive survey of these results, including tabulations of computer searches is given in [37].

One particularly good result will be derived from the following lemmas.

LEMMA 5.2. *Let*

$$r = 2^k + 1, \qquad e = \gcd(2k, L), \tag{5.248}$$

and assume e divides k. Then $\Delta_r(y)$ takes on three values: 0, $2^{(L+e)/2}$, and $-2^{(L+e)/2}$, with 0 occurring for $2^L - 2^{L-e}$ different values of y.

Proof. Squaring and simplifying (5.247) gives

$$[\Delta_r(y)]^2 = \sum_{x,z \in \mathrm{GF}(2^L)} (-1)^{\mathrm{Tr}[y(x+z)+x^r+z^r]}$$

$$\overset{(1)}{=} \sum_{x,w \in \mathrm{GF}(2^L)} (-1)^{\mathrm{Tr}[yw+x^r+(w+x)(w^{2^k}+x^{2^k})]}$$

$$\overset{(2)}{=} \sum_{w \in \mathrm{GF}(2^L)} (-1)^{\mathrm{Tr}(yw+w^r)} \sum_{x \in \mathrm{GF}(2^L)} (-1)^{\mathrm{Tr}[x(w^{2^k}+w^{2^{L-k}})]}$$

$$\tag{5.249}$$

Equality (1) results from the substitution of $z = w + x$ and the form of r, and equality (2) employs trace linearity and the fact that wx^{2^k} and $xw^{2^{L-k}}$ have the same trace since they are conjugates.

As x varies over $\mathrm{GF}(2^L)$, the trace in the inner sum is 0 and 1 equally often (see trace Property T-4) provided that $w^{2^k} + w^{2^{L-k}}$ is not zero. Hence, the sum on x is 0 or 2^L, and (5.249) reduces to

$$[\Delta_r(y)]^2 = 2^L \sum_{w \in \Omega} (-1)^{\text{Tr}(yw + w^r)}, \qquad (5.250)$$

where

$$\Omega = \left\{ w: w^{2^k} + w^{2^{L-k}} = 0, w \in \text{GF}(2^L) \right\}$$
$$= \left\{ w: w^{2^{2k}} = w, w \in \text{GF}(2^L) \right\}$$
$$= \left\{ w: w \in \text{GF}(2^{2k}), w \in \text{GF}(2^L) \right\}. \qquad (5.251)$$

The intersection of $\text{GF}(2^{2k})$ and $\text{GF}(2^L)$ is $\text{GF}(2^e)$, where e is the gcd $(2k, L)$.

The order of every w in $\text{GF}(2^e)$ divides $2^e - 1$, and, therefore, assuming e divides k,

$$w^r = w^{(2^e)^c} \cdot w = w^2, \qquad (5.252)$$

where c is an integer. Since the traces of w and w^2 are identical, (5.250) reduces to

$$[\Delta_r(y)]^2 = 2^L \sum_{w \in \text{GF}(2^e)} (-1)^{\text{Tr}_2^{2^e}\{[\text{Tr}_{2^e}^{2^L}(y) + \text{Tr}_{2^e}^{2^L}(1)]w\}} \qquad (5.253)$$

by (5.125). Since $\text{Tr}_{2^e}^{2^L}(y + 1)$ is zero for 2^{L-e} values of y by trace property T-4,

$$[\Delta_r(y)]^2 - 2^{L+e} \quad \text{for } 2^{L-e} \text{ values of } y. \qquad (5.254)$$

When $\text{Tr}_{2^e}^{2^L}(y + 1)$ is non-zero, the outer trace in (5.253) is zero and one equally often as w scans through $\text{GF}(2^e)$ and, therefore,

$$[\Delta_r(y)]^2 = 0 \quad \text{for } 2^L - 2^{L-e} \text{ values of } y. \qquad (5.255)$$

■

Lemma 5.2 provides the mathematical background for computing the periodic cross-correlation between an m-sequence and its r-th decimation, when r is of the form $2^k + 1$. The application of this result to the cross-correlation of m-sequences requires the following lemma to determine when the r-th decimation of an m-sequence is itself an m-sequence.

LEMMA 5.3. *For all integers m and n,*

$$\gcd(2^m - 1, 2^n - 1) = 2^{\gcd(m, n)} - 1 \qquad (5.256)$$
$$\gcd(2^m + 1, 2^n - 1) = 1 \text{ iff } n/\gcd(m, n) \text{ is odd.} \qquad (5.257)$$

Proof. By Euclidean division,

$$m = an + r, \quad 0 \le r < n, \qquad (5.258)$$

and, using binary representations of $2^m - 1$ and $2^n - 1$ as m- and n-tuples of ones, Euclidean division of $2^m - 1$ by $2^n - 1$ gives

$$(2^m - 1) = b(2^n - 1) + (2^r - 1), \quad 0 \le 2^r - 1 < 2^n - 1, \quad (5.259)$$

for some integer b. Hence, there is a correspondence between terms in Euclid's algorithm (see Appendix 5A.2) for determining $\gcd(m, n)$ and that for finding $\gcd(2^m - 1, 2^n - 1)$, which easily yields a proof of (5.256).

A proof of (5.257) begins by applying (5.256) to give

$$\gcd((2^m - 1)(2^m + 1), 2^n - 1) = 2^{\gcd(2m, n)} - 1, \quad (5.260)$$

and noting that

$$\gcd(2^m - 1, 2^m + 1) = 1, \quad (5.261)$$

since these numbers differ by two and are odd. Therefore,

$$\gcd(2^m + 1, 2^n - 1) = 1$$

$$\Leftrightarrow \gcd(2m, n) \mid m$$

$$\Leftrightarrow n/\gcd(m, n) \text{ is odd.} \quad (5.262)$$

∎

When the decimation coefficient r is relatively prime to $2^L - 1$, Theorem 5.8 indicates that the resulting decimation of an m-sequence is also an m-sequence. Placing Lemma 5.2 in correlation terms and restricting the decimation by r to produce an m-sequence, gives the following theorem.

THEOREM 5.17. *Let $\{a_n\}$ be a roots-of-unity m-sequence with period $2^L - 1$, and let $P_{aa'}(\tau)$ be the periodic cross-correlation of $\{a_n\}$ with $\{a'_n\}$, its decimation by r. Let r and e be defined by (5.248), with $L/\gcd(k, L)$ odd. Then*

$$P_{aa'}(\tau) = \begin{cases} -1 & \text{for } 2^L - 2^{L-e} - 1 \text{ values of } \tau, \\ -1 + 2^{(L+e)/2} & \text{for } 2^{L-e-1} + 2^{(L-e-2)/2} \text{ values of } \tau, \\ -1 - 2^{(L+e)/2} & \text{for } 2^{L-e-1} - 2^{(L-e-2)/2} \text{ values of } \tau \end{cases}$$

$$(5.263)$$

where $0 \le \tau < 2^L - 1$.

Final determination of the counts shown in this theorem can be made after showing that $\sum_\tau P_{aa'}(\tau) = 1$.

Information concerning other pairs of m-sequences with good periodic cross-correlation is given in Table 5.16. It is assumed in this listing that r is relatively prime to $2^L - 1$, and e is the greatest common divisor of L and $2k$. Note that while Welch's proof of the second listed result is unpublished, it is the $m = 3$ case of the third listed result. Each of the entries in Table 5.16 gives rise to peak cross-correlation magnitudes on the order of $2^{(L+1)/2}$ when L is odd, and $2^{(L+2)/2}$ when L is even. However, this table does not list all such cases.

Table 5.16

Maximum absolute value of the cross-correlation between m-sequences with
characteristic polynomials $m_\alpha(z)$ and $m_\alpha^r(z)$, α a primitive element of GF(2^L).

r	$\max|P_{aa'}(\tau)|$	Comments
$2^k + 1$	$2^{(L+e)/2} + 1$	[30]
$2^{2k} - 2^k + 1$	$2^{(L+e)/2} + 1$	(Welch, unpublished)
$(2^{mk} + 1)/(2^k + 1)$	$2^{(L+e)/2} + 1$	m odd, [37]
$2^{(L+2)/2} - 1$	$2^{(L+2)/2} - 1$	$L = 0 \bmod 4$, [37]
$2^{L-1} - 1$	$\leq 2^{(L+2)/2}$	[38]

Periodic cross-correlation calculations can be carried out analytically on
m-sequences of differing periods, sometimes with good results. This is
illustrated in the following theorem and proof.

THEOREM 5.18. *Let α be a primitive element of* GF(2^L), *L even, and let*

$$r = 2^{L/2} + 1, \tag{5.264}$$

α^r being a primitive element of GF($2^{L/2}$). *Then the roots-of-unity m-sequences of periods $2^L - 1$ and $2^{L/2} - 1$, with elements defined by*

$$a_n = (-1)^{\mathrm{Tr}_2^{2^L}(\alpha^n)} \tag{5.265}$$

and

$$a_n' = (-1)^{\mathrm{Tr}_2^{2^{L/2}}(\alpha^{rn})} \tag{5.266}$$

respectively, have periodic cross-correlation (over $2^L - 1$ sequence elements) values $-1 \pm 2^{L/2}$ at all shifts.

Proof. The cross-correlation between $\{a_n\}$ and $\{a_n'\}$ at shift τ is given by

$$P_{aa'}(\tau) = \sum_{n-1}^{2^L-1} (-1)^{\mathrm{Tr}_2^{2^L}(\alpha^{n+\tau}) + \mathrm{Tr}_2^{2^{L/2}}(\alpha^{rn})}$$

$$= -1 + \sum_{x \in \mathrm{GF}(2^L)} (-1)^{\mathrm{Tr}_2^{2^L}(yx) + \mathrm{Tr}_2^{2^{L/2}}(x^r)} \tag{5.267}$$

where $x = \alpha^n$, $y = \alpha^\tau$, and the field element 0 has been added to the sum
range. The elements of GF(2^L) can be represented in the form

$$x = u\beta + v \tag{5.268}$$

where u and v are elements of GF($2^{L/2}$) and β is the root of an irreducible
quadratic polynomial over GF($2^{L/2}$) (see Appendix 5A.5, Theorem 5A.9).

Then, using trace properties (5.123) and (5.125) on (5.268) gives

$$P_{aa'}(\tau) = -1 + \sum\sum_{u,v \in GF(2^{L/2})} (-1)^{Tr_2^{2L/2}[u\,Tr_2^L/2(y\beta) + v\,Tr_2^L/2(y) + (u\beta+v)(u\beta^{2L/2}+v)]}$$

$$= -1 + \sum_{u \in GF(2^{L/2})} (-1)^{Tr_2^{2L/2}[Tr_2^L/2(y\beta)u + \beta'u^2]}$$

$$\times \sum_{v \in GF(2^{L/2})} (-1)^{Tr_2^{2L/2}[(Tr_2^L/2(y) + u\,Tr_2^L/2(\beta) + 1)v]}, \qquad (5.269)$$

the latter step using the fact that v and v^2 have the same trace. The interior sum over v in (5.269) is zero by trace property T-4 unless the coefficient of v is zero, in which case the sum is $2^{L/2}$. Furthermore, the coefficient of v in (5.265) is zero for exactly one value of u, namely

$$u_y = \left[Tr_2^{2L}_{L/2}(\beta)\right]^{-1}\left[1 + Tr_2^{2L}_{L/2}(y)\right], \qquad (5.270)$$

and it follows that

$$P_{aa'}(\tau) = -1 + 2^{L/2}(-1)^{Tr_2^{2L}(y\beta u_y) + Tr_2^{L/2}(\beta'u_y^2)}. \qquad (5.271)$$

∎

Theorems 5.17 and 5.18 will now be applied to the design of large sets of sequences with good periodic correlation properties.

5.7.4 Linear Designs

Two roots-of-unity m-sequences, $\{a_n\}$ and $\{a_n'\}$, with corresponding sequences $\{b_n\}$ over $GF(p)$ and $\{b_n'\}$ over $GF(p')$, possessing known cross-correlation properties, can be used to construct a larger set of sequences with easily determined correlation properties. In this general situation, the periodic cross-correlation of the roots-of-unity m-sequences is defined as

$$P_{aa'}(\tau) = \sum_{n=1}^{N_m} \rho^{b_{n+\tau}}\rho'^{-b_n'}, \qquad (5.272)$$

where ρ and ρ' are primitive complex p-th and p'-th roots of unity respectively, and

$$N_m = lcm(p^L - 1, p'^{L'} - 1), \qquad (5.273)$$

L and L' being the linear spans of the corresponding m-sequences. The following theorem relates the cross-correlation properties defined in (5.272) to those of a larger collection of sequences.

THEOREM 5.19. *Let $\{b_n\}$ and $\{b_n'\}$ be the m-sequences described above and let*

$$a_n^{(j)} = \rho^{b_{n+j}}\rho'^{-b_n'}. \qquad (5.274)$$

Then the set of N_d sequences, namely $\{a_n^{(0)}\}$, $\{a_n^{(1)}\}, \ldots, \{a_n^{(N_d-1)}\}$, where

$$N_d = \gcd(p^L - 1, p'^{L'} - 1), \qquad (5.275)$$

have periodic cross-correlation (over N_m symbols) taking on values $-N_m/(p'^{L'} - 1)$, $-N_m/(p^L - 1)$, and the cross-correlation values calculated in (5.272).

Proof. The cross-correlation of $\{a_n^{(j)}\}$ and $\{a_n^{(k)}\}$ can be simplified immediately, using the shift-and-add property of m-sequences.

$$P_{jk}(\tau) = \sum_{n=1}^{N_m} \rho^{b_{n+\tau+j} - b_{n+k}} \rho'^{-b'_{n+\tau} + b'_n}. \qquad (5.276)$$

Case (a): Assume $\tau + j - k \neq 0 \bmod p^L - 1$, and $\tau \neq 0 \bmod p'^{L'} - 1$. Then from (5.63),

$$P_{jk}(\tau) = \sum_{n=1}^{N_m} \rho^{b_{n+k+\tau'(\tau+j-k)}} \rho'^{-b'_{n+\tau''(\tau)}}$$

$$= P_{aa'}(k + \tau'(\tau + j - k) - \tau''(\tau)). \qquad (5.277)$$

Case (b): Assume $\tau + j - k \neq 0 \bmod p^L - 1$, and $\tau = 0 \bmod p'^{L'} - 1$. In this case, b'_n cancels in the exponent of (5.276) and

$$P_{jk}(\tau) = \sum_{n=1}^{N_m} \rho^{b_{n+k+\tau'(\tau+j-k)}} = -N_m/(p^L - 1). \qquad (5.278)$$

Case (c): Assume $\tau + j - k = 0 \bmod p^L - 1$, and $\tau \neq 0 \bmod p'^{L'} - 1$. As in case (c), (5.276) reduces to

$$P_{jk}(\tau) = \sum_{n=1}^{N_m} \rho'^{-b_{n+\tau''(\tau)}} = -N_m/(p'^{L'} - 1). \qquad (5.279)$$

Case (d): Assume $\tau + j - k = 0 \bmod p^L - 1$, and $\tau = 0 \bmod p'^{L'} - 1$. Then

$$P_{jk}(\tau) = N_m. \qquad (5.280)$$

This last case can occur only if

$$\tau = C_1(p'^{L'} - 1), \quad \tau + j - k = C_2(p^L - 1), \qquad (5.281)$$

whence

$$j - k = C_2(p^L - 1) - C_1(p'^{L'} - 1). \qquad (5.282)$$

Since the right side of (5.282) is a multiple of N_d, the allowed range of j and k excludes the occurrence of case (d), except when $j = k$ and $\tau = 0$. ∎

When $p = p'$, there is a common mathematical structure available for describing the sequences in Theorem 5.19. The characteristic polynomial of

$\{b_n\}$ is a minimum polynomial $m_\alpha(z)$ of degree L over GF(p) with α a primitive element of GF(p^L).

$$m_\alpha(z) = z^L + \sum_{j=1}^{L} q_j z^{L-j}. \tag{5.283}$$

Likewise, $m_\beta(z)$ of degree L' over GF(p) is the characteristic polynomial of $\{b'_n\}$, where β is a primitive element of GF($p^{L'}$), and we assume that $m_\alpha(z) \neq m_\beta(z)$.

$$m_\beta(z) = z^{L'} + \sum_{j=1}^{L'} s_j z^{L'-j}. \tag{5.284}$$

The generator for a sequence $\{a_n^{(j)}\}$ can be organized with p-ary shift registers as shown in Figure 5.15. This generator form is equivalent to a single $L + L'$ stage shift-register generator producing a sequence over GF(p) with characteristic polynomial $m_\alpha(z)m_\beta(z)$.

Let $\boldsymbol{b}_k$ denote an L-tuple of $\{b_n\}$ beginning at the k-th symbol, and similarly denote an L'-tuple from $\{b'_n\}$, beginning at the k-th symbol, by $\boldsymbol{b}'_k$. Hence, the states of the two Fibonacci generators in Figure 5.15 can be specified by one such L-tuple and one such L'-tuple from the component sequences, provided the registers are not loaded with all-zeroes vectors 0. The results of various register initializations on the generator's GF(p) output sequence $\{c_n\}$ are shown in Table 5.17. The notation $\{b_n^{(j)}\}$ has been used here to describe the GF(p) form of the output sequence, and, hence, when neither register is initialized with a zero vector,

$$a_n^{(j)} = a_{n+j} a'_n = \rho^{b_{n+j} - b'_n} = \rho^{b_n^{(j)}} \tag{5.285}$$

for all j and n where ρ is a complex primitive p-th root of unity.

The set of sequences $\{c_n\}$ for all i and k, enumerated in Table 5.17 forms a linear space. Specifically, if $\{c_n\}$ and $\{c'_n\}$ both satisfy the linear recursion having characteristic polynomial $m_\alpha(z)m_\beta(z)$, then so do sequences of the form $\{dc_n + ec'_n\}$ for any choice of d, e GF(p). Since Table 5.17 enumerates the $p^{L+L'}$ distinct sequences which satisfy such a recursion, the sequence $\{dc_n + ec'_n\}$ must correspond to one of the tabulated sequences. Hence, the initial N_m-tuples of the GF(p) sequences $\{c_n\}$ listed in Table 5.17 and generated by the structure in Figure 5.15, form a *linear cyclic code* [6], [7], [40], [41], [48] and we shall refer to the sequence set as a *linear design*.

By proper initialization of the register contents, the generator of Figure 5.15 can be used to produce any of the N_d possible sequences $\{a_n^{(j)}\}$, $j = 0, \ldots, N_d - 1$, the component sequences $\{a_n\}$ and $\{a'_n\}$, and the all-ones sequence. The latter sequence, definitely not pseudorandom in nature, is simply avoided by not initializing both registers with all zeroes. A purely random non-zero initialization of the registers is equally likely to produce any one of the sequences of interest. To load the registers for the production

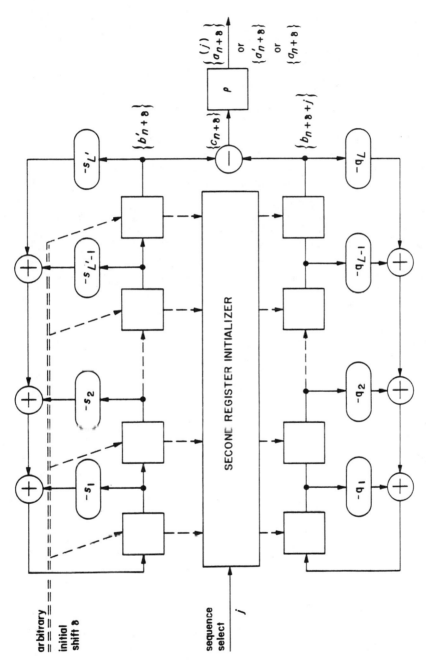

Figure 5.15. Fibonacci form of a linear generator structure for $\{a_n^{(j)}\}$ when both $\{b_n\}$ and $\{b_n'\}$ are over GF(p).

Table 5.17

Initialization data for the linear generator structure of Figure 5.15.

Register Initializations		Output	Size of	Output
Lower	Upper	$\{c_n\}$	Index Set	Linear Span
b_k	b_i'	$\{b_{n+i-1}^{(k-i)}\}$	$(p^L - 1)(p^{L'} - 1)$	$L + L'$
0	b_i'	$\{b_{n+i-1}'\}$	$p^{L'} - 1$	L'
b_k	0	$\{b_{n+k-1}\}$	$p^L - 1$	L
0	0	$\{0\}$	1	1

of $\{a_{n+\delta}^{(j)}\}$ for a particular j at a random shift δ, one register of the sequence generator in Figure 5.15 may be initialized randomly with non-zero contents, and then the allowed choice(s) for the initial contents of the other register must be determined mathematically. This calculation could be done off line, thereby complicating the use of the system by requiring that operators have tables of allowed register contents, or it could be mechanized in hardware, as suggested in Figure 5.15.

If the second register initialization is mechanized in hardware and the computation can be carried out in a fraction of the registers' clock time with a unique answer, then the second register can be eliminated and the initializing hardware can be used to calculate the second register's output symbol for each successive contents of the first register. This calculating hardware, and the adder which combines the two component sequences, form an NLFFL function of the contents of the first register to produce the output sequence over GF(p). Note that the structure of this NLFFL is dependent on the sequence to be generated, i.e., the index j, and is generally rather complex.

Insight into the balance properties of a sequence $\{b_n^{(j)}\}$ in a linear design can be obtained by relating balance to correlation. Let N_b be the number of occurrences of the symbol b in one period of $\{b_n^{(j)}\}$, and note that

$$\sum_{b \in \mathrm{GF}(p)} N_b \rho^b = \sum_{n=1}^{N_m} \rho^{b_n^{(j)}} = \sum_{n=1}^{N_m} \rho^{b_{n+j} - b_n'} = P_{aa'}(j). \tag{5.286}$$

Knowledge of the periodic cross-correlation values on the left side of (5.286) imposes constraints on the possible values of N_b, and vice versa. The clearest example of this relation occurs when $p = 2$, and is summarized in the following.

THEOREM 5.20. *Let* $\{b_n^{(j)}\}$, $j = 0, 1, \ldots, N_d - 1$, *be the compound sequences of a binary linear design, based on component m-sequences* $\{b_n\}$ *and* $\{b_n'\}$ *with corresponding roots-of-unity m-sequences* $\{a_n\}$ *and* $\{a_n'\}$ *respectively. Then the number of compound sequences in the binary linear design, having* N_1 *ones in one period, is given by the number of shifts j,*

$0 \leq j < N$, *for which*

$$P_{aa'}(j) = N_m - 2N_1,$$

where N_m and N_1 are defined in (5.273) and (5.275) with $p = p' = 2$, and L and L' are the linear spans of the component m-sequences.

Note that $N_m - 2N_1$ is the number of zeroes in the binary sequence $\{b_n^{(j)}\}$ minus the number of ones, and, hence, $P_{aa'}(j)$ can be interpreted as the imbalance in $\{b_n^{(j)}\}$.

Example 5.14 (Gold sequences). In a 1967 paper [35], Gold suggested combining two binary m-sequences with identical periods to produce a set of SSMA codes with good periodic correlation properties. Gold considered a linear design in which two component sequences, each with linear span L, were employed, one with characteristic polynomial $m_\alpha(z)$, and the other with characteristic polynomial $m_{\alpha^r}(z)$, where

$$r = 2^{\lfloor (L+2)/2 \rfloor} + 1, \tag{5.287}$$

with $\lfloor x \rfloor$ denoting the integer part of x. This choice of r satisfies the conditions of the cross-correlation Theorem 5.17 with $e - 1$ for L odd and $e = 2$ for L even. The combining process of Theorem 5.19, in this case with $p = p' = 2$ and $N_m = N_d = 2^L - 1$, yields a binary linear design with all cross-correlation and out-of-phase autocorrelation values taken from the set $-1, -1 + 2^{\lfloor (L+2)/2 \rfloor}, -1 - 2^{\lfloor (L+2)/2 \rfloor}$.

Including the component sequences, Gold's design contains $2^L + 1$ sequences, a sufficient number for Sidelnikov's bound (5.245) to apply. Specializing the bound to this case gives

$$P_{max} > \left[2(2^r - 1) \quad 2\right]^{1/2} - (2^{L+1} - 1)^{1/2}. \tag{5.288}$$

where P_{max} is defined in (5.232)–(5.235). For $L \geq 3$, this implies that P_{max} is greater than $-1 + 2^{(L+1)/2}$, and, since the binary sequence set's period is odd, P_{max} must be odd, yielding

$$P_{max} \geq 1 + 2^{(L+1)/2}. \tag{5.289}$$

Gold sequences with L odd have P_{max} satisfying this bound with equality, and, hence, are an optimal design from the viewpoint of minimizing the absolute periodic correlation parameter P_{max}.

Theorem 5.20 indicates that only $2^L - 2^{L-e} - 1$ of the compound sequences in the Gold set are balanced, with the imbalance in the remaining 2^{L-e} sequences being on the order of $2^{\lfloor (L+2)/2 \rfloor}$.

Example 5.15 (the small set of Kasami sequences). Kasami [30] in 1966 reported results on the enumeration of weights of linear cyclic codes over GF(2). He noted that this problem is equivalent (as in Theorem 5.20) to the evaluation of cross-correlation or out-of-phase autocorrelation between two binary (± 1) sequences, and produced forms of the results described in Theorems 5.16 and 5.18.

The small set of Kasami sequences is a binary linear design whose component sequences are (1) an m-sequence with period $2^L - 1$, L even, and characteristic polynomial $m_\alpha(z)$, $\alpha \in GF(2^L)$, and (2) an m-sequence with shorter period $2^{L/2} - 1$ and characteristic polynomial $m_{\alpha^r}(z)$, where

$$r = 2^{L/2} + 1, \tag{5.290}$$

α^r being an element of $GF(2^{L/2})$. The period of the latter m-sequence divides the longer period of the first m-sequence (see Lemma 5.3), and Theorem 5.18 states that the periodic cross-correlation of these two m-sequences, over the longer period, is $-1 \pm 2^{L/2}$ at all relative phase shifts.

Application of Theorem 5.19 to these m-sequences, with $p = p' = 2$, $N_m = 2^L - 1$, and $N_d = 2^{L/2} - 1$, yields a binary linear design now referred to as the small set of Kasami sequences, with all cross-correlation and out-of-phase autocorrelation values from the set -1, $-1 + 2^{L/2}$, and $-1 - 2^{L/2}$. Adding the long-period m-sequence to this set gives $2^{L/2}$ sequences with a $P_{\max}$ of $1 + 2^{L/2}$.

Evaluation of the Welch bound (5.241) at $J = 2^{L/2}$ and $N = 2^L - 1$, gives

$$P_{\max} \geq \left(\frac{N^2(J - 1)}{NJ - 1} \right)^{1/2} = (2^L - 2^{L/2} - 1 + r)^{1/2}$$

$$> (2^L - 2 \cdot 2^{L/2} + 1)^{1/2} = 2^{L/2} - 1 \tag{5.291a}$$

for some r, $0 < r < 1$. Since the inequality is strict and since $P_{\max}$ must be odd for binary sequences with odd periods, the Welch bound for this binary case reduces to

$$P_{\max} \geq 2^{L/2} + 1. \tag{5.291b}$$

The small set of Kasami sequences achieves this bound with equality, and, therefore, is optimal in the sense of minimizing the maximum absolute correlation parameter $P_{\max}$.

Theorem 5.20 can be used to verify that none of the compound sequences in the Kasami set is balanced.

5.7.5 A Transform-Domain Design Philosophy

The linear designs of the previous section have excellent periodic correlation properties, but short linear spans. A transform domain design approach will now be described which results in a NLFFL design producing balanced binary sequences with optimal correlation properties and longer linear span [32], [42].

Consider two functions, $r(x)$ and $s(x)$, mapping the elements of $GF(2^d)$ into $+1$ and -1. These functions can be viewed as representing NLFFL functions operating on a Galois m-sequence generator with characteristic polynomial $m_\alpha(z)$, whose d-stage register contents $r_1^{(n)}, \ldots, r_d^{(n)}$ represent α^n (see (5.127)). The resulting binary ($+1$ and -1) sequences, $\{r(\alpha^n)\}$ and $\{s(\alpha^n)\}$, have periodic cross-correlation at shift τ given by

$$R_{rs}(\tau) = \sum_{n=1}^{2^d-1} r(\alpha^{n+\tau}) s(\alpha^n)$$

$$= -r(0)s(0) + \sum_{x \in \mathrm{GF}(2^d)} r(yx) s(x). \qquad (5.292)$$

where $y = \alpha^\tau$.

The inner product of $r(yx)$ and $s(x)$ over $x \in \mathrm{GF}(2^d)$ in (5.292) can be studied in a transform domain using any inner product-preserving transform. Two closely related transforms will be considered for this purpose, the first being the *trace transform* which is defined as

$$\hat{r}(\lambda) = \frac{1}{2^{d/2}} \sum_{x \in \mathrm{GF}(2^d)} r(x)(-1)^{\mathrm{Tr}(\lambda x)} \qquad (5.293)$$

for all $\lambda \in \mathrm{GF}(2^d)$. The following properties of the trace transform can be verified:

(a) Inversion theorem:

$$r(x) = \frac{1}{2^{d/2}} \sum_{\lambda \in \mathrm{GF}(2^d)} \hat{r}(\lambda)(-1)^{\mathrm{Tr}(\lambda x)}. \qquad (5.294)$$

(b) Multiplicative shifting theorem: For $y \neq 0$,

$$s(x) = r(yx) \text{ for all } x \text{ iff } \hat{s}(\lambda) = \hat{r}(y^{-1}\lambda) \text{ for all } \lambda. \qquad (5.295)$$

(c) Parseval's relation:

$$\sum_{x \in \mathrm{GF}(2^d)} r(x)s(x) = \sum_{\lambda \in \mathrm{GF}(2^d)} \hat{r}(\lambda)\hat{s}(\lambda). \qquad (5.296)$$

(d) Additive shifting theorem:

$$s(x) = r(x + y) \text{ for all } x \text{ iff } \hat{s}(\lambda) = \hat{r}(\lambda)(-1)^{\mathrm{Tr}(\lambda y)} \text{ for all } \lambda, \qquad (5.297)$$

$$\hat{s}(\lambda) = \hat{r}(\lambda + y) \text{ for all } \lambda \text{ iff } s(x) = r(x)(-1)^{\mathrm{Tr}(xy)} \text{ for all } x. \qquad (5.298)$$

Another equivalent inner-product-preserving transform is defined after a one-to-one linear mapping of $\mathrm{GF}(2^d)$ onto the space $\mathcal{V}_d$ of d-tuples with elements in $\mathrm{GF}(2)$. Specifically, let $\beta_1, \ldots, \beta_d$ be an arbitrary basis for $\mathrm{GF}(2^d)$, and let $\gamma_1, \ldots, \gamma_d$ be another basis for $\mathrm{GF}(2^d)$ with the property that

$$\mathrm{Tr}(\beta_i \gamma_j) = \begin{cases} 1, & i = j \\ 0, & i \neq j. \end{cases} \qquad (5.299)$$

Two bases satisfying (5.299) are called *complementary*. It can be verified [42] that every basis for $GF(2^d)$ has a complementary basis. Let x and λ be arbitrary elements of $GF(2^d)$, represented in complementary bases by

$$x = \sum_{i=1}^{d} x_i \beta_i, \qquad \lambda = \sum_{i=1}^{d} \lambda_i \gamma_i. \qquad (5.300)$$

Coordinate values can be determined by applying (5.299) and trace linearity to (5.301) to yield

$$x_i = \mathrm{Tr}(\gamma_i x), \qquad \lambda_i = \mathrm{Tr}(\beta_i \lambda) \qquad (5.301)$$

for all i. These same relations allow $\mathrm{Tr}(\lambda x)$ to be interpreted as an inner product of vectors of coefficients from the expansions (5.300).

$$\mathrm{Tr}(\lambda x) = x^t \lambda. \qquad (5.302)$$

The mapping from $x \in GF(2^d)$ to $x \in \mathcal{V}_d$, can be used to define

$$R(x) \triangleq r(x) \qquad (5.303)$$

for all x in $GF(2^d)$, substitution of (5.302) and (5.303) into the trace transform definition (5.293) gives

$$\hat{r}(\lambda) = \frac{1}{2^{d/2}} \sum_{x \in \mathcal{V}_d} R(x)(-1)^{\lambda^t x} \triangleq \tilde{R}(\lambda). \qquad (5.304)$$

The right side of (5.304) defines the *Fourier transform* $\tilde{R}(\lambda)$ of the function $R(x)$. The inversion theorem, Parseval's relation, and additive shifting theorem for the Fourier transform are easily derived from the corresponding results for trace transforms, in view of (5.304). Both transforms will be used in the design approach to follow.

Returning to the periodic cross-correlation computation of (5.292), we now develop a bound on cross-correlation. Applying Parseval's relation (5.296) gives

$$R_{rs}(\tau) = -r(0)s(0) + \sum_{\lambda \in GF(2^d)} \hat{r}(y^{-1}\lambda)\hat{s}(\lambda). \qquad (5.305)$$

Therefore,

$$|R_{rs}(\tau)| \leq 1 + \sum_{\lambda \in y\mathcal{S}_r \cap \mathcal{S}_s} |\hat{r}(y^{-1}\lambda)||\hat{s}(\lambda)|, \qquad (5.306)$$

where the sum over λ in (5.306) has been limited to those λ for which the corresponding term is non-zero, i.e.,

$$\mathcal{S}_s \triangleq \{\lambda: \hat{s}(\lambda) \neq 0\} \qquad (5.307)$$

and

$$y\mathcal{S}_r = \{\lambda: \hat{r}(y^{-1}\lambda) \neq 0\} = \{\lambda y: \lambda \in \mathcal{S}_r\}. \qquad (5.308)$$

Parseval's relation also implies that

$$\sum_{\lambda \in \mathcal{S}_r} |\hat{r}(\lambda)|^2 = \sum_{\lambda \in GF(2^d)} |\hat{r}(\lambda)|^2 = \sum_{x \in GF(2^d)} |r(x)|^2 = 2^d, \qquad (5.309)$$

which yields a bound on the trace transform, namely

$$B_r \triangleq \max_{\lambda} |\hat{r}(\lambda)|^2 \geq \frac{2^d}{|\mathscr{S}_r|}. \tag{5.310}$$

Applying the definition on the left side of (5.310) to (5.306) gives a general bound on correlation magnitude.

$$|R_{rs}(\tau)| \leq 1 + \sqrt{B_r B_s} |y\mathscr{S}_r \cap \mathscr{S}_s|. \tag{5.311}$$

The above development suggests the following procedure for designing a set $\mathscr{B}$ of distinct NLFFL functions operating on identical m-sequence shift registers, to produce sequences with good periodic auto- and cross-correlation properties:

Property P-1.

Assume that the transforms of all functions in $\mathscr{B}$ are non-zero on the same subset $\mathscr{S}$ of $GF(2^d)$ and are all zero outside $\mathscr{S}$. Notice that

$$0 \notin \mathscr{S} \Leftrightarrow \hat{r}(0) = 0 = \sum_{x \in GF(2^d)} r(x) \Leftrightarrow \{r(\alpha^n)\} \text{ is balanced,}$$

$$\tag{5.312}$$

for all $r \in \mathscr{B}$. Hence, a set of balanced sequences will be achieved in this design if and only if 0 is not in $\mathscr{S}$.

Property P-2.

Choose the set $\mathscr{S}$ (with $0 \notin \mathscr{S}$) so that $y\mathscr{S} \cap \mathscr{S}$ is as small as possible for all non-integer y in $GF(2^d)$. To calculate the average size of the intersection, let $a_i = 1$ for each i for which $\alpha^i \in \mathscr{S}$, and let $a_i = 0$ otherwise. Then for $y = \alpha^\tau$, we have

$$|y\mathscr{S} \cap \mathscr{S}| - \sum_{i=0}^{2^d-2} a_{i+\tau}a_i, \tag{5.313}$$

and bounding the maximum intersection by the average intersection gives

$$\max_{\substack{y \in GF(2^d) \\ y \notin GF(2)}} |y\mathscr{S} \cap \mathscr{S}| \geq \frac{1}{2^d - 2} \sum_{\tau=1}^{2^d-2} \sum_{i=0}^{2^d-2} a_{i+\tau}a_i$$

$$\geq \frac{|\mathscr{S}|(|\mathscr{S}| - 1)}{2^d - 2}. \tag{5.314}$$

Property P-3.

Design the set $\mathscr{B}$ of functions so that for each r and s in $\mathscr{B}$, $r \neq s$,

$$\sum_{x \in GF(2^d)} r(x)s(x) = 0 = \sum_{\lambda \in GF(2^d)} \hat{r}(\lambda)\hat{s}(\lambda). \tag{5.315}$$

This orthogonality guarantees that $|R_{rs}(0)|$ is 1. Since in the transform domain $\hat{r}(\lambda)$ can be viewed as a $|\mathcal{S}|$-tuple, the achievable size of $\mathcal{B}$ is limited by $|\mathcal{S}|$, i.e.,

$$|\mathcal{B}| \leq |\mathcal{S}|. \tag{5.316}$$

Property P-4.

Design each function in $\mathcal{B}$ so that its trace transform has constant magnitude, namely $2^d/|\mathcal{S}|$ on $\mathcal{S}$ by (5.309). If this objective can be achieved and if the bounds (5.314) and (5.316) can be achieved with equality, then by (5.311) it is possible to find orthogonal NLFFL functions such that

$$|R_{rs}(\tau)| \leq 1 + \frac{2^d|\mathcal{S}|}{2^d - 2}, \tag{5.317}$$

for all $\tau \neq 0 \bmod 2^d - 1$.

The final parameter, namely $|\mathcal{S}|$, can be chosen with the aid of the Welch bound, which states that $|R_{rs}(\tau)|$ must be on the order of $2^{d/2}$ when $|\mathcal{B}|$ is large. Assuming that (5.317) is valid and tight, the Welch bound (5.241) indicates that $|\mathcal{S}|$ should be on the order of $2^{d/2}$, with larger values of $|\mathcal{S}|$ resulting in designs which cannot meet that bound with equality. The above design approach, with $|\mathcal{S}| = 2^{d/2}$, will be successfully demonstrated in the next section.

5.7.6 Bent Sequences

Rothaus [43] defined a function $f(x)$ mapping the space $\mathcal{V}_k$ of binary k-tuples over GF(2) into $\mathcal{V}_1$ (GF(2)) to be *bent* if the Fourier transform of $(-1)^{f(x)}$, namely

$$\tilde{F}(\lambda) \triangleq \frac{1}{2^{k/2}} \sum_{x \in \mathcal{V}_k} (-1)^{f(x)+x^t\lambda}, \tag{5.318}$$

is ± 1 for all λ in $\mathcal{V}_k$. Although there are many families of bent functions [43]–[45], only one bent function form will be described here.

THEOREM 5.21 (Rothaus [43]). *Let*

$$x = \begin{bmatrix} x_1 \\ x_2 \end{bmatrix}, \quad x_1, x_2 \in \mathcal{V}_j. \tag{5.319}$$

Let the function $f_z(x)$, *mapping* $\mathcal{V}_{2j}$ *onto* $\mathcal{V}_1$, *be defined as*

$$f_z(x) = x_1^t x_2 + g(x_2) + z^t x, \tag{5.320}$$

where $g(x_2)$ *is an arbitrary function mapping* $\mathcal{V}_j$ *into* $\mathcal{V}_1$. *Then* $f_z(x)$ *is bent for every* z *in* $\mathcal{V}_{2j}$.

Proof. The Fourier transform of $f_z(x)$ is given by

$$\tilde{F}_z(\lambda) = 2^{-j} \sum_{x_1, x_2 \in \mathcal{V}_j} (-1)^{x_1'x_2 + g(x_2) + x'z + x'\lambda}$$

$$= 2^{-j} \sum_{x_2 \in \mathcal{V}_j} (-1)^{g(x_2) + x_2'(z_2 + \lambda_2)}$$

$$\times \sum_{x_1 \in \mathcal{V}_j} (-1)^{x_1'(x_2 + z_1 + \lambda_1)}, \tag{5.321}$$

where, in the latter step, z and λ have been appropriately partitioned into vectors of dimension j. The sum over x_1 in (5.321) is zero if and only if $x_2 + z_1 + \lambda_1$ is not zero, its alternative value being 2^j. Hence,

$$\tilde{F}_z(\lambda) = (-1)^{g(z_1 + \lambda_1) + (z_1 + \lambda_1)'(z_2 + \lambda_2)}. \tag{5.322}$$

∎

With regard to dimensionality, this design will now fix

$$d = 2k = 4j, \tag{5.323}$$

with $g(\cdot)$ an arbitrary function from $\mathcal{V}_j$ to $\mathcal{V}_1$, and $f_z(\cdot)$ a bent function mapping $\mathcal{V}_k$ into $\mathcal{V}_1$. Let $L(x)$ define a linear mapping from $GF(2^d)$ to $\mathcal{V}_k$. The set $\mathcal{B}$ of NLFFL functions used to produce bent sequences is now defined to contain all functions of the form

$$r_z(x) = (-1)^{f_z(L(x)) + \text{Tr}(\sigma x)}, \tag{5.324}$$

where σ is fixed and $z \in \mathcal{V}_k$.

The trace transform of $r_z(x)$ is given by

$$\hat{r}_z(\lambda) = 2^{-k} \sum_{x \in GF(2^d)} (-1)^{f_z(L(x)) + \text{Tr}(\sigma x) + \text{Tr}(\lambda x)}. \tag{5.325}$$

Representing $(-1)^{f_z}$ in terms of its inverse transform on $\mathcal{V}_k$ gives

$$\hat{r}_z(\lambda) = 2^{-k} \sum_{x \in GF(2^d)} (-1)^{\text{Tr}((\sigma + \lambda)x)} 2^{-j} \sum_{\lambda' \in \mathcal{V}_k} \tilde{F}_z(\lambda')(-1)^{\lambda''L(x)}. \tag{5.326}$$

The linear mapping $L(x)$ can be represented as

$$L(x) = Mx = M \begin{bmatrix} \text{Tr}(\gamma_1 x) \\ \vdots \\ \text{Tr}(\gamma_d x) \end{bmatrix}, \tag{5.327}$$

where M is a $k \times d$ matrix over $GF(2)$, of rank k, and x is the vector representation of the $GF(2^d)$ element x (see (5.300)–(5.301)). The inner product $\lambda''L(x)$ in (5.326) can be rewritten using (5.327) and the linearity of the trace function, so that

$$\lambda''L(x) = \text{Tr}(\ell(\lambda')x), \tag{5.328}$$

where the *adjoint* $\ell(\lambda')$ of $L(x)$ is

$$\ell(\lambda') = \lambda''M\gamma, \tag{5.329}$$

and γ is the vector of basis elements $\gamma_1, \ldots, \gamma_d$. Since the rank of M is $k, \ell(\lambda')$ is a one-to-one mapping from $\mathscr{V}_k$ onto a k-dimensional linear subspace of $GF(2^d)$ with basis $M\gamma$.

Substituting (5.328) into (5.326) and interchanging the order of summation gives

$$\hat{r}_z(\lambda) = 2^{-j} \sum_{\lambda' \in \mathscr{V}_k} \tilde{F}_z(\lambda')$$

$$\cdot 2^{-k} \sum_{x \in GF(2^d)} (-1)^{\text{Tr}\{[\sigma + \lambda + \ell(\lambda')]x\}} \tag{5.330}$$

The sum over x is non-zero if and only if the coefficient of x (in the trace function) is zero. Hence, $\hat{r}_z(\lambda)$ is non-zero if and only if $\sigma + \lambda$ is in the range of $\ell(\cdot)$. Denoting the inverse of ℓ on the range of ℓ by ℓ^{-1}, it follows that

$$\hat{r}_z(\lambda) = \begin{cases} 2^j \tilde{F}_z(\ell^{-1}(\sigma + \lambda)), & \text{for } \sigma + \lambda \in \text{range}(\ell) \\ 0, & \text{otherwise.} \end{cases} \tag{5.331}$$

Since $\tilde{F}_z$ is the transform of a bent function, $|\hat{r}_z(\lambda)|$ takes on two values, 0 and 2^j, as λ varies over $GF(2^d)$, thereby satisfying design properties P-1 and P-4 of the previous section. One purpose of the added term $\text{Tr}(\sigma x)$ in the design of $f_z(x)$ now is apparent: The element σ can be chosen to make $\hat{r}_z(0) = 0$, thereby guaranteeing that the bent sequences are balanced.

Two applications of Parseval's relation can be used to verify the orthogonality of functions in $\mathscr{B}$.

$$\sum_{\lambda \in GF(2^d)} \hat{r}_z(\lambda)\hat{r}_w(\lambda) = 2^k \sum_{\lambda \in \mathscr{V}_k} \tilde{F}_z(\lambda)\tilde{F}_w(\lambda)$$

$$= 2^k \sum_{x \in \mathscr{V}_k} (-1)^{(z-w)'x}$$

$$= 0 \text{ for } z \neq w. \tag{5.332}$$

Thus, design objective (Property P-3) is satisfied, guaranteeing good "in-phase" cross-correlation properties of bent sequences.

All requirements of the design procedure (Properties P-1 to P-4) have been satisfied except Property P-2, i.e., the minimization of $|y\mathscr{S} \cap \mathscr{S}|$. In this design, the set $\mathscr{S}$ is defined from (5.331) as

$$\mathscr{S} = \{\lambda: \sigma + \lambda \in \text{range}(\ell)\}. \tag{5.333}$$

Since σ is a constant and the range of ℓ contains 2^k points, $\mathscr{S}$ contains 2^k points, and (5.314) indicates that a good design may at best achieve

$$|y\mathscr{S} \cap \mathscr{S}| \leq 1 \quad \text{for all } y \notin GF(2). \tag{5.334}$$

The final design step is performed by choosing an appropriate basis for the range of $\ell(\cdot)$ so that (5.334) is satisfied. Let

$$\text{range}(\ell) = \{\delta\varepsilon_0 : \delta \in GF(2^k)\}, \tag{5.335}$$

where ε_0 is any element of $GF(2^d)$ that is not contained in a smaller field. Then $\mathscr{S}$ in (5.333) may be represented as

$$\mathscr{S} = \{\lambda : \sigma + \lambda = \delta\varepsilon_0, \delta \in GF(2^k)\}$$
$$= \{\delta\varepsilon_0 + \sigma : \delta \in GF(2^k)\}. \tag{5.336}$$

We will now demonstrate that (5.334) is satisfied. Viewing $GF(2^d)$ as an extension field of $GF(2^k)$ which can be represented as a two-dimensional vector space over $GF(2^k)$ (see Appendix 5A.4), an arbitrary element y of $GF(2^d)$ can be represented as

$$y = y_1\varepsilon_0 + y_0, \tag{5.337}$$

where the basis for the vector space is 1 and ε_0, and the coefficients y_1 and y_0 are in $GF(2^k)$. Then a point in the intersection of $y\mathscr{S}$ and $\mathscr{S}$ corresponds to a solution of the equation

$$(y_1\varepsilon_0 + y_0)(\delta\varepsilon_0 + \sigma) = \delta'\varepsilon_0 + \sigma \tag{5.338}$$

for the unknowns δ and δ' in $GF(2^k)$. The element ε_0^2 can also be represented as $e_1\varepsilon_0 + e_0$, where e_1 and e_0 are in $GF(2^k)$ and e_0 is not 0. A similar representation may be used for the element σ. This reduces the solution of (5.338) to the solution of a pair of linear equations over $GF(2^k)$, by setting the coefficients of the basis elements ε_0 and 1 to zero, giving

$$(y_1e_1 + y_0)\delta + \delta' = (y_0 + y_1e_1 + 1)\sigma_1 + y_1\sigma_0$$
$$y_1e_0\delta = y_1e_0\sigma_1 + (y_0 + 1)\sigma_0. \tag{5.339}$$

These equations are linearly independent and have a unique solution, except when $y_1 = 0$ and $y_0 = 1$, corresponding to the uninteresting case when $y = 1$. Therefore, (5.334) is satisfied, there being a single point in the intersection $y\mathscr{S} \cap \mathscr{S}$ when $y \notin GF(2)$, and hence, design objective P-2 has been achieved. The choice of σ can now be made so that $\hat{r}_z(0)$ is zero, by requiring that σ not be of the form $\delta\varepsilon_0$ for any δ in $GF(2^k)$, i.e., by requiring that σ not be in range (ℓ).

The specification of range (ℓ) in (5.335) may be used to determine a suitable choice for M. Let $\phi_1, \ldots, \phi_k$ be a basis for $GF(2^k)$, and let ϕ be the vector composed of these basis elements. Then, equating

$$\varepsilon_0\phi = M\gamma \tag{5.340}$$

guarantees by (5.329) that range (ℓ) matches the designed range specified in (5.335). Multiplying the i-th entry of the vector in (5.340) by β_j, taking the trace of the result, and using the fact that γ is a vector of elements from a complementary basis for $(\beta_j : j = 1, \ldots, d)$, gives the ij-th element of M to be

$$m_{ij} = \text{Tr}(\varepsilon_0\phi_i\beta_j). \tag{5.341}$$

Thus, the elements of M can be determined by trace calculations. The design can be completed by choosing a vector s so that

$$\text{Tr}(\sigma x) = s' x, \tag{5.342}$$

the requirement that σ not be in the range (ℓ) corresponding to s' not being in the row space of M.

It is worth noting that both transforms were useful at certain points in the derivation, but neither transform need be mechanized. Furthermore, completion of a design does not require the determination of a complementary basis. A compact description of the preceding development is given in the following result.

THEOREM 5.22. *Let α be a primitive element of $\text{GF}(2^d)$, d divisible by 4, and let x represent the contents of a Galois-configured m-sequence generator having the minimum polynomial of α over $\text{GF}(2)$ as its characteristic polynomial, i.e.,*

$$x = \sum_{i=0}^{d-1} x_i \alpha^i. \tag{5.343a}$$

Let ε_0 be any element of $\text{GF}(2^d)$ that is not contained in a smaller field, and let $\{\phi_1, \ldots, \phi_{d/2}\}$ be a basis for $\text{GF}(2^{d/2})$ over $\text{GF}(2)$. Let M be a $d/2 \times d$ matrix whose i,j-th entry m_{ij} is given by

$$m_{ij} = \text{Tr}_2^{2^d}\left(\varepsilon_0 \phi_i \alpha^{j-1}\right) \tag{5.343b}$$

and let s' be any d-dimensional vector not contained in the linear subspace spanned by the rows of M. Then the $2^{d/2}$ NLFFL functions of the form

$$r_z(x) = (-1)^{f_z(Mx) + s' x} \tag{5.343c}$$

indexed by $z \in \mathcal{V}_{d/2}$, where $f_z(\cdot)$ is a bent function of the form (5.320), produce balanced sequences with periodic cross-correlation and out-of-phase autocorrelation bounded in magnitude by $1 + 2^{d/2}$.

Example 5.16. Consider a primitive polynomial $z^{12} + z^6 + z^4 + z + 1$ with root α in $\text{GF}(4096)$, which serves as the characteristic polynomial of a Galois LFSR. This register's state sequence has period 4095, and its contents x represent the field element x in the basis $\{\alpha^i : i = 0, 1, \ldots, 11\}$ as indicated in (5.343). In this notation, the set $(\alpha^{65i} : i = 0, 1, \ldots, 5)$ is composed of elements having order dividing 63, α^{65} being primitive in $\text{GF}(64)$, and, hence, these elements form a basis for the subfield $\text{GF}(64)$. One possible choice for the element ε_0, which must be outside $\text{GF}(64)$, is α. With the choices just stated, the 6×12 basis reduction matrix M has ij-th entry

$$m_{ij} = \text{Tr}_2^{4096}\left(\alpha^{65(i-1) + (j-1) + 1}\right), \tag{5.344}$$

resulting in the matrix

$$M = \begin{bmatrix} 000000000010 \\ 001100010111 \\ 100001110000 \\ 101001100100 \\ 110010100001 \\ 011001001001 \end{bmatrix} \tag{5.345}$$

A suitable choice for the vector s^t, which must be outside the row space of M, is

$$s^t = (000000000001). \tag{5.346}$$

Employing a bent function of the form described in Theorem 5.21, operating on Mx, with the arbitrary function $g(\ )$ being a three-input AND gate, one bent-sequence generator design is shown in Figure. 5.16. The set of bent sequences produced by this generator has an absolute correlation measure P_{max} of 65.

The linear span L of a bent sequence can be bounded by Key's technique (5.172) to give

$$L \le \sum_{i=1}^{D} \binom{d}{i} \tag{5.347}$$

where D is the order of the nonlinearity employed in a d-stage register design, D presumably taking on the maximum allowed value $d/4$ (when $d \ge 8$) as determined by the choice of arbitrary nonlinearity $g(\cdot)$ (see (5.320)). Key's bound is not tight in this case, the intuitive reason for this being the fact that the nonlinearity is imposed on a lower dimensional subspace. The following theorem provides upper and lower bounds on the linear span of a well-designed bent sequence.

THEOREM 5.23 [46]. *Let L denote the maximum-achievable linear span of bent sequences produced by maximal-order $(d/4)$ NLFFL functions operating on a d-stage LFSR with d at least 8 and divisible by 4. Then L can be upper- and lower-bounded as follows:*

$$L \le \sum_{i=1}^{\frac{d}{4}-1} \binom{d}{i} + \binom{d/2}{d/4} 2^{d/4} - \sum_{i=1}^{\left\lfloor \frac{\frac{d}{4}-1}{2} \right\rfloor} \binom{d/2}{i}, \tag{5.348a}$$

$$L \ge \begin{cases} 20, & d = 8 \\ \binom{d/2}{d/4} 2^{d/4} + d + \frac{1}{2} \sum_{i=2}^{\frac{d}{4}-1} \binom{d/2}{i} 2^i, & d > 8. \end{cases} \tag{5.348b}$$

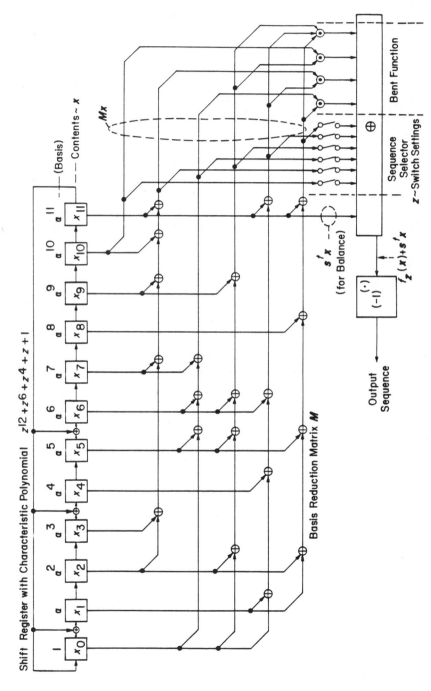

Figure 5.16. Typical bent sequence generator. Connections correspond to Example 5.16.

Table 5.18
Bounds on the maximum linear span of bent function sequences employing
d-stage generators.

d	lower bound	upper bound
8	20	32
12	202	232
16	1416	1808
20	10334	14204
24	76804	114512
28	577098	938004

The term $\binom{d/2}{d/4}2^{d/4}$ tends to dominate the lower bound and, therefore, may be used as a conservative estimate of the achievable linear span. These bounds on the linear span of bent sequences are evaluated in Table 5.18 for several values of register length d. A comparison of the properties of various DS-SSMA signal sets is shown in Table 5.19.

In summary, bent sequences and their generators have several desirable features:

(a) optimal periodic cross-correlation properties for multiple access uses (although the choice of short periods is limited).
(b) Large linear span and balance for AJ applications.
(c) Easy sequence selection in hardware.
(d) Efficient use of memory in the receiver.

Thus, these sequences have potential applications in both the multiple access and jamming environments.

Table 5.19
Properties of sequence sets with period $2^d - 1$.

Family	d	size	P_{max}	maximum achievable linear span	maximum achievable linear span for $d \leq 20$
Gold	odd	$2^d + 1$	$1 + 2^{(d+1)/2}$	$2d$	38
Gold	2 (mod 4)	$2^d + 1$	$1 + 2^{(d+2)/2}$	$2d$	36
Kasami (small set)	even	$2^{d/2}$	$1 + 2^{d/2}$	$3d/2$	30
Kasami (large set)	even	$2^{d/2}(2^d + 1)$	$1 + 2^{(d+2)/2}$	$5d/2$	50
Bent Sequences	0 (mod 4)	$2^{d/2}$	$1 + 2^{d/2}$	$\geq \binom{d/2}{d/4}2^{d/4}$	≥ 8064

5.8 FREQUENCY-HOPPING MULTIPLE-ACCESS DESIGNS

5.8.1 Design Criteria

The design of PN sequences to be used in frequency-hopping multiple-access (FHMA) systems differs from DSMA design in several ways:

1. The sequences' alphabet size is larger in an FHMA design, with one alphabetical character for each synthesizable frequency. In the DSMA case, that alphabet size usually corresponds to the number of carrier phase shifts used to modulate the sequence, the number typically being two (BPSK) or four (QPSK).
2. Interference occurs in a FHMA system when two distinct transmitters use the same frequency simultaneously. Hence, the Hamming distance between the FH sequences being used by the transmitters is a measure of the quality of the sequence design, the ideal design possessing the maximum possible distance between FH sequences. In the DSMA case, optimal designs were based on minimum absolute correlation between sequences, which corresponded, in the binary case under ideal conditions, to a design in which all Hamming distances between sequences were equal to half the number of symbols over which the distances were being calculated.
3. The one-to-one mapping of a PN generator's output symbols into frequencies in a FHMA system is arbitrary and preserves Hamming distance. If this mapping can be changed by the system operators, e.g., used as a code-of-the-day, this may make the system less susceptible to jammers attempting to predict the hopping pattern, but should not be relied upon, as a substitute for large linear span, to baffle intelligent jammers. The predetermined mapping of PN generator output to signal phase (or amplitude), required in a DSMA system to preserve the correlation properties designed into the PN generator outputs, allows the intelligent jammer to observe the PN generator output directly; hence, design characteristics such as linear span are, to some extent, more important in the DSMA case.

The Hamming distance design criterion for FHMA schemes will now be formalized.

Let $\{ f_n^{(k)} \}$ represent the k-th in a set of K periodic sequences of symbols, each symbol being in one-to-one correspondence with a transmitted frequency, and denote the common period by N. Then the maximum self-interference which a sequence in this set can encounter corresponds to the minimum Hamming distance H_A defined by

$$H_A = \min_k \ \min_{0 < \tau < N} H(f_{k,\tau}, f_{k,0}), \qquad (5.349)$$

where $H(x, y)$ denotes the Hamming distance between the vectors x and y, and $f_{k,t}$ denotes the vector of N consecutive symbols from the sequence

$\{f_n^{(k)}\}$, beginning with the t-th symbol. Similarly the maximum interference between a pair of FH signals in the design corresponds to the minimum Hamming distance H_C, given by

$$H_C = \min_{i \neq k} \min_{0 \leq \tau < N} H(f_{i,\tau}, f_{k,0}), \qquad (5.350)$$

and, therefore, a reasonable design criterion for FHMA PN generators consists of minimizing

$$H_{\min} = \min(H_A, H_C). \qquad (5.351)$$

This criterion corresponds to designing cyclic codes (over large alphabets) with large minimum distance between code words.

5.8.2 A Bound on Hamming Distance

Theorem 5.4 provides an upper bound on H which depends only on the sequence period and alphabet size. The approach used in deriving Theorem 5.4 can be modified to give the following bound on H_C.

THEOREM 5.24 [11]. *Let* $\{f_n^{(1)}\}$ *and* $\{f_n^{(2)}\}$ *denote sequences of period N over an alphabet $\mathscr{A}$, and let $f_{i,n}$ denote the N-tuple of consecutive elements from $\{f_n^{(i)}\}$ beginning at element n. Let*

$$H_C = \min_{0 \leq \tau < N} H(f_{1,\tau}, f_{2,0}) \qquad (5.352)$$

and determine α and β from

$$N = \alpha|\mathscr{A}| + \beta, \qquad 0 \leq \beta < |\mathscr{A}|. \qquad (5.353)$$

Then

$$H_C \leq N - \frac{\alpha+1}{N}(N + \beta - |\mathscr{A}|). \qquad (5.354)$$

Proof. Following the approach of Theorem 5.4, it is easily verified that

$$H_C \leq \max_{\{N_{1,a}\}, \{N_{2,a}\}} \frac{1}{N}\left(N^2 - \sum_{a \in \mathscr{A}} N_{1,a}N_{2,a}\right), \qquad (5.355)$$

where $N_{i,a}$ is the number of occurrences of the symbol a in one period of $\{f_n^{(i)}\}$. Assuming for simplicity that $\mathscr{A}$ is the set of integers, $\{1, 2, \ldots, |\mathscr{A}|\}$, the maximizing choices for the symbol counts are

$$N_{1,a} = \begin{cases} \alpha + 1, & a = 1, 2, \ldots, \beta \\ \alpha, & a = \beta + 1, \ldots, |\mathscr{A}| \end{cases} \qquad (5.356)$$

$$N_{2,a} = \begin{cases} \alpha, & a = 1, 2, \ldots, |\mathscr{A}| - \beta \\ \alpha + 1, & a = |\mathscr{A}| - \beta + 1, \ldots, |\mathscr{A}|. \end{cases} \qquad (5.357)$$

Note that these maximizing counts correspond to balanced sequences. Substitution of (5.356) and (5.357) into (5.355) yields the final result. ∎

This bound on H_C for a two-sequence set also must apply for larger sets of sequences. Surprisingly, significantly larger sets can achieve this two-sequence bound with equality.

This section includes two well-known FHMA PN generator designs which achieve the bound of Theorem 5.24 with equality. One approach, analyzed by Lempel and Greenberger [11], adapts an m-sequence generator for FH purposes. The second approach described here is based on the use of Reed-Solomon code words as hopping patterns [47]. (Further material on the structure of Reed-Solomon codes is available in most major texts on error-correcting codes [6], [7], [40], [41], [48].)

5.8.3 An FHMA Design Employing an M-Sequence Generator

Let $\{b_n\}$ denote an m-sequence of period $q^L - 1$ with elements in GF(q), and let $\boldsymbol{b}_n$ denote a J-tuple of consecutive elements from $\{b_n\}$, beginning with the n-th element. Furthermore let $s(\boldsymbol{x})$ denote an arbitrary one-to-one mapping of J-tuples over GF(q) onto a set of q frequencies. A set of q^J sequences $\{f_n^{(v)}\}$ of frequencies, the set indexed by the J-tuple $\boldsymbol{v}$ of elements from GF(q), can now be defined by the relation

$$f_n^{(v)} = s(\boldsymbol{b}_n + \boldsymbol{v}). \qquad (5.358)$$

One possible generator mechanization is shown in Figure 5.17. Notice that indexing into a particular sequence is easily accommodated in this generator mechanization, and is independent of shift register initialization.

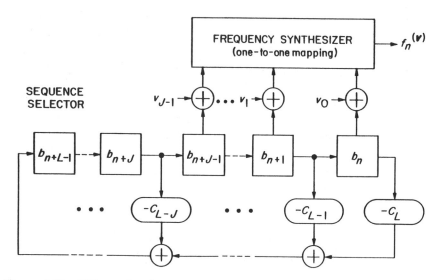

Figure 5.17. FH pseudonoise generator for SSMA use, based on an m-sequence design employing a q-ary shift register. All operations are in GF(q).

The Hamming distance between two of the sequences defined by (5.358) and calculated over a full period at a relative shift τ, can be derived as follows:

$$H(f_{v,\tau}, f_{w,0}) = \sum_{n=1}^{q^L-1} h\left(s(b_{n+\tau} + v), s(b_n + w)\right)$$

$$\overset{(1)}{=} \sum_{n=1}^{q^L-1} h'(b_{n+\tau} + v, b_n + w)$$

$$\overset{(2)}{=} \sum_{n=1}^{q^L-1} h'\left(b_{n+\tau'(\tau)}, w - v\right)$$

$$\overset{(3)}{=} \begin{cases} q^L - 1 - q^{L-J}, & w \neq v \\ q^L - q^{L-J}, & w = v \end{cases} \qquad (5.359)$$

Both $h(\ ,\ )$ and $h'(\ ,\ )$ are zero if their two arguments (scalars and vectors respectively) are equal, and are one otherwise. At (1) in (5.359), the Hamming distance-preserving property of one-to-one mappings is used; (2) is based on the shift-and-add property (5.63) of m-sequences; and (3) results from the balance property (5.61) of m-sequences.

The minimum Hamming distance $H_{\min}$ for this design is $q^L - 1 - q^{L-J}$, indicating that at most q^{L-J} matching frequencies appear in a comparison of two sequences over one period, at any relative shift. The quantities β and α of (5.353) are $q^J - 1$ and $q^{L-J} - 1$ respectively, and, hence, using the fact that H_A and H_C must be integer, it can be verified that these Hamming measures for this design satisfy the bounds of Theorem 5.4 and Theorem 5.24 with equality.

5.8.4 Reed-Solomon Sequences

This algebraic design is based on mechanizing calculations in $GF(q)$, and employing a one-to-one mapping $s(\cdot)$ from elements of $GF(q)$ to a set of q output frequency values. Let γ denote a primitive element of $GF(q)$, and define the polynomial $P_c(z)$ of degree $t - 1$, $t > 1$, over $GF(q)$ to be

$$P_c(z) = \sum_{i=0}^{t-1} c_i z^i. \qquad (5.360)$$

Elements of the sequences in this set are defined by

$$f_n^{(c)} = s(P_c(\gamma^n)) \qquad (5.361)$$

for all n, where c is any coefficient vector $(c_0, \ldots, c_{t-1})$ over $GF(q)$ with $c_1 = 1$. Hence, there are q^{t-1} sequences in this set. Each of these has period $q - 1$, the multiplicative order of γ.

Two sequences, $\{f_n^{(c)}\}$ and $\{f_n^{(d)}\}$, are cyclically distinct if

$$P_c(\gamma^{n+\tau}) \neq P_d(\gamma^n) \tag{5.362}$$

for some value of n at each possible value of τ. This condition (5.362) reduces by (5.360) to

$$\sum_{i=0}^{t-1} \left(c_i \gamma^{\tau i}\right) \gamma^{ni} = \sum_{i=0}^{t-1} c_i \gamma^{(\tau+n)i} \neq \sum_{i=0}^{t-1} d_i \gamma^{ni}. \tag{5.363}$$

Since the t sequences $\{\gamma^{ni}\}$, $i = 0, 1, \ldots, t - 1$, are linearly independent, the condition (5.363) can be satisfied if and only if $c_i \gamma^{\tau i} \neq d_i$ for some i at each value of τ. Fixing $c_1 = d_1 = 1$ for all sequences guarantees that this inequality is satisfied for $i = 1$ at all $\tau \neq 0 \mod q - 1$. When $\tau = 0 \mod q - 1$, the condition (5.363) reduces to $c \neq d$. Hence, the set of sequences with $c_1 = 1$ are cyclically distinct.

Let $e_i = c_i \gamma^{\tau i}$ and denote the vector $(e_0, \ldots, e_{t-1})$ by e. Based on the left side of (5.363),

$$P_c(\gamma^{n+\tau}) = P_e(\gamma^n). \tag{5.364}$$

The Hamming distance over one period between $\{f_n^{(c)}\}$ and $\{f_n^{(d)}\}$ at relative shift τ is given by

$$H(f_{c,\tau}, f_{d,0}) = \sum_{n=1}^{q-1} h\left(P_c(\gamma^{n+\tau}), P_d(\gamma^n)\right)$$

$$= \sum_{n=1}^{q-1} h\left(P_e(\gamma^n), P_d(\gamma^n)\right). \tag{5.365}$$

Notice that

$$P_e(\gamma^n) = P_d(\gamma^n)$$

$$\Leftrightarrow P_{e-d}(\gamma^n) = 0$$

$$\Leftrightarrow \gamma^n \text{ is a root of } P_{e-d}(z). \tag{5.366}$$

The number of roots in GF(q) of such a polynomial is bounded by $t - 1$, its maximum possible degree. Use of this fact in (5.365) leads to the distance bound for Reed-Solomon sequences, namely

$$H_{\min} \geq q - t. \tag{5.367}$$

Since the sequences' alphabet size is q and their period is $q - 1$, the bound on $H_{\min}$ based on Theorem 5.24 is

$$H_{\min} \leq q - 2. \tag{5.368}$$

In view of (5.367) and (5.368), the Reed-Solomon design with $t = 2$ produces a set of q sequences over GF(q) with $H_{\min} = q - 2$, indicating that in a comparison of two sequences, at any relative shift τ, the $t = 2$ design guarantees at most one frequency match.

A conceptual block diagram of a $t = 3$ Reed-Solomon sequence generator is shown in Figure 5.18. In this form, a squaring device is required to randomly initialize this generator within a particular sequence. Another configuration can be obtained by combining the two single-stage q-ary LFSRs into a single two-stage LFSR. Still another possibility is to square the contents γ^n of the first single-stage q-ary register to directly determine the input to the c_2 multiplier, thereby eliminating the second register and the need for a squarer during initialization. Notice that the generator for a $t = 2$ design is contained within that shown in Figure 5.18, and is de-

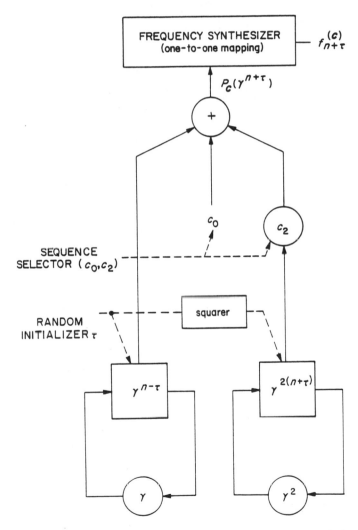

Figure 5.18. Mathematical block diagram of a Reed-Solomon FH PN generator. All arithmetic operations are in GF(q).

termined by setting c_2 to zero permanently, thereby eliminating the need for the right-hand (γ^2) shift register.

Example 5.17. A Reed-Solomon design with q being a power of 2 results in a relatively simple implementation. In this example let $q = 32$, $t = 3$, and let γ be a root of the primitive polynomial $z^5 + z^2 + 1$ over GF(2). Then γ'' can be generated by a Galois-configured, binary LFSR having characteristic polynomial $z^5 + z^2 + 1$, as noted in Section 5.5. Hence, let

$$\gamma' = \left(1, \gamma, \gamma^2, \gamma^3, \gamma^4\right), \quad \gamma'' = \gamma's_n \tag{5.369}$$

where s_n is a state vector over GF(2) representing the contents of the five stages of the LFSR at time n. The companion matrix for this register is (see (5.36))

$$A = \begin{bmatrix} 0 & 0 & 0 & 0 & 1 \\ 1 & 0 & 0 & 0 & 0 \\ 0 & 1 & 0 & 0 & 1 \\ 0 & 0 & 1 & 0 & 0 \\ 0 & 0 & 0 & 1 & 0 \end{bmatrix} \tag{5.370a}$$

and from the development of Section 5.8.2,

$$s_{n+1} = As_n. \tag{5.370b}$$

Squaring γ'' as represented in (5.369) and using the fact that $\gamma^5 = \gamma^2 + 1$ to reduce the resultant powers of γ, gives

$$s_{2n} = Ss_n \tag{5.371}$$

where

$$S = \begin{bmatrix} 1 & 0 & 0 & 0 & 1 \\ 0 & 0 & 0 & 1 & 0 \\ 0 & 1 & 0 & 1 & 0 \\ 0 & 0 & 0 & 1 & 1 \\ 0 & 0 & 1 & 0 & 0 \end{bmatrix}. \tag{5.372}$$

Let y_n and k_i be coefficient vectors for the representations

$$P_c(\gamma'') = \gamma'y_n, \quad c_i = \gamma'k_i. \tag{5.373}$$

Then it is easily verified that

$$y_n = k_0 + \left[I + \left(k_{20}I + k_{21}A + k_{22}A^2 + k_{23}A^3 + k_{24}A^4 \right)S \right]s_n.$$

A block diagram of the generator in matrix/vector terms is shown in Figure 5.19. The complexity of this mechanization comes from the requirement that the value of k_2 be changeable, as part of the sequence selection process. Were it not for this specification, the quantity in brackets in (5.374) could be reduced to a single binary matrix, making the generator structure quite simple.

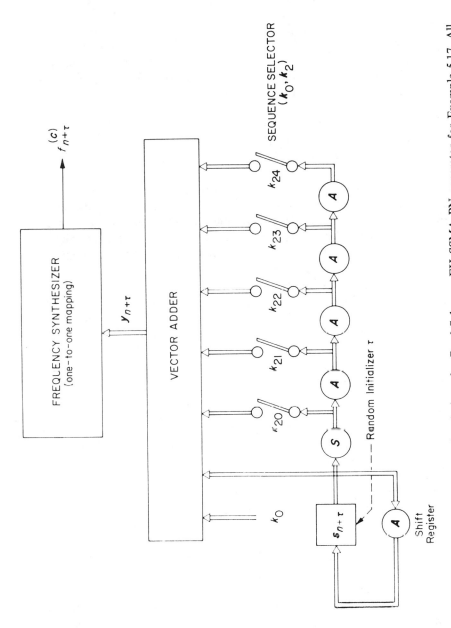

Figure 5.19. Matrix/vector description of a Reed-Solomon FH SSMA PN generator for Example 5.17. All arithmetic operations are in GF(2).

There is a simple tradeoff in the design shown in Figure 5.19, involving generator complexity and the size of the sequence set. If the switches $k_{20}, \ldots, k_{24}$ are left permanently open, thereby eliminating the need for five matrix multipliers, then the input to the vector adder is $k_0 + s_{n+\tau}$, the thirty-two possible choices of k_0 corresponding to the thirty-two period-31 sequences of a $t = 2$ Reed-Solomon design with $H_{\min} = 30$. If instead k_0 and only the k_{21} switch are enabled, then the matrix multiply S is required, and there are sixty-four possible sequences that can be generated, a subset of the $H_{\min} = 29$, $t = 3$ design. Each additional multiplication by A and k_{2i} switch enablement doubles the number of possible FH sequences, and maintains $H_{\min}$ at 29. When the full generator in Figure 5.19 is employed, any of a set of 1024 period-31 sequences can be generated.

The FHMA design of Lempel and Greenberger, based on an m-sequence generator, is in fact a $t = 2$ Reed-Solomon sequence generator with several elements of the vector adder output disabled. Or equivalently, the Lempel-Greenberger design with $J = L$ is a $t = 2$ Reed-Solomon sequence generator. In making this comparison, the reader should note that the definitions of q differ in the discussions of these two designs.

Both of the sequence sets discussed in this section are generated linearly; hence, the sequences driving the frequency synthesizers have short linear span. Therefore, these designs, while optimal for FHMA applications, are potentially weak when the threat of intelligent jamming exists.

5.9 A LOOK AT THE LITERATURE

While we have covered considerable ground in this chapter and presented the fundamentals of PN generator design, much more is available in the open literature. As starting points for literature searches, we refer the reader to Golomb's comprehensive bibliography (406 references) in the revised edition of [12], to the "Cumulative Index 1953–1981," of the *IEEE Transactions on Information Theory* published with the September 1982 issue (especially papers listed under the headings: sequences, shift register sequences, shift registers, and pseudonoise sequences), and to Sarwate and Pursley's survey article [31] (136 references) on pseudonoise sequences. Also of interest may be Levine's book [51] on cryptographic patents, many being SS related.

Additional references on bent sequences [52], spectral analysis of sequences [53], [54], [55], correlation bounds [56], complex sequences meeting the Welch bound [65], [66], frequency-hopping designs [57], [58], [59], [67], and linear span bounds [60], [61] are given below. A special DSMA PN sequence design criterion, which takes into account the effects of data modulation on sequence cross-correlations, has been studied [31], [62], [63], [64], and is useful in short period designs.

The core material of this chapter has been based solely on the LFSR which provides a long state sequence from which the output sequence is derived. The design of DeBruijn sequences, i.e., sequences of period 2^m generated by m-stage binary registers with nonlinear feedback, are surveyed in [69].

5.10 REFERENCES

[1] W. V. Quine, "The problem of simplifying truth functions," *Am. Math. Monthly*, vol. 59, no. 8, pp. 521–531, October 1952.

[2] E. J. McCluskey, "Minimization of Boolean functions," *Bell Syst. Tech. J.*, vol. 35, no. 5, pp. 1417–1444, November 1956.

[3] F. J. Hill and G. R. Peterson, *Introduction to Switching Theory and Logical Design*, New York: John Wiley, 1968.

[4] I. S. Reed, "A class of multiple error-correcting codes and the decoding scheme," *IRE Trans. Inform. Theory*, vol. 4, pp. 38–49, 1954.

[5] D. E. Muller, "Application of Boolean algebra to switching circuit design and to error correction," *IRE Trans. Electron. Comput.*, vol. 3, pp. 6–12, 1954.

[6] W. W. Peterson and E. J. Weldon, Jr., *Error-Correcting Codes*, 2nd ed., Cambridge, MA: M.I.T. Press, 1972.

[7] E. R. Berlekamp, *Algebraic Coding Theory*, New York: McGraw-Hill, 1968.

[8] J. L. Massey, "Shift register synthesis and BCH decoding," *IEEE Trans. Inform. Theory*, IT-15, pp. 122–127, January 1969.

[9] W. H. Mills, "Continued fractions and linear recurrences," *Math. Comput.*, vol. 29, pp. 173–180, January 1975.

[10] R. A. Scholtz and L. R. Welch, "Continued fractions and Berlekamp's Algorithm," *IEEE Trans. Inform. Theory*, IT-25, no. 1, pp. 19–27, January 1979.

[11] A. Lempel and H. Greenberger, "Families of sequences with optimal Hamming correlation properties," *IEEE Trans. Inform. Theory*, IT-15, pp. 90–94, January 1974.

[12] S. W. Golomb, *Shift Register Sequences*, San Francisco, CA: Holden-Day, 1967; revised edition, Aegean Park Press, Laguna Hills, CA, 1982.

[13] T. C. Bartee and P. W. Wood, "Coding for Tracking Radar Ranging," M.I.T. Lincoln Lab., Lexington, MA, Tech. Rept. 318, 1963.

[14] J. H. Lindholm, "An analysis of the pseudo-randomness properties of subsequences of long m-sequences," *IEEE Trans. Inform. Theory*, IT-14, pp. 569–576, July 1968.

[15] A. B. Cooper and P. H. Lord, "Subsequence Correlation Analysis," Research and Technology Div., Rome Air Development Center, USAF Systems Command, Griffiss AFB, Rome, NY, Tech. Rept. RADC-TR-67-591, Dec. 1967.

[16] H. F. Mattson, Jr., and R. J. Turyn, "On correlation by subsequences," Sylvania Applied Research Lab., Sylvania Electronic Systems, Waltham, MA, Res. Note 692, Feb. 1967.

[17] S. Wainberg and J. K. Wolf, "Subsequences of pseudo-random sequences," *IEEE Trans. Commun. Tech.*, COM-18, pp. 606–612, October 1970.

[18] S. A. Fredricsson, "Pseudo-randomness properties of binary shift register sequences," *IEEE Trans. Inform. Theory*, pp. 115–120, January 1975.

[19] V. Pless, "Power moment identities on weight distributions in error correcting codes," *Inform. Contr.*, vol. 6, pp. 147–152, 1963.

[20] N. E. Bekir, "Bounds on the Distribution of Partial Correlation for PN and Gold Sequences," Ph.D. Dissertation, Electrical Engineering Dept., Univ. of Southern California, Los Angeles, CA, Jan. 1978.

[21] N. I. Akhiezer, *The Classical Moment Problem*, New York: Hafner, 1965.

[22] J. A. Shohat and J. D. Tamarkin, *The Problems of Moments*, New York: American Math. Society, 1943.

[23] E. L. Key, "An analysis of the structure and complexity of nonlinear binary sequence generators," *IEEE Trans. Inform. Theory*, IT-22, pp. 732–736, November 1976.

[24] W. Diffie and M. E. Hellman, "Privacy and authentication: an introduction to cryptography," *Proc. IEEE*, vol. 67, pp. 397–427, March 1979.

[25] E. J. Groth, "Generation of binary sequences with controllable complexity," *IEEE Trans. Inform. Theory*, IT-17, pp. 288–296, May 1971.

[26] L. D. Baumert, *Cyclic Difference Sets*, Lecture Notes in Math. No. 182, New York: Springer Verlag, 1971.

[27] J. Singer, "A theorem in finite projective geometry and some applications to number theory," *Trans. Amer. Math. Soc.*, vol. 43, pp. 377–385, 1938.

[28] B. Gordon, W. H. Mills, and L. R. Welch, "Some new difference sets," *Canad. J. Math.*, vol. 14, pp. 614–625, 1962.

[29] L. R. Welch, "Lower bounds on the maximum cross correlation of signals," *IEEE Trans. Inform. Theory*, IT-20, pp. 397–399, May 1974.

[30] T. Kasami, "Weight distribution formula for some class of cyclic codes," Coordinated Science Lab., Univ. Illinois, Urbana. Tech. Rep. R-285, April 1966 (AD 632574).

[31] D. V. Sarwate and M. B. Pursley, "Crosscorrelation properties of pseudorandom and related sequences," *Proc. IEEE*, vol. 68, pp. 593–619, May 1980.

[32] J. D. Olsen, R. A. Scholtz, and L. R. Welch, "Bent-function sequences," *IEEE Trans. Inform. Theory*, IT-28, pp. 858–864, November, 1982.

[33] R. A. Scholtz and L. R. Welch, "Group characters: sequences with good correlation properties," *IEEE Trans. Inform. Theory*, IT-24, pp. 537–545, September 1978.

[34] R. A. Scholtz, "Optimal CDMA codes," *IEEE National Telecommunications Conference Record*, November 1979, pp. 54.2.1–54.2.4.

[35] R. Gold, "Optimum binary sequences for spread spectrum multiplexing," *IEEE Trans. Inform. Theory*, IT-13, pp. 619–621, October 1967.

[36] R. Gold, "Maximal recursive sequences with 3-valued recursive cross-correlation functions," *IEEE Trans. Inform. Theory*, IT-14, pp. 154–156, January 1968.

[37] Y. Niho, "Multi-valued Cross-correlation Functions between Two Maximal Linear Recursive Sequences," Ph.D. dissertation, Electrical Engineering Department, Univ. Southern California, 1972.

[38] T. A. Dowling and R. J. McEliece, "Cross-correlation of reverse maximal-length shift-register sequences," JPL Space Programs Summary 37-53, vol. 3, pp. 192–193, 1968.

[39] V. M. Sidelnikov, "On the mutual correlation of sequences," *Soviet Math. Dokl.*, vol. 12, pp. 197–201, 1971.

[40] F. J. MacWilliams and N. J. A. Sloane, *The Theory of Error-Correcting Codes*, New York: North Holland Publishing Co., 1977.

[41] S. Lin and D. J. Costello, Jr., *Error Control Coding: Fundamentals and Applications*, Englewood Cliffs, NJ: Prentice-Hall, 1983.

[42] J. D. Olsen, "Nonlinear Binary Sequences with Asymptotically Optimum Periodic Cross-Correlation," Ph.D. dissertation, Electrical Engineering Department, Univ. Southern California, December 1977.

[43] O. S. Rothaus, "On 'bent' functions," *J. Comb. Theory*, Series A20, pp. 300–305, 1976.

[44] J. F. Dillon, "Elementary Hadamard Difference Sets," Ph.D. dissertation, University of Maryland, 1974.

[45] J. F. Dillon, "Elementary Hadamard difference sets," *Proc. 6th S.E. Conf. Combinatorics, Graph Theory, and Computing* (Utilitas Math., Winnepeg, 1975), pp. 237–249.

[46] P. V. Kumar and R. A. Scholtz, "Bounds on the linear span of bent sequences," *IEEE Trans. Inform. Theory*, IT-29, pp. 854–862, November 1983.

[47] G. Solomon, "Optimal frequency hopping for multiple access," *Proc. of the 1977 Symposium on Spread Spectrum Communications*, Naval Electronics Laboratory Center, San Diego, CA, 13–16 March 1973, pp. 33–35.

[48] R. E. Blahut, *Theory and Practice of Error Control Codes*, Reading, MA: Addison-Wesley, 1983.

[49] R. A. Scholtz and L. R. Welch, "GMW sequences," *IEEE Trans. Inform. Theory*, IT-30, pp. 548–553, May 1984.

[50] P. H. R. Scholefield, "Shift registers generating maximum-length sequences," *Electronic Technology*, vol. 37, pp. 389–394, October 1960.

[51] J. Levine, *United States Cryptographic Patents 1861–1981*. Terre Haute, IN: Cryptologia, Inc., February 1983.

[52] A. Lempel and M. Cohn, "Maximal families of bent sequences," *IEEE Trans. Inform. Theory*, IT-28, pp. 865–868, November 1982.

[53] R. A. Scholtz, "The spread spectrum concept," *IEEE Trans. Commun.*, COM-25, pp. 748–755, August 1977.

[54] S. G. Glisic, "Power density spectrum of the product of two time displaced versions of a maximum length binary pseudonoise signal," *IEEE Trans. Commun.*, COM-31, pp. 281–286, February 1983.

[55] R. C. Titsworth and L. R. Welch, "Modulation by Random and Pseudorandom Sequences," JPL Laboratory Report No. 20-387, 1959.

[56] J. E. Mazo, "Some theoretical observations on spread-spectrum communications," *Bell Syst. Tech. J.*, vol. 58, pp. 2013–2023, 1979.

[57] R. M. Marsareau and T. S. Seay, "Multiple access frequency hopping patterns with low ambiguity," *IEEE Trans. Aerosp. Electron. Syst.*, AES-17, pp. 571–578, July 1981.

[58] G. Einarsson, "Address assignment for a time-frequency-coded spread spectrum system," *Bell Syst. Tech. J.*, vol. 59, pp. 1241–1255, September 1980.

[59] R. C. Singleton, "Maximum distance q-nary codes," *IEEE Trans. Inform. Theory*, IT-10, pp. 116–118, April 1964.

[60] M. P. Ristenbatt and J. L. Daws, Jr., "Performance criteria for spread spectrum communications," *IEEE Trans. Commun.*, COM-25, pp. 756–763, August 1977.

[61] R. A. Games and A. H. Chan, "A fast algorithm for determining the complexity of a binary sequence with period 2^n," *IEEE Trans. Inform. Theory*, IT-29, pp. 144–146, January 1983.

[62] J. L. Massey and J. J. Urhan, Jr., "Sub-baud decoding," *Proc. 13th Annu.*

Allerton Conf. Circuit and System Theory, pp. 539–547, 1975.

[63] M. B. Pursley and D. V. Sarwate, "Evaluation of correlation parameters for periodic sequences," *IEEE Trans. Inform. Theory*, IT-23, pp. 508–513, July 1977.

[64] D. V. Sarwate, "Bounds on crosscorrelation and autocorrelation of sequences," *IEEE Trans. Inform. Theory*, IT-25, pp. 720–724, November 1979.

[65] W. O. Alltop, "Complex sequences with low periodic correlations," *IEEE Trans. Inform. Theory*, IT-26, pp. 350–354, May 1980.

[66] R. A. Scholtz and L. R. Welch, "Group characters: signals with good correlation properties," *IEEE Trans. Inform. Theory*, IT-24, pp. 537–545, September 1978.

[67] D. V. Sarwate and M. B. Pursley, "Hopping patterns for frequency-hopped multiple-access communication," *IEEE Int. Conf. Commun.*, June 1978, pp. 7.4.1–7.4.3.

[68] R. Turyn, "Sequences with small correlation," *Error Correcting Codes*, H. B. Mann, ed., New York: John Wiley, 1968.

[69] H. Fredricksen, "A survey of full length nonlinear shift register cycle algorithms," *SIAM Review*, vol. 24, pp. 195–221, April 1982.

APPENDIX 5A. FINITE FIELD ARITHMETIC

5A.1. COMMUTATIVE GROUPS

DEFINITION 5A.1. *A* commutative (Abelian) group *is a collection $\mathcal{G}$ of elements and a rule of combination, &, satisfying the following axioms*:

1. *Closure: If $g_1 \in \mathcal{G}$ and $g_2 \in \mathcal{G}$, then $g_1 \& g_2 \in \mathcal{G}$.*
2. *Associative law: For any g_1, g_2, and g_3 in $\mathcal{G}$, $(g_1 \& g_2) \& g_3 = g_1 \& (g_2 \& g_3)$.*
3. *Identity: There exists an element e in $\mathcal{G}$ such that for all $g \in \mathcal{G}$, $g \& e = e \& g = g$.*
4. *Inverse: For each element $g \in \mathcal{G}$ there exists an element $\bar{g}$ such that $g \& \bar{g} = e = \bar{g} \& g$.*
5. *Commutative law: If $g_1 \in \mathcal{G}$ and $g_2 \in \mathcal{G}$, then $g_1 \& g_2 = g_2 \& g_1$.*

It can be verified that the identity element e is unique as is each element's inverse. The identity element is its own inverse; if $\bar{h}$ is the inverse of g, then $\bar{g}$ is the inverse of h.

Obviously the integers under ordinary addition are a simple example of a commutative group. All N-dimensional vectors of real numbers form a commutative group under addition. All complex numbers of unit magnitude form a commutative group under multiplication.

DEFINITION 5A.2. *A subgroup $\mathcal{S}$ of the group $\mathcal{G}$ is a collection of elements from $\mathcal{G}$ which form a group. A subgroup $\mathcal{S}$ of $\mathcal{G}$, which is not equal to $\mathcal{G}$, is called a* proper *subgroup.*

The identification of a subgroup $\mathcal{S}$ within $\mathcal{G}$ permits the following procedure to be performed.

Coset Construction Algorithm:

(a) Let $\mathcal{C}_0 = \mathcal{S}$. $n = 0$.

(b) Construct $\mathcal{A}_{n+1} = \mathcal{G} - \bigcup\limits_{i=0}^{n} \mathcal{C}_i$

(c) If $\mathcal{A}_{n+1}$ is empty, then STOP.

(d) Increase n by 1.

(e) Select an element g_n from $\mathcal{A}_n$.

(f) Construct $\mathcal{C}_n = \{ g : g = g_n \& s, \ s \in \mathcal{S} \}$

(g) Go to (b).

The sets $\mathcal{C}_i$, $i > 0$, are called *cosets* of the subgroup $\mathcal{S}$. It can be shown that

(a) Two elements g and g' are in the same coset if and only if $g \& \bar{g}' \in \mathcal{S}$.

(b) There is a one-to-one correspondence between the elements of $\mathcal{S}$ and the elements of each coset.

(c) $\mathcal{C}_i$ and $\mathcal{C}_j$ are disjoint for $i \neq j$, and unique, and hence, when $\mathcal{G}$ is finite,

$$|\mathcal{G}| = N|\mathcal{S}|, \qquad (5A.1)$$

where $|\mathcal{A}|$ denotes the number of elements in $\mathcal{A}$, and $N - 1$ is the number of cosets generated in the above construction algorithm. The number $|\mathcal{G}|$ of elements in a group $\mathcal{G}$ is called the *order* of $\mathcal{G}$. If the order of $\mathcal{G}$ is prime, then by (5A.1), $\mathcal{G}$ contains no proper subgroups.

Example 5A.1. The integers $0, 1, \ldots, 11$ under addition modulo 12 form a group $\mathcal{G}$ (The term "modulo" is described in Section 5A.2 for those not familiar with it.) One proper subgroup $\mathcal{S}$ of $\mathcal{G}$ is

$$\mathcal{S} = \{0, 3, 6, 9\},$$

and the cosets of this subgroup are

$$\{1, 4, 7, 10\} \text{ and } \{2, 5, 8, 11\}.$$

Suppose that g is an element of a finite-order commutative group. An element g^m can be constructed by combining m of the elements g according to the rule of combination used in $\mathcal{G}$. Thus,

$$g^1 = g$$
$$g^2 = g \& g$$
$$g^3 = g \& g \& g, \text{ etc.}$$

DEFINITION 5A.3. *Let e be the identity element of $\mathscr{G}$ and let $g \in \mathscr{G}$. Then the smallest positive integer k such that*

$$g^k = e \tag{5A.2}$$

is called the exponent *of g.*

It can be shown that:

(a) The collection of elements defined by

$$\mathscr{S}_g = \{ g^j \colon 1 \le j \le k \} \tag{5A.3}$$

is a subgroup of $\mathscr{G}$. $\mathscr{S}_g$ is called the *cyclic group generated by g*. If there exists a $g \in \mathscr{G}$ such that $\mathscr{S}_g = \mathscr{G}$, then $\mathscr{G}$ is cyclic and g is called a *primitive element* of $\mathscr{G}$.

(b) The exponent of g is the order of $\mathscr{S}_g$. (This has led to the interchangeable use of exponent and order in describing an element g.)

(c) The exponent of g divides the order of $\mathscr{G}$, i.e.,

$$|\mathscr{S}_g| \| |\mathscr{G}|. \tag{5A.4}$$

(The vertical bar between two quantities should be read "divides.")

Example 5A.2. Consider the integers relatively prime to 9 under multiplication mod 9. (The term "relatively prime" is defined in Section 2 for those not familiar with it.) The elements of this collection are 1, 2, 4, 5, 7, and 8. This group has the operation table:

×	1	2	4	5	7	8
1	1	2	4	5	7	8
2	2	4	8	1	5	7
4	4	8	7	2	1	5
5	5	1	2	7	8	4
7	7	5	1	8	4	2
8	8	7	5	4	2	1

A tabulation of the cyclic subgroup generated by each element gives

$$\mathscr{S}_1 = \{1\}$$
$$\mathscr{S}_2 = \{1, 2, 4, 8, 7, 5\}$$
$$\mathscr{S}_4 = \{1, 4, 7\}$$
$$\mathscr{S}_5 = \{1, 5, 7, 8, 4, 2\}$$
$$\mathscr{S}_7 = \{1, 7, 4\}$$
$$\mathscr{S}_8 = \{1, 8\}.$$

Clearly this six element group is cyclic, the elements 2 and 5 are primitive, 7 and 4 have exponent 3, and 8 has exponent 2.

A cyclic group $\mathcal{G}$ of order k is *isomorphic* to the integers $0, 1, \ldots, k - 1$ under addition modulo k. Specifically, using the convention $g^0 = e$ and assuming g generates $\mathcal{G}$, the correspondence

$$g^m \leftrightarrow m, \qquad m = 0, 1, \ldots, k - 1, \tag{5A.5}$$

implies that the group operation tables for $\mathcal{G}$ and for modulo k addition are identical. That is,

$$g^m g^n = g^h \text{ if and only if } h = (m + n) \bmod k,$$

for all m and n. This isomorphism can be used to find the exponent of each element of $\mathcal{G}$. Let k_m denote the order of g^m. Then k_m is by definition the smallest integer such that

$$g^{mk_m} = e, \tag{5A.6}$$

or, under the isomorphism, the smallest integer k_m such that mk_m is a multiple of k. Thus, k_m can be determined directly from k and m as

$$mk_m = \mathrm{lcm}(k, m), \tag{5A.7}$$

where lcm( ) denotes the *least common (integer) multiple* of the integers in parentheses.

Example 5A.3. The cyclic group of Example 5A.2 has six elements and has primitive element 2. Solving (5A.7) leads to the following table, and verifies the calculations of Example 5A.2.

m	element 2^m	exponent k_m
1	2	6
2	4	3
3	8	2
4	7	3
5	5	6
6	1	1

5A.2. RINGS AND FIELDS

In the real number system we have learned to think in terms of two fundamental operations: addition and multiplication. The abstract description of this is a field.

DEFINITION 5A.4. *A field is a non-empty collection $\mathcal{F}$ of elements which has the following properties*:

1. *$\mathcal{F}$ is a commutative group under an operation, $+$ (addition), with the identity element under $+$ denoted by 0.*

2. *Excluding* 0, *the remaining elements of* $\mathscr{F}$ *are a commutative group under a second operation,* $\cdot$ *(multiplication), with the identity element under* $\cdot$ *denoted by* 1.
3. *Distributive law:* *If* f_1, f_2 *and* f_3 *are in* $\mathscr{F}$, *then*

$$f_1 \cdot (f_2 + f_3) = (f_1 \cdot f_2) + (f_1 \cdot f_3)$$
$$(f_2 + f_3) \cdot f_1 = (f_2 \cdot f_1) + (f_3 \cdot f_1).$$

Additive and multiplicative inverses of a field element will be denoted by $-f$ and f^{-1}, respectively.

Many algebraic structures have all but one or two of the properties of a field. Two structures, the integers under addition and multiplication, and the polynomials under addition and multiplication, have all but one of these properties; neither has multiplicative inverses for its elements. These are examples of structures, with addition and multiplication operations, which are called *commutative rings*. A commutative ring is an Abelian group under addition, is closed, associative, and commutative under multiplication, and satisfies the distributive law.

Two properties of rings and fields are the following:

Property R1. Let r be an element of a ring $\mathscr{R}$ with additive identity 0. Then

$$r \cdot 0 = 0 \cdot r = 0 \quad \text{for all } r \in \mathscr{R}.$$

Property R2. Let r_1 and r_2 be elements of a ring $\mathscr{R}$. Then

$$(-r_1) \cdot r_2 = r_1 \cdot (-r_2) = -(r_1 \cdot r_2) = (-1) \cdot (r_1 \cdot r_2) \quad \text{for all } r_1, r_2 \in \mathscr{R}.$$

Henceforth, the following notational simplifications will be observed:

$$r_1 \cdot r_2 = r_1 r_2, \tag{5A.8}$$

$$r_1 + (-r_2) = r_1 - r_2 \tag{5A.9}$$

Lack of multiplicative inverses precludes use of the cancellation law. That is, if r_1 is a ring element which does not have a multiplicative inverse, then $r_1 r_2 = r_1 r_3$ does *not* imply $r_2 = r_3$.

Example 5A.4. The ring of integers $0, 1, \ldots, 7$ under addition and multiplication mod 8 exhibits the following relation:

$$4 \cdot 1 = 4 \cdot 3 = 4 \cdot 5 = 4 \cdot 7.$$

Yet 1, 3, 5, and 7 are distinct elements of the ring.

In the ring of integers it is possible to define a process called *division*, which is akin to multiplication by a multiplicative inverse. The result of the *Euclidean division algorithm* is: If n_1 and n_2 are any two real integers, $n_2 \neq 0$, then n_1 can be represented by

$$n_1 = qn_2 + r, \tag{5A.10}$$

where q and r are integers and $0 \leq r < |n_2|$. If $r = 0$, we say that n_2 divides n_1, and denote this by $n_2|n_1$. The *remainder* r is often called (the value of) n_1 *modulo* n_2, or in abbreviated fashion $n_1 \bmod n_2$.

The *greatest common divisor* $\gcd(m, n)$ of two integers m and n, defined as the largest integer dividing both m and n, can be determined by repeated application of the Euclidean division algorithm. Furthermore, it can be shown that

$$\gcd(m, n) = am + bn \qquad (5A.11)$$

where a and b are also elements of the ring of integers. If $\gcd(m, n) = 1$, then m and n are said to be *relatively prime*.

The relation (5A.11) is extremely important in the sections to follow so we will review it here and indicate an algorithm for determining it. Reviewing briefly, Euclid's division algorithm states that

$$m_i = a_i n_i + r_i, \quad 0 \leq r_i < |n_i|. \qquad (5A.12)$$

From this relation, it follows that if g divides any two of the three quantities, $m_i, a_i n_i,$ and r_i, then it divides the third also. Hence,

$$\gcd(m_i, n_i) = \gcd(n_i, r_i). \qquad (5A.13)$$

Let's suppose we wish to compute $\gcd(m, n)$, $m > n$. Then the algorithm's structure is as follows.

Euclid's Greatest Common Divisor Algorithm

(a) Let $m_1 = m$, $n_1 = n$, $i = 1$.
(b) Compute r_i.
(c) If $r_i = 0$ then [$\gcd(m, n) = n_i$, STOP].
(d) If $r_i \neq 0$, set $m_{i+1} = n_i$, $n_{i+1} = r_i$.
(e) Increase i by one and go to (b).

Suppose that we also wish to determine the integer coefficients a and b which are used to combine m and n to give the $\gcd(m, n)$. This can be done by keeping track of how m_i and n_i can be written in terms of m and n as the algorithm progresses. Let's define

$$A^{(i)} = \begin{bmatrix} A_1^{(i)} \\ A_2^{(i)} \end{bmatrix} \quad \text{where } m_i = A_1^{(i)}m + A_2^{(i)}n. \qquad (5A.14)$$

$$B^{(i)} = \begin{bmatrix} B_1^{(i)} \\ B_2^{(i)} \end{bmatrix} \quad \text{where } n_i = B_1^{(i)}m + B_2^{(i)}n \qquad (5A.15)$$

As the algorithm is set up,

$$A^{(1)} = \begin{bmatrix} 1 \\ 0 \end{bmatrix}, \quad B^{(1)} = \begin{bmatrix} 0 \\ 1 \end{bmatrix}, \qquad (5A.16)$$

and since $m_{i+1} = n_i$, it follows that

$$A^{(i+1)} = B^{(i)} \qquad (5A.17)$$

A recursion for $B^{(i+1)}$ can also be determined, using the fact that

$$n_{i+1} = r_i = m_i - a_i n_i. \qquad (5A.18)$$

Hence,

$$B^{(i+1)} = A^{(i)} - a_i B^{(i)}. \qquad (5A.19)$$

Hence, the modified algorithm for determining $gcd(m, n)$, $m > n$, is:

Revised GCD Algorithm:

(a) Let $m_1 = m$, $n_1 = n$, $i = 1$, $A^{(i)} = \begin{bmatrix} 1 \\ 0 \end{bmatrix}$, $B^{(1)} = \begin{bmatrix} 0 \\ 1 \end{bmatrix}$.

(b) Compute a_i and r_i by Euclid's Division Algorithm.

(c) If $r_i = 0$, the $[gcd(m, n) = n_i = B_1^{(i)} m + B_2^{(i)} n$. STOP].

(d) If $r_i \neq 0$, then compute m_{i+1}, n_{i+1}, $A^{(i+1)}$ and $B^{(i+1)}$, using the proper recursions.

(e) Increase i by 1 and go to (b).

Further insight into this algorithm can be seen by comparing it to continued fraction computations.

Example 5A.5. Find the $gcd(46410, 36310300)$.

i	m_i	n_i	a_i	r_i	$(A^{(i)})^t$	$(B^{(i)})^t$
1	36310300	46410	782	17680	$(1, 0)$	$(0, 1)$
2	46410	17680	2	11050	$(0, 1)$	$(1, -782)$
3	17680	11050	1	6630	$(1, -782)$	$(-2, 1565)$
4	11050	6630	1	4420	$(-2, 1565)$	$(3, -2347)$
5	6630	4420	1	2210	$(3, -2347)$	$(-5, 3912)$
6	4420	2210	2	0	$(-5, 3912)$	$(8, -6259)$

Therefore $gcd(46410, 36310300) = 2210$.

Check: $46410 = 21 \cdot 2210$

$36310300 = 6430 \cdot 2210$

$2210 = 8 \cdot 36310300 - 6259 \cdot 46410$

5A.3. POLYNOMIALS

A *polynomial* $P(z)$ of *degree* n over a commutative ring $\mathscr{R}$ (i.e., with coefficients in $\mathscr{R}$) can be written as

$$P(z) = \sum_{i=0}^{n} r_i z^i, \quad r_n \neq 0. \tag{5A.20}$$

The coefficients, r_i, $i = 0, \ldots, n$, of the polynomial are elements of $\mathscr{R}$. The *indeterminate* z obeys all the properties of a commutative ring element with respect to the coefficients; hence z can be thought of as an element in $\mathscr{R}$ or a larger ring containing $\mathscr{R}$. The collection of all polynomials over a ring is itself a ring $\mathscr{R}_P$. Our main interest in polynomial rings is restricted to the case where the coefficients come from a field.

A Euclidean division algorithm also exists for polynomials: If $P_1(z)$ and $P_2(z)$ are polynomials over a field $\mathscr{F}$ (or over a ring $\mathscr{R}$ with the highest degree coefficient r_n of $P_2(z)$ being the multiplicative identity of $\mathscr{R}$), then there exists polynomials $Q(z)$ and $R(z)$ over $\mathscr{F}$ such that

$$P_1(z) = Q(z)P_2(z) + R(z), \tag{5A.21}$$

where $0 \leq \deg R(z) < \deg P_2(z)$. The polynomial $R(z)$ is often referred to as the *value* of $P_1(z)$ *modulo* $P_2(z)$ or simply $P_1(z)$ mod $P_2(z)$. Again, if $R(z) = 0$, we say that $P_2(z)$ divides $P_1(z)$. We can also define the *greatest common divisor* of two polynomials $P_1(z)$ and $P_2(z)$ over a field $\mathscr{F}$ as the highest degree polynomial dividing both $P_1(z)$ and $P_2(z)$. By an algorithm identical to that used in the ring of integers (see Section 5A.2), the $\gcd(P_1(z), P_2(z))$ can be found and shown to be of the form

$$\gcd(P_1(z), P_2(z)) = C_1(z)P_1(z) + C_2(z)P_2(z) \tag{5A.22}$$

where $C_1(z)$ and $C_2(z)$ are polynomials over the same field as $P_1(z)$ and $P_2(z)$.

Example 5A.6. Find $\gcd(z^{16} + z^{15} + z^{14} + z^{13} + z^{12} + z^{11} + z^{10} + z^9 + z^7 + z^4 + z + 1, z^{13} + z^{12} + z^{11} + z^9 + z^7 + z^3 + z)$. These polynomials are over GF(2) and only polynomial coefficients are listed in the following table. The notation is identical to that used for integers earlier.

	m_i	n_i	a_i	r_i	$(B^{(i)})^t$
1	(1111111010010011)	(11101010001010)	(1001)	(1000001001001)	(0, 1)
2	(11101010001010)	(1000001001001)	(11)	(101001010001)	(1, 1001)
3	(1000001001001)	(101001010001)	(10)	(10011101011)	(11, 11010)
4	(101001010001)	(10011101011)	(10)	(1110000111)	(111, 11101)
5	(10011101011)	(1110000111)	(11)	(1100010)	(1101, 1100000)
6	(1110000111)	(1100010)	(1011)	(110001)	(100000, 10011101)
7	(1100010)	(110001)	(10)	---	(10111101, 10100101111)

$\gcd = z^5 + z^4 + 1$

Check: $z^{16} + z^{15} + z^{14} + z^{13} + z^{12} + z^{11} + z^{10} + z^9 + z^7 + z^4 + z + 1 = (z^5 + z^4 + 1)$
$\times (z^{11} + z^9 + z^7 + z^6 + z + 1)$

$z^{13} + z^{12} + z^{11} + z^9 + z^7 + z^3 + z = (z^5 + z^4 + 1)(z^8 + z^6 + z^5 + z^3 + z)$

$(z^5 + z^4 + 1) = (z^{16} + z^{15} + z^{14} + z^{13} + z^{12} + z^{11} + z^{10} + z^9 + z^7 + z^4 + z + 1)$
$\times (z^7 + z^5 + z^4 + z^3 + z^2 + 1) + (z^{13} + z^{12} + z^{11} + z^9 + z^7 + z^3 + z)$
$\times (z^{10} + z^8 + z^5 + z^3 + z^2 + z + 1)$

Suppose α is an element of a ring which contains the ring $\mathscr{R}$ of polynomial coefficients. We say that α is *root of multiplicity K*, of the polynomial $P(z)$ if $(z - \alpha)^K$ divides $P(z)$. To determine if α is a root of $P(z)$, long division can be used, generally resulting in

$$P(z) = (z - \alpha)Q(z) + P(\alpha), \qquad (5A.23)$$

where the polynomial $Q(z)$ has degree one less than $P(z)$. Hence, we have the following theorem:

THEOREM 5A.1 (Factorization Theorem). *The element α is a root of $P(z)$ if and only if $P(\alpha) = 0$. If $P(\alpha) = 0$, we can determine if $(z - \alpha)^2$ divides $P(z)$ by determining if $(z - \alpha)$ divides $Q(z)$. This operation can be iterated until the number of $(z - \alpha)$ factors of $P(z)$ is determined.*

The student should observe that "strange" things can happen in the factorization of polynomials *over rings*: (a) The number of roots of a polynomial over a ring, which are in the ring, may possibly be greater than or less than the degree of the polynomial. (b) Factorizations of polynomials over rings are not unique.

Example 5A.7. Consider the following polynomials over the ring of integers $0, 1, \ldots, 7$ under addition and multiplication modulo 8.

(a) $z^2 + 7$ has four roots in the ring, namely $1, 3, 5, 7$.
(b) $z^2 + z + 1$ has no roots in the ring.
(c) z^2 has two roots in the ring, namely 0 and 4, each of multiplicity 2.
(d) Furthermore with an excess of roots available, factorization is certainly not unique.

$$z^2 + 7 = (z + 1)(z + 7) = (z + 5)(z + 3)$$

$$z^2 = (z)(z) = (z + 4)(z + 4)$$

(e) There can be more factors on one side of an equality than the degree of the other side:

$$(4z + 2)(2z + 4) = 4z.$$

These unusual properties disappear when we require that the polynomial be over a field.

Polynomials over any field satisfy certain basic factorization properties. These properties are most easily described in terms of a special kind of polynomial.

DEFINITION 5A.5. *A polynomial of degree n in an indeterminate z is said to be* monic *if the (non-zero) coefficient of z^n is 1.*

The reader can easily verify that if a polynomial has coefficients in a *field*, then the following are true:

(a) Every polynomial over a finite field can be uniquely written as a monic polynomial times an element of the coefficient field.

(b) The product of two monic polynomials of degrees m and n is a monic polynomial of degree $m + n$.

(c) If a monic polynomial is the product of two polynomials, then it is the product of two monic polynomials.

(d) If $F(z)$ and $H(z)$ are two polynomials over a field, which have greatest common divisor 1, then $F(z)|H(z)J(z)$ implies $F(z)|J(z)$.

(e) A monic polynomial over a field $\mathscr{F}$ can always be uniquely factored into a product of irreducible monic polynomials over $\mathscr{F}$.

(f) Let $P(z)$ be a polynomial of degree d over a field $\mathscr{F}$, and let $a_1, \ldots, a_n$ be the distinct roots of $P(z)$ in $\mathscr{F}$ with multiplicities $m_1, \ldots, m_n$, respectively. Then

$$\sum_{i=1}^{n} m_i \le d. \tag{5A.24}$$

If a polynomial $P(z)$ over a field $\mathscr{F}$ can be written as a product of two or more polynomials (each with degree greater than 0) over the same field $\mathscr{F}$, we say that $P(z)$ is *factorable over $\mathscr{F}$*. On the other hand, if $P(z)$ cannot be factored in this manner, $P(z)$ is said to be *irreducible over $\mathscr{F}$*. In specifying whether a polynomial is irreducible or factorable, the coefficient field be specified. Notice that the greatest common divisor of an irreducible polynomial $F(z)$ and any other polynomial $G(z)$ over the same field must be 1 or $F(z)$.

Example 5A.8. $z^2 + 1$ is irreducible over the real numbers. It is factorable over the complex numbers since

$$z^2 + 1 = (z + i)(z - i)$$

where i is a name for a root of the polynomial $z^2 + 1$ over the real numbers.

5A.4. ANALYZING THE STRUCTURE OF GALOIS FIELDS

A field with a finite number of elements q is called a Galois field, and is denoted by GF(q). The addition and multiplication tables of GF(16) are shown in Figure 5A.1, as are the tables for GF(2) (i.e., $\{0, 1\}$) and GF(4) ($\{0, 1, a, b\}$). The reader may want to refer to GF(16) as a typical example of the abstract fields in the discussions to follow.

Assuming that the addition and multiplication tables of a finite field GF(q) exist, let's explore the structures which GF(q) must contain. Suppose that the cyclic subgroup, $\mathscr{I}$, generated by 1 under addition has p elements,

$$\mathscr{I} = \{0, 1, \ldots, p - 1\}. \tag{5A.25}$$

$\mathscr{I}$ is the set of *integers* of GF(q), and it can be verified that addition and multiplication of elements of $\mathscr{I}$ corresponds to arithmetic modulo p.

It can be shown that subsets of an arbitrary field GF(q), which are closed under addition and multiplication and contain the identities 0 and 1, are in fact fields in their own right. Hence, the following can be verified:

THEOREM 5A.2. *The integers* $0, 1, 2, \ldots, p - 1$, *of a finite field form a subfield under addition and multiplication modulo p where p is the number of distinct integer elements of the field.*

Example 5A.9. In the addition and multiplication tables of GF(16) (see Figure 5A.1), $1 + 1 = 0$ and hence $\{0, 1\} = $ GF(2).

Now consider the case in which a subfield GF(q) (not necessarily an integer subfield) exists within a larger field, e.g., as GF(2) or GF(4) exist within GF(16) in Figure 5A.1. Let x be a field element in the larger field but *not* in GF(q). Find all elements in the larger field which are equal to linear combinations of powers of x with coefficients in GF(q), i.e., elements of the form

$$v_a = \sum_{i=0}^{d-1} a_i x^i, \tag{5A.26}$$

where $a = (a_0, a_1, \ldots, a_{d-1})$ is a coefficient vector of elements from GF(q). The right side above will be referred to as a *pseudopolynomial* in x, since x is not an indeterminate and the sum is simply a field element. For a given value of d, the number of possible coefficient vectors a is q^d. If the larger finite field contains n elements, we shall choose d so that $q^d > n$. This means that

$$v_a = v_b \tag{5A.27}$$

for some $a \neq b$. Hence,

$$\sum_{i=0}^{d-1} a_i x^i = \sum_{i=0}^{d-1} b_i x^i \tag{5A.28}$$

or

$$\sum_{i=0}^{d-1} (a_i - b_i) x^i = 0 \tag{5A.29}$$

and the coefficients $(a_i - b_i)$ cannot all be zero, though the non-zero coefficient with largest subscript need not be $(a_{d-1} - b_{d-1})$. Replacing x by an indeterminate z above, it is apparent that the set of polynomials over GF(q) which has x as a root, is not empty.

+	0	1	a	b	c	d	e	f	g	h	j	k	m	n	p	q
0	0	1	a	b	c	d	e	f	g	h	j	k	m	n	p	q
1	1	0	b	a	f	j	q	c	p	k	d	h	n	m	g	e
a	a	b	0	1	d	c	m	j	k	p	f	g	e	q	h	n
b	b	a	1	0	j	f	n	d	h	g	c	p	q	e	k	m
c	c	f	d	j	0	a	k	1	m	q	b	e	g	p	n	h
d	d	j	c	f	a	0	g	b	e	n	1	m	k	h	q	p
e	e	q	m	n	k	g	0	h	d	f	p	c	a	b	j	1
f	f	c	j	d	1	b	h	0	n	e	a	q	p	g	m	k
g	g	p	k	h	m	e	d	n	0	b	q	a	c	f	1	j
h	h	k	p	g	q	n	f	e	b	0	m	1	j	d	a	c
j	j	d	f	c	b	1	p	a	q	m	0	n	h	k	e	g
k	k	h	g	p	e	m	c	q	a	1	n	0	d	j	b	f
m	m	n	e	q	g	k	a	p	c	j	h	d	0	1	f	b
n	n	m	q	e	p	h	b	g	f	d	k	j	1	0	c	a
p	p	g	h	k	n	q	j	m	1	a	e	b	f	c	0	d
q	q	e	n	m	h	p	1	k	j	c	g	f	b	a	d	0

·	0	1	a	b	c	d	e	f	g	h	j	k	m	n	p	q
0	0	0	0	0	0	0	0	0	0	0	0	0	0	0	0	0
1	0	1	a	b	c	d	e	f	g	h	j	k	m	n	p	q
a	0	a	b	1	g	h	j	k	m	n	p	q	c	d	e	f
b	0	b	1	a	m	n	p	q	c	d	e	f	g	h	j	k
c	0	c	g	m	d	e	f	a	h	j	k	b	n	p	q	1
d	0	d	h	n	e	f	a	g	j	k	b	m	p	q	1	c
e	0	e	j	p	f	a	g	h	k	b	m	n	q	1	c	d
f	0	f	k	q	a	g	h	j	b	m	n	p	1	c	d	e
g	0	g	m	c	h	j	k	b	n	p	q	1	d	e	f	a
h	0	h	n	d	j	k	b	m	p	q	1	c	e	f	a	g
j	0	j	p	e	k	b	m	n	q	1	c	d	f	a	g	h
k	0	k	q	f	b	m	n	p	1	c	d	e	a	g	h	j
m	0	m	c	g	n	p	q	1	d	e	f	a	h	j	k	b
n	0	n	d	h	p	q	1	c	e	f	a	g	j	k	b	m
p	0	p	e	j	q	1	c	d	f	a	g	h	k	b	m	n
q	0	q	f	k	1	c	d	e	a	g	h	j	b	m	n	p

Figure 5A.1. Addition and multiplication in GF(16).

One polynomial totally characterizes the properties of x and the way it interacts with elements of GF(q).

DEFINITION 5A.6. *The monic polynomial $m_x(z)$ over GF(q), of minimum degree d_m with respect to all polynomials over GF(q) having x as a root, is called the* minimum polynomial $m_x(z)$ *of x over GF(q).*

The minimum polynomial of a field element x over a subfield GF(q) must exist since it has already been demonstrated that at least one polynomial over GF(q) having x as a root, exists. It can also be shown that the minimum polynomial of x is unique and irreducible over GF(q).

Suppose we now investigate the collection $\mathcal{V}$ of all elements of the form

$$v_a = \sum_{i=0}^{d_m - 1} a_i x^i \tag{5A.30}$$

$$a = (a_0, a_1, \ldots, a_{d_m - 1}). \tag{5A.31}$$

where the coefficients a_i are in GF(q), x is a field element not in GF(q), and d_m is the degree of the minimum polynomial of x over GF(q).

(a) For two distinct vectors a and b of dimensions d_m,

$$v_a \neq v_b.$$

Proof. Assume $v_a = v_b$. Then v_{a-b} is a pseudopolynomial of degree $d_m - 1$ or less in x with coefficients in GF(q) which is equal to 0. But the minimum polynomial of x has degree d_m, and, hence, $v_a \neq v_b$. ∎

(b) $\mathcal{V}$ is closed under addition.

Proof.

$$v_a + v_b = v_{a+b} \tag{5A.32}$$

∎

(c) $\mathcal{V}$ is closed under multiplication.

Proof.

$$v_a v_b = \left(\sum_{i=0}^{d_m - 1} a_i x^i \right) \left(\sum_{i=0}^{d_m - 1} b_i x^i \right)$$

$$= \sum_{i=0}^{2(d_m - 1)} c_i x^i = C(x) \tag{5A.33}$$

where c_i is in GF(q). By the Euclidean division algorithm for polynomials in an indeterminate, $C(z) = Q(z)m_x(z) + R(z)$, deg $R(z) <$ deg $m_x(z) = d_m$. Now since $m_x(x) = 0$, substitution of x for z gives $C(x) = R(x)$, a pseudopolynomial in the field element x of degree less than d_m with coefficients in GF(q). ∎

Since this subset $\mathcal{V}$ of the field is closed under addition and multiplication, we have the following result:

THEOREM 5A.3. *Let GF(q) be a subfield of a larger field and let x be a field element not in GF(q), the minimum polynomial $m_x(z)$ of x over GF(q) having degree d_m. Then the pseudopolynomials in x over GF(q), having degree less than d_m, are the distinct elements of a field GF(q^{d_m}).*

Example 5A.10. Consider the addition and multiplication tables of GF(16) shown in figure 5A.1. Forming pseudopolynomials in a over GF(2), and evaluating them using Figure 5A.1 gives

$$a + 1 = b, \; a^2 = b, \; a^2 + 1 = a, \; a^2 + a = 1, \; a^2 + a + 1 = 0.$$

Hence, $a^2 + a + 1$ is the lowest degree pseudopolynomial in a which is zero, and the minimum polynomial of a is

$$m_a(z) = z^2 + z + 1.$$

The smallest subfield containing a is GF(4), whose elements are represented as

element	a-representation
0	0
1	1
a	a
b	$a + 1$

and whose addition and multiplication tables are characterized completely by the GF(2) addition and multiplication tables and the fact that $a^2 + a + 1 = 0$.

The element c of Figure 5A.1 is not in GF(4). To determine the minimum polynomial of c over GF(4), evaluate the monic pseudopolynomials in c over GF(4).

$$c + 1 = f, \; c + a = d, \; c + b = j, \; c^2 = d.$$

Note that

$$c + a = d = c^2 \Rightarrow c^2 + c + a = 0.$$

(Addition and subtraction are identical operations in fields containing GF(2).) Hence,

$$m_c(z) = z^2 + z + a \; (\text{over GF}(4)).$$

If instead the minimum polynomial of c over GF(2) had been determined, it would be seen that

$$m_c(z) = z^4 + z + 1 \; (\text{over GF}(2)).$$

Representations of GF(16) elements as pseudopolynomials in c over GF(4) and GF(2) are shown in Table 5A.1. It is obvious from the above pseudopolynomial representations that the elements of GF(16) may be viewed as a 2-dimensional vector space over GF(4) or a 4-dimensional vector space over GF(2). In either case, vector addition corresponds to field element addition. For example, the field element h can be viewed as the vector (a, b) of pseudopolynomial coefficients over GF(4), or as $(1, 0, 1, 1)$ over GF(2).

Using the c-representations over GF(4) of GF(16) elements, only the GF(4) arithmetic tables and the fact that $c^2 + c + a = 0$ are necessary to do calculations. For example:

$$e(m + k) = (bc + a)((bc) + (ac + a)) \qquad \text{(representation)}$$
$$= (bc + a)(c + a) \qquad \text{(addition in GF(4))}$$
$$= bc^2 + (a + ab)c + a^2 \qquad \text{(distributive law)}$$
$$= b(c + a) + (a + ab)c + a^2 \qquad (c^2 = c + a)$$
$$= a \qquad \text{(GF(4) arithmetic)}$$

The same calculations using c-representations over GF(2) along with the fact that $c^4 + c + 1 = 0$ is:

$$e(m + k) = c^3((c^3 + c^2 + c) + (c^3 + c)) \qquad \text{(representation)}$$
$$= c^5 \qquad \text{(GF(2) arithmetic)}$$
$$= c(c + 1) \qquad (c^4 = c + 1)$$
$$= c^2 + c \qquad \text{(distributive law)}$$
$$= a \qquad \text{(representation)}$$

The above agree with the direct calculation

$$e(m + k) = ed = a$$

from Figure 5A.1.

Clearly the cumbersome tables of Figure 5A.1 can be totally eliminated, and the elements of GF(16) viewed as the pseudopolynomials in c of degree

Table 5A.1
Representations of elements in GF(16) as pseudopolynomials in c over GF(4) and GF(2).

element	c-representation over GF(4)	c-representation over GF(2)
0	0	0
1	1	1
a	a	$c^2 + c$
b	b	$c^2 + c + 1$
c	c	c
d	$c + a$	c^2
e	$bc + a$	c^3
f	$c + 1$	$c + 1$
g	ac	$c^3 + c^2$
h	$ac + b$	$c^3 + c + 1$
j	$c + b$	$c^2 + 1$
k	$ac + a$	$c^3 + c$
m	bc	$c^3 + c^2 + c$
n	$bc + 1$	$c^3 + c^2 + c + 1$
p	$ac + 1$	$c^3 + c^2 + 1$
q	$bc + b$	$c^3 + 1$

three or less. The structure of GF(16) arithmetic then is dictated by GF(2) arithmetic and the fact that $c^4 + c + 1 = 0$.

5A.5. THE CONSTRUCTION OF FINITE FIELDS

Using the structures uncovered in the previous section, it now is possible to construct finite fields. To first construct the integer subfield, note that the integers $0, 1, \ldots, p - 1$ are a commutative ring under addition and multiplication modulo p. Hence, $0, 1, 2, \ldots, p - 1$ form a field under addition and multiplication mod p if the integers $1, 2, \ldots, p - 1$ have inverses under multiplication mod p.

THEOREM 5A.4. *If p is prime, then $1, 2, \ldots, p - 1$ all have inverses under multiplication* mod p.

Proof. Suppose $1 \le j < p$. Then, since p is prime, the greatest common divisor of j and p is 1. Hence, there exist integers n_1 and n_2 such that

$$1 = \gcd(j, p) = n_1 j + n_2 p. \tag{5A.34}$$

Reducing both sides of the equation mod p gives

$$1 = ((n_1 \bmod p) j) \bmod p \tag{5A.35}$$

Hence, $n_1 \bmod p$ is the inverse of j.   ∎

THEOREM 5A.5. *If p is not prime, then at least one element in the collection $1, 2, \ldots, p - 1$ does not have an inverse under multiplication* mod p, *and the collection is not closed under multiplication* mod p.

Proof. The integer p not prime means $p = p_1 p_2$ for some p_1 and p_2 in the range $1 < p_1 < p, 1 < p_2 < p$. Hence, under multiplication mod p

$$0 = p_1 p_2 \bmod p. \tag{5A.36}$$

If p_2 had an inverse under multiplication mod p, the above equation would imply

$$0 = p_1 \bmod p. \tag{5A.37}$$

Since p_1 is a non-zero element of the ring, p_2 cannot have an inverse under multiplication.   ∎

Summarizing these results gives:

THEOREM 5A.6. *The integers $0, 1, \ldots, p - 1$ form a field under addition and multiplication* mod p *if and only if p is a prime number.*

The next step in the field synthesis problem is to assume the existence of a field GF(q), and attempt to construct a larger field containing it. Emulating the structures of the previous section, let $P(z)$ be a polynomial of degree d_m over $GF(q)$ in an indeterminate z. It is easily demonstrated that the collection of pseudopolynomials in x of degree $d_m - 1$ or less forms a commutative ring under multiplication modulo $P(x)$. Again, the question of field existence resolves to determining when multiplicative inverses exist.

THEOREM 5A.7. *If $P(z)$ is an irreducible polynomial over $GF(q)$, having x as a root, then the nonzero pseudopolynomials in x over $GF(q)$ of degree less than the degree of $P(z)$ all have multiplicative inverses under multiplication mod $P(x)$.*

Proof. Let $J(z)$ be a polynomial in $z, 1 \leq \deg J(z) < \deg P(z)$. Since $P(z)$ is irreducible, the greatest common divisor of $J(z)$ and $P(z)$ is 1. Hence, there exist polynomials $N_1(z)$ and $N_2(z)$ such that

$$1 = \gcd(J(z), P(z)) = N_1(z)J(z) + N_2(z)P(z) \qquad (5A.38)$$

Reducing this equation mod $P(z)$ gives

$$1 = ((N_1(z) \bmod P(z))J(z)) \bmod P(z) \qquad (5A.39)$$

Hence, by setting the indeterminate z equal to x, it follows that $N_1(x)$ mod $P(x)$ is the inverse of $J(x)$ under multiplication mod $P(x)$. ∎

By now the reader should recognize the similarity of the roles played by the prime numbers within the ring of integers and the irreducible polynomials within the ring of polynomials. The proof of the following theorem follows exactly the proof of the equivalent theorem for the integers.

THEOREM 5A.8. *If $P(z)$ is not irreducible over the field $GF(q)$ then at least one non-zero pseudopolynomial in x over $GF(q)$ with degree less than the degree of $P(x)$ does not have an inverse under multiplication mod $P(x)$.*

With the background developed thus far, it is possible to conclude the following:

THEOREM 5A.9. *Let $GF(q)$ be a known field. A larger field $GF(q^m)$ containing $GF(q)$ exists if, and only if, a polynomial $P(z)$ of degree m over $GF(q)$, irreducible over $GF(q)$, exists. The distinct members of $GF(q^m)$ are all pseudopolynomials in x of degree less than m over $GF(q)$, and addition and multiplication in $GF(q^m)$ is polynomial addition and multiplication mod $P(x)$, where x is a root of $P(z)$.*

In the method of construction just outlined, $GF(q)$ is called the *ground field* and $GF(q^m)$ is termed the (algebraic) *extension field* of $GF(q)$. The term "extension field of $GF(q)$" may be used to describe any larger field containing $GF(q)$; the modifier "algebraic" refers to the method of construction by which the field GF(q) is augmented with the root of an algebraic expression to construct the larger field.

Since each finite field contains an integer field having a prime number of elements, and since any field containing the integer subfield can be constructed by the method stated in the above theorem, a finite field GF(q) containing q elements cannot exist unless q is a power of a prime number p. The prime p is called the *characteristic* of the field GF(q) when $q = p^n$. Hence, finite fields containing 6, 10, 12, 14, 15, 18, (etc.) elements do not exist. The existence of fields containing 2, 3, 5, 7, 11, 13, 17, (etc.) elements has already been demonstrated, since these are prime fields (integer fields mod p). The existence of fields containing 4, 8, 9, 16, 25, 27, 32 (etc.) elements can be demonstrated; this is equivalent to showing the existence of an irreducible polynomial of the appropriate degree over the appropriate integer fields.

The number $N_I(d)$ of irreducible polynomials over GF(q) of degree d can be shown to be

$$N_I(d) = \frac{1}{d} \sum_{m \mid d} \mu\left(\frac{d}{m}\right) q^m, \qquad (5A.40)$$

where the sum is over all integer divisors of the degree d and $\mu(\ )$ is the Möbius function:

$$\mu(k) = \begin{cases} 1 & k - 1 \\ (-1)^j & k = \text{product of } j \text{ distinct primes} \\ 0 & k \text{ divisible by a square.} \end{cases} \qquad (5A.41)$$

It is easily verified that $N_I(d)$ is a positive integer for all degrees d.

5A.6. PRIMITIVE ELEMENTS

The non-zero elements of a Galois field GF(q) form a multiplicative group of order $q - 1$; hence, the exponent e of the non-zero elements of GF(q) under multiplication must divide $q - 1$ (see Section 5A.1). If the exponent e of an element x equals the order $q - 1$ of the multiplicative group of GF(q), then x is termed a *primitive* field element. (It can be shown that the multiplicative group of a finite field is cyclic.)

Example 5A.11. In the Galois field of sixteen elements shown in Figure 5A.1, the element c is primitive. This can be verified from Table 5A.2.

The addition and multiplication tables for GF(16), shown in Figure 5A.1, were constructed from Table 5A.2 in the following manner. An irreducible polynomial over GF(2) of degree 4, namely $z^4 + z + 1$ was known to have a primitive element as its root, which we named c. The powers of c were reduced to pseudopolynomials of degree 3 or less in c, since $c^4 + c + 1$

Table 5A.2
Representations of GF(16)

Element name	Representations	
	Power of c	Pseudopolynomial in c
1	c^0	1
c	c^1	c
d	c^2	c^2
e	c^3	c^3
f	c^4	$1 + c$
a	c^5	$c + c^2$
g	c^6	$c^2 + c^3$
h	c^7	$1 + c \quad\;\; + c^3$
j	c^8	$1 + \quad\; c^2$
k	c^9	$c + \quad\; c^3$
b	c^{10}	$1 + c + c^2$
m	c^{11}	$c + c^2 + c^3$
n	c^{12}	$1 + c + c^2 + c^3$
p	c^{13}	$1 + \quad\; c^2 + c^3$
q	c^{14}	$1 + \quad\quad\quad\;\; c^3$

$= 0$, giving the representation columns in the table. Names were assigned as in the first column. The addition table of Figure 5A.1 was constructed using the pseudopolynomial representations and coefficient addition in GF(2); the multiplication table was constructed using the power representations and the fact that $c^{15} = 1$. Obviously the table embodies the additive and multiplicative properties of the field.

The representation columns of the table can be generated by shift registers storing the coefficients of c. Suppose a binary shift register is constructed as shown below, with the coefficient of c^i shifting into the coefficient of c^{i+1}.

COEFFICIENT OF

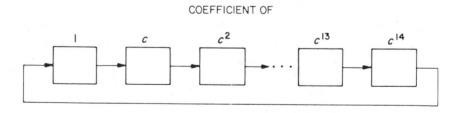

If 011010110000001 is present in the register, this represents $c + c^2 + c^4 + c^6 + c^7 + c^{14}$. A single shift of the contents of the register corresponds to a multiplication by c. Hence, in the above example, 101101011000000 represents $c(c + c^2 + c^4 + c^6 + c^7 + c^{14}) = 1 + c^2 + c^3 + c^5 + c^7 + c^8$. If we had started the register with 100000000000000, the successive contents of the register would have represented $1, c, c^2, c^3, c^4, \ldots, c^{14}, c^{15}, c^{16}, \ldots$. The above representation for multiplication by c only uses the fact that c has

order 15. A similar shift register using the additional fact that $c^4 = c + 1$, is
shown below.

COEFFICIENT OF

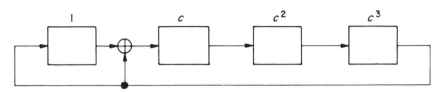

A shift (multiplication by c) of the c^3 coefficient results in $c^4 = c + 1$ being
added to the contents of the register. If this register is started with 1000, the
register contents after each shift will represent $1, c, c^2, \ldots, c^{13}, c^{14}$,
$c^{15}, c^{16}, \ldots$. The representations will be pseudopolynomials of degree 3 or
less in c, i.e., the pseudopolynomials in Table 5A.2.

The multiplicative order of each element in a field $GF(q^d)$ can be
determined from the representation table. To determine the order of c^i
where c is primitive, we first compute the gcd $(i, q^d - 1)$. Then the smallest
number e for which e is a multiple of $q^d - 1$ is

$$e = \frac{lcm(i, q^d - 1)}{i} = \frac{q^d - 1}{\gcd(i, q^d - 1)}. \tag{5A.42}$$

Hence, e is the order of c^i.

It can be shown that all the roots of an irreducible polynomial have the
same order. Hence, an irreducible polynomial can be labelled with the order
of its roots. An irreducible polynomial with a primitive field element as its
root is called a *primitive polynomial*. The number of primitive polynomials
$N_p(d)$ of degree d over $GF(q)$ is given by

$$N_p(d) = \frac{q^d - 1}{d} \prod_{i=1}^{J} \frac{p_i - 1}{p_i} \tag{5A.43}$$

where p_i, $i = 1, \ldots, J$ are the distinct primes which divide $q^d - 1$. Certainly
$N_p(d)$ is greater than zero for all powers of a prime as represented by q, and
all degrees d. As a result every finite field can be constructed using a
primitive irreducible polynomial.

Example 5A.12. In GF(16) the orders of the elements are as follows:

1	has order 1
c^5, c^{10}	have order 3
c^3, c^6, c^9, c^{12}	have order 5
$c, c^2, c^4, c^7, c^8, c^{11}, c^{13}, c^{14}$	have order 15.

Excluding the elements of GF(4), namely 0, 1, c^5, and c^{10}, any of the remaining twelve elements could be used to generate GF(16) as c did in Section 5A.5, and, hence, all must be roots of irreducible polynomials of degree 4. From the previous section, the number $N_I(4)$ irreducible polynomials of degree 4 over GF(2) is

$$N_I(4) = \tfrac{1}{4}\left[\mu(1)2^4 + \mu(2)2^2 + \mu(4)2\right]$$
$$= \tfrac{1}{4}\left[2^4 - 2^2\right] = 3,$$

while the number $N_p(4)$ of primitive irreducible polynomials of degree 4 over GF(2) is

$$N_p(4) = \frac{2^4 - 1}{4}\left(\frac{5-1}{5}\right)\left(\frac{3-1}{3}\right) = 2$$

Hence, the nonprimitive elements c^3, c^6, c^9, and c^{12} must be roots of the same irreducible degree-4 polynomial. The eight elements of order 15 must be roots of the two degree-4 primitive polynomials.

5A.7. FINDING IRREDUCIBLE AND PRIMITIVE POLYNOMIALS

Some fundamental ideas are necessary before further progress can be made. The following are stated without proof:

(a) A finite field of specified size is unique. This means that different irreducible polynomials of the same degree over the same ground field must lead to the same extension field structure.
(b) $GF(q^{d'}) \subset GF(q^d) \Leftrightarrow d'|d$. This is a dimensionality constraint that comes from viewing $GF(q^d)$ as the result of a construction employing a polynomial of degree d/d' over $GF(q^{d'})$.
(c) All roots of an irreducible polynomial lie in the same field. If this were not the case, the construction of Section 5 might yield different extension fields using the same polynomial.
(d) The elements of $GF(q^d)$ are the roots of the polynomial $z^{q^d} - z$. A proof of this fact follows directly from noting that the non-zero elements of $GF(q^d)$ have order dividing $q^d - 1$ and, therefore, are roots of $z^{q^d-1} - 1$.

As a result of (a)–(d), it follows that the polynomial containing all elements of $GF(q^d)$ as roots can be factored as follows:

$$z^{q^d} - z = \prod_{d'|d} \prod_{m(z) \in P_{d'}} m(z) \tag{5A.44}$$

where P_d denotes the set of all monic irreducible polynomials of degree d over GF(q).

Let $A(z)$ be a polynomial of degree d over GF(q). The following algorithm can be used to determine if $A(z)$ is irreducible.

Irreducibility Test:

(a) $d' = 1$.
(b) $N_{d'}(z) = \gcd(A(z), z^{q^{d'}} - z)$.
(c) If $N_{d'}(z) \neq 1$, then $[A(z)$ has a factor $N_{d'}(z)$, STOP].
(d) If $d' < \lfloor d/2 \rfloor$, then $[d' = d' + 1,$ Go to (b)].
(e) $A(z)$ is irreducible. STOP.

Step (b) will recover any polynomial factors over $GF(q)$ that are common to $A(z)$ and $z^{q^{d'}} - z$. If the test runs its full course for all d' in the range $1 \leq d' \leq d/2$, $A(z)$ will have been checked for all possible irreducible factors of degree at most $d/2$. Since if $A(z)$ is factorable, it must have a factor in this degree range, the test shows $A(z)$ to be irreducible if and only if no factors are uncovered.

Euclid's algorithm of Section 5A.2 can be mechanized to determine the greatest common divisor in step (b). However, the first part of this calculation may be difficult since the degree $q^{d'}$ may be a very large number. The solution to this problem is to note that the first step in Euclid's algorithm will be to evaluate $z^{q^{d'}} - z$ mod $A(z)$. This can be done with the following routine.

Evaluation of $R(z) = z^{q^d} - z \bmod A(z)$

(a) $R_0(z) = z$, $e = 1$.
(b) $R_e(z) = [R_{e-1}(z)]^q \bmod A(z)$.
(c) If $e < d$, then $[e = e + 1,$ Go to (b)].
(d) $R(z) = [R_d(z) - z] \bmod A(z)$. STOP.

The degrees of the polynomials in the calculation never exceed qd, because of the judicious insertions of mod $A(z)$ reductions into the calculations.

The operation of raising a polynomial $R(z)$ over $GF(q)$ to the q-th power is a simple procedure as indicated by the following theorem.

THEOREM 5A.10. *Let $R(z)$ be a polynomial over* $GF(q)$. *Then*

$$(R(z))^q = R(z^q) \tag{5A.45}$$

Proof. Consider a polynomial $R(z)$ with coefficients in the ground field $GF(q)$, where $q = p^n$, p being a prime number.

$$R(z) = \sum_{i=0}^{d} a_i z^i, \quad a_i \in GF(q). \tag{5A.46}$$

If we raise $R(z)$ to a power equal to the characteristic of the field $GF(q)$ we get

$$(R(z))^p = \sum_{j=0}^{p} \binom{p}{j}(a_d z^d)^j \left(\sum_{i=0}^{d-1} a_i z^i \right)^{p-j} \tag{5A.47}$$

Here we have used the binomial expansion treating the d-th term as one element and the sum of the remaining terms as the other element in the expansion. The binomial coefficient $\binom{p}{j}$ counts the number of times a term appears and, hence, can be considered to be the element of GF(q) corresponding to the sun of $\binom{p}{j}$ unit elements. In GF(q), integer elements are added mod p. Since p is prime, $\binom{p}{j}$ is divisible by p unless $j = p$ or $j = 0$. Hence,

$$\binom{p}{j} \bmod p = \begin{cases} 1 & \text{if } j = p \text{ or } j = 0 \\ 0 & \text{otherwise.} \end{cases} \tag{5A.48}$$

Thus,

$$(R(z))^p = \left(a_d z^d\right)^p + \left(\sum_{i=0}^{d-1} a_i z^i\right)^p, \tag{5A.49}$$

and iterating this result on the remaining polynomial of degree $(d-1)p$ gives

$$(R(z))^p = \sum_{i=0}^{d} (a_i)^p (z^p)^i. \tag{5A.50}$$

If this procedure is repeated n times we have

$$(R(z))^{p^n} = \sum_{i=0}^{d} (a_i)^{p^n} (z^{p^n})^i. \tag{5A.51}$$

Now a_i is in GF(q) where $q = p^n$. Hence, the order of a_i divides $q - 1$ and $a_i^q = a_i$. Therefore,

$$(R(z))^q = \sum_{i=0}^{d} a_i z^{qi} = R(z^q). \tag{5A.52}$$

∎

This theorem makes the computation of $R(z)$ equivalent to simply changing the powers of z in the polynomial. Hence, the evaluation algorithm for $z^{q^{d'}} - z \bmod A(z)$ is extremely efficient numerically.

Highly structured sieve methods which employ efficient exhaustive search techniques could be used to construct a table of irreducible polynomials. However, this approach bogs down at a relatively low degree, since a large number of polynomials is involved. A much faster approach for finding a single irreducible polynomial of high degree is simply to select a polynomial $A(z)$ of degree d over GF(q) at random, and apply the irreducibility test outlined above. The probability of success on any given test is $N_I(d)/q^d$, a quantity which generally decreases as $1/d$.

The random selection process by which a trial polynomial $A(z)$ is selected can be improved by not choosing polynomials which are easily shown to be factorable. For example, when $q = 2$, note the following guidelines.

(a) $A(z)$ has a factor z if the coefficient of 1 in the polynomial $A(z)$ is zero. (This obviously works for all q.)

(b) $A(z)$ has a root 1 if $A(z)$ has an even number of non-zero coefficients.

Observance of these facts by selecting trial polynomials, which do not have roots in GF(2), has the effect of improving the probability of test success by approximately a factor of 4. Many other results, e.g., Theorem 5A.10, can be used to further eliminate poor choices.

After determining that a polynomial $A(z)$ is irreducible, it is also possible to determine the order of its roots. For example, if one wishes to test an irreducible polynomial $A(z)$ of degree d over GF(q) to see if it is primitive, first determine the prime decomposition of $q^d - 1$.

$$q^d - 1 = \prod_{i=1}^{J} p_i^{e_i}$$

where p_i, $i = 1, \ldots, J$ are distinct primes. Then carry out the following algorithm.

Primitivity Test

(a) $i = 1$.

(b) $N_i(z) = z^{(q^d-1)/p_i} \bmod A(z)$.

(c) If $N_i(z) = 1$, then [$A(z)$ is not primitive. STOP].

(d) If $i < J$, then [$i = i + 1$, Go to (b)].

(e) $A(z)$ is primitive. STOP.

Step (b) in this test checks to see if the exponent of $A(z)$ lacks the prime factor p_i, which, if this is the case, results in $N_i(z) = 1$. The difficult part of this test is to determine the prime decomposition of $q^d - 1$. A decomposition table for the case $q = 2$, for $1 < d \le 120$ is included in Appendix 5B. A selected table of primitive polynomials over this range of parameters is also included in Appendix 5B, along with references to more extensive listings.

5A.8. PROPERTIES OF MINIMUM POLYNOMIALS

Once a primitive polynomial $m_x(z)$ of degree d over GF(q) has been determined, the elements of GF(q^d) (excluding zero), can be represented as pseudopolynomials in x or as powers of x and addition and multiplication operations defined. The minimum polynomials of the remaining field elements of GF(q^d) can be calculated directly.

Suppose that y is another element of GF(q^d) and its minimum polynomial is desired. Knowledge of y as a power of x allows the exponent of y

to be determined as in Section 5A.6. This provides the following additional information.

THEOREM 5A.11. *Let y be an element of exponent e in an extension field of* GF(q). *Then the following are equivalent statements:*

(a) *d is the smallest integer for which $e|q^d - 1$.*
(b) *GF(q^d) is the smallest field containing y.*
(c) *d is the degree of the minimum polynomial of y over GF(q).*

Proof. Let a be the smallest integer such that $e|q^a - 1$; let b be the smallest integer such that $y \in$ GF(q^b); let c be the degree of the minimum polynomial of y over GF(q). The results of previous sections justify the following relations:

(a) $e|q^a - 1 \Rightarrow y$ is a root of $z^{q^a} - z \Rightarrow y \in$ GF(q^a) $\Rightarrow b \le a$.
(b) $y \in$ GF(q^b) $\Rightarrow y$ is a root of $z^{q^b} - z \Rightarrow c \le b$.
(c) $m_y(z)$ irreducible of degree $c \Rightarrow m_y(z)|z^{q^c} - z \Rightarrow y^{q^c} = y \Rightarrow e|q^c - 1$
  $\Rightarrow a \le c$.
(d) Since $a \le c \le b \le a$, $a = b = c$.   ∎

It is now possible to find the minimum polynomial over GF(q) of a field element y.

Linear Equation Algorithm for Determining $m_y(z)$ over GF(q)

(a) Determine the exponent of y.
(b) Determine the degree d' of $m_y(z)$ over GF(q).
(c) Represent y^i as a pseudopolynomial over GF(q) in x, x being a primitive element of the field GF(q^d) containing y, and the minimum polynomial of x being known. Hence, determine a_{ij} in GF(q) such that

$$y^i = \sum_{j=0}^{d-1} a_{ij}x^j, \quad i = 0,1,\ldots,d'. \tag{5A.53}$$

(d) Find the coefficients $b_0,\ldots,b_{d'-1}$ in the minimum polynomial $m_y(z)$, such that

$$m_y(y) = \sum_{i=0}^{d'-1} b_i y^i + y^{d'} = 0, \tag{5A.54}$$

or equivalently, solve the following equations in GF(q) for $b_0,\ldots,b_{d'-1}$:

$$\sum_{i=0}^{d'-1} b_i a_{ij} = -a_{d'j}, \quad j = 0,1,\ldots,d-1. \tag{5A.55}$$

(e) The minimum polynomial of y over GF(q) is

$$m_y(z) = z^{d'} + b_{d'-1}z^{d'-1} + \cdots + b_1 z + b_0. \tag{5A.56}$$

We shall illustrate this algorithm with an example using the description of GF(16) in Section 5A.6.

Example 5A.13. Consider the element h in GF(16).

(a) $h = c^7$, c primitive, so the order of h is $15/\gcd(7, 15) = 15$

(b) $m_h(z)$ over GF(2) has degree $d_m = 4$ since

15 does not divide $(2 - 1)$

15 does not divide $(2^2 - 1)$

15 does not divide $(2^3 - 1)$

15 divides $(2^4 - 1)$.

(c) $h^0 = 1 = 1$
$h^1 = c^7 = 1 + c + c^3$
$h^2 = c^{14} = 1 + c^3$
$h^3 = c^6 = c^2 + c^3$
$h^4 = c^{13} = 1 + c^2 + c^3$

(d) $h^4 + b_3 h^3 + b_2 h^2 + b_1 h + b_0 = 0$
gives for the coefficients of c^i:
$i = 0$: $1 + b_2 + b_1 + b_0 = 0$
$i = 1$: $b_1 = 0$
$i = 2$: $1 + b_3 = 0$
$i = 3$: $1 + b_3 + b_2 + b_1 = 0$
which are easily solved over GF(2) to give $b_0 = b_3 = 1$, $b_1 = b_2 = 0$.

(e) The minimum polynomial of h over GF(2) is $z^4 + z^3 + 1$.

Instead of specifying a monic polynomial by its coefficients, it can be described by listing its roots.

DEFINITION 5A.7. *The q-conjugates of x are the d − 1 roots of the minimum polynomial $m_x(z)$ of x over* GF(q), *after x has been excluded, d being the degree of $m_x(z)$.*

The q-conjugates of x are easily derived from x via the following theorem.

THEOREM 5A.12. *Let $m_x(z)$, the minimum polynomial of x over* GF(q), *have degree d. Then $x, x^q, x^{q^2}, \ldots, x^{q^{d-1}}$, are all distinct and are the d roots of $m_x(z)$.*

Proof. Using Theorem 5A.10, we have that

$$m_x(z) = 0 \Rightarrow [m_x(z)]^q = 0 \Rightarrow m_x(z^q) = 0.$$

Hence, if x is a non-zero root of a polynomial over $GF(q)$, then so is x^q. Iterating this line of reason we see that $x, x^q, x^{q^2}, x^{q^3}, \ldots$, must all be roots of $m_x(z)$.

We now show that $x, x^q, x^{q^2}, \ldots, x^{q^{d-1}}$ are all distinct. Suppose $x^{q^j} = x^{q^i}$ for $0 \le i < j$. Multiplying both sides of this equation by the inverse of x^{q^i} in $GF(q^d)$ gives

$$x^{q^j - q^i} = 1 \tag{5A.57}$$

which implies that the order e of x divides $q^j - q^i$. Hence,

$$e \mid q^i(q^{j-i} - 1).$$

Since $e \mid q^d - 1$, it follows that $\gcd(e, q) = 1$,

$$e \mid (q^{j-i} - 1),$$

and, therefore, by Theorem 5A.11, $d \le j - i$. Since j and i are both non-negative integers, $x, x^2, \ldots, x^{q^{d-1}}$ are all distinct. Because a polynomial over a field can have no more roots than its degree, these elements are all the roots of $m_x(z)$.   ∎

We now have an alternative algorithm for determining the roots of $m_y(z)$ over $GF(q)$ whenever the arithmetic for any extension field of $GF(q)$, containing y, is specified.

q-Conjugate Algorithm for Determining $m_y(z)$ over $GF(q)$.

(a) Compute $y, y^q, y^{q^2}, \ldots$, until an integer d is reached for which $y^{q^d} = y$.
(b) Form the polynomial

$$m_y(z) = \prod_{i=0}^{d-1} \left(z - y^{q^i} \right) \tag{5A.58}$$

(c) Evaluate the coefficients of $m_y(z)$ in (b) using the arithmetic specified for the field.

We shall again provide an example from $GF(16)$. This time, for the sake of variety, we shall use the notation of the addition and multiplication tables of Figure 5A.1.

Example 5A.14. As a check, let us again consider the element h of $GF(16)$, and try to find $m_h(z)$ over $GF(2)$ using the q-conjugate method.

(a) $h^{2^0} = h$
$\quad h^{2^1} = h^2 = q$
$\quad h^{2^2} = q^2 = p$
$\quad h^{2^3} = p^2 = m$
$\quad h^{2^4} = m^2 = h$

(b) $m_h(z) = (z - h)(z - h^2)(z - h^4)(z - h^8)$
$m_h(z) = (z - h)(z - q)(z - p)(z - m)$
$\quad = (z^2 - cz + g)(z^2 - fz + k)$
$\quad = z^4 - (c + f)z^3 + (k + q + fc)z^2 - (ck + fg)z + gk$
$\quad = z^4 + z^3 + 1$ (over GF(2))

(c) This is a check on the previous computation.

Using Theorem 5A.12, the entire collection of non-zero elements of $GF(q^d)$ can be separated into disjoint sets, each set containing all the roots of a single irreducible polynomial over $GF(q)$. Suppose that each field element in this construction is represented as a power of a primitive element x of $GF(q^d)$, and to simplify notation, represent x^i by its base-x logarithm i. Hence, one logarithm set in the construction will be

$$\mathscr{S}_1 = \{1, q, q^2, \ldots, q^{d-1}\}, \tag{5A.59}$$

representing the roots $x, x^q, x^{q^2}, \ldots, x^{q^{d-1}}$ of $m_x(z)$. Another set in the construction, representing the minimum polynomial $m_{x^e}(z)$ of degree d', is

$$S_e = \{e, eq, eq^2, \ldots, eq^{d'-1}\}. \tag{5A.60}$$

The logarithms in S_e may be calculated by a coset-like construction. In fact, if e is relatively prime to $q^d - 1$, then $\mathscr{S}_e$ is a coset of the subgroup $\mathscr{S}_1$ in the group of integers relatively prime to $q^d - 1$ under multiplication modulo $q^d - 1$, and represents the roots of another primitive polynomial. If e is not relatively prime to $q^d - 1$, then $\mathscr{S}_e$ is not the result of a formal coset construction; however, all the distinct sets $\mathscr{S}_e$ are called *cyclotomic cosets*. Note that the d symbol, base q expansion of e and eq^i modulo $q^d - 1$ are related for all i by a cyclic shift, making the determination of the sets $\mathscr{S}_e$ a simple matter in a base q number system.

Example 5A.15. In GF(16), the root logarithm sets are given by $\mathscr{S}_1$, $\mathscr{S}_3$, $\mathscr{S}_5$, and $\mathscr{S}_7$, with $\mathscr{S}_1$ being the basic subgroup, $\mathscr{S}_7$ representing the one formal coset of $\mathscr{S}_1$, and $\mathscr{S}_3$ and $\mathscr{S}_5$ representing the remaining cyclotomic cosets. Using the element c of Table 5A.2 as the primitive element, the q-conjugate algorithm can be used to construct the polynomials corresponding to the above sets:

$$\mathscr{S}_1 = \{1, 2, 4, 8\} \rightarrow z^4 + z + 1$$

$$\mathscr{S}_3 = \{3, 6, 12, 9\} \rightarrow z^4 + z^3 + z^2 + z + 1$$

$$\mathscr{S}_5 = \{5, 10\} \rightarrow z^2 + z + 1$$

$$\mathscr{S}_7 = \{7, 14, 13, 11\} \rightarrow z^4 + z^3 + 1$$

$$\mathscr{S}_0 = \{0\} \rightarrow z + 1$$

The following relation reduces the amount of work necessary to evaluate the irreducible polynomials associated with the root sets. Let

$$A(z) = \sum_{i=0}^{d} a_i z^i = \prod_{i=1}^{d} (z - r_i) \qquad (5A.61)$$

be an irreducible polynomial whose roots have exponent e. Then the corresponding *reciprocal polynomial*, defined by

$$B(z) = a_0^{-1} z^d A\left(\frac{1}{z}\right)$$

$$= \sum_{i=0}^{d} a_0^{-1} a_{d-i} z^i = \prod_{i=1}^{d} (z - r_i^{-1}), \qquad (5A.62)$$

is easily determined by reversing the coefficient order (and scaling if necessary). The roots of $B(z)$ are the reciprocals of the roots of $A(z)$ and have the same order. Hence, in Example 5A.15, knowing that c and c^{14} are reciprocals and that $m_c(z)$ is $z^4 + z + 1$ is enough to conclude directly that $m_{c^{14}}(z)$ is $z^4 + z^3 + 1$.

5A.9. THE TRACE FUNCTION

The *trace polynomial* from $GF(q^n)$ to $GF(q)$ is defined as

$$\mathrm{Tr}_q^{q^n}(z) = \sum_{j=0}^{n-1} z^{q^j}, \qquad (5A.63)$$

where the superscript and subscript on Tr have been added to clarify the fields under consideration. The *trace* in $GF(q)$ of an element α in $GF(q^n)$ is then defined as $\mathrm{Tr}_q^{q^n}(\alpha)$, i.e., the trace polynomial evaluated at α. For a given element α which is in several fields, $\mathrm{Tr}_q^{q^n}(\alpha)$ will vary as the choice of q and/or q^n varies. We shall drop the superscripts and subscripts on $\mathrm{Tr}_q^{q^n}(\alpha)$ only when n and q are obvious in the context of the discussion.

Example 5A.16. In GF(16) model of Figure 5A.1,

$$\mathrm{Tr}_2^{16}(a) = a + a^2 + a^4 + a^8 = a + a^2 + a + a^2 = 0$$

Since a is in GF(4), we may also compute

$$\mathrm{Tr}_2^{4}(a) = a + a^2 = 1.$$

Notice that the variations of trace value with the parameter n are possible. To illustrate variations with the range field of the trace function, consider the element e in GF(16),

$$\mathrm{Tr}_{16}^{16}(e) = e$$

$$\mathrm{Tr}_{4}^{16}(e) = e + e^4 = b$$

$$\mathrm{Tr}_{2}^{16}(e) = e + e^2 + e^4 + e^8 = 1$$

We now will prove useful properties of the trace function.

Property 1. When α is in $GF(q^n)$, $Tr_q^{q^n}(\alpha)$ has values in $GF(q)$.

Proof.

$$\left[Tr_q^{q^n}(\alpha)\right]^q = \left(\sum_{j=0}^{n-1} \alpha^{q^j}\right)^q \qquad \text{(Definition)}$$

$$= \sum_{j=0}^{n-1} \alpha^{q^{j+1}} \qquad \text{(Theorem 5A.10)}$$

$$= \sum_{j=1}^{n-1} \alpha^{q^j} + \alpha \qquad \left(\alpha \in GF(q^n) \Rightarrow \alpha^{q^n} = \alpha\right)$$

$$= Tr_q^{q^n}(\alpha). \qquad \text{(Definition)}$$

Therefore, $Tr_q^{q^n}(\alpha)$ is a field element whose q-th power equals itself. Only elements of $GF(q)$ have this property. ∎

Property 2. Conjugate field elements have the same trace, i.e.,

$$Tr_q^{q^n}(\alpha^q) = Tr_q^{q^n}(\alpha) \text{ for } \alpha \in GF(q^n). \qquad (5A.64)$$

The proof of this property follows from the previous property and the fact (Theorem 5A.10) that $P(z^q) = [P(z)]^q$.

Property 3. The trace is linear: for $a, b \in GF(q)$ and $\alpha, \beta \in GF(q^n)$,

$$Tr_q^{q^n}(a\alpha + b\beta) = a\,Tr_q^{q^n}(\alpha) + b\,Tr_q^{q^n}(\beta) \qquad (5A.65)$$

Proof.

$$Tr(a\alpha + b\beta) = \sum_{j=0}^{n-1} (a\alpha + b\beta)^{q^j} \qquad \text{(Definition)}$$

$$= \sum_{j=0}^{n-1} \left(a^{q^j}\alpha^{q^j} + b^{q^j}\beta^{q^j}\right) \qquad \text{(Theorem 5A.10)}$$

$$= \sum_{j=0}^{n-1} a\alpha^{q^j} + b\beta^{q^j} \qquad \left(a \in GF(q) \Rightarrow a^q = a\right)$$

$$= a\,Tr_q^{q^n}(\alpha) + b\,Tr_q^{q^n}(\beta) \qquad \text{(Definition)}$$

∎

Property 4. There are q^{n-1} elements in $GF(q^n)$ which have trace value a for each a in $GF(q)$.

Proof. An element α in $\mathrm{GF}(q^n)$ has

$$\mathrm{Tr}_q^{q^n}(\alpha) = a \qquad (5\mathrm{A}.66)$$

if and only if α is a root of the trace equation

$$z + z^q + z^{q^2} + \cdots + z^{q^{n-1}} - a = 0, \; a \in \mathrm{GF}(q).$$

Every element of $\mathrm{GF}(q^n)$ must be the root of exactly one such equation, and each of the q equations has exactly q^{n-1} roots. Since all roots of the trace equations are accounted for by elements of $\mathrm{GF}(q^n)$, there must be exactly q^{n-1} elements of $\mathrm{GF}(q^n)$ with trace a in $\mathrm{GF}(q)$. ∎

Property 5. If $\mathrm{GF}(q) \subset \mathrm{GF}(q^k) \subset \mathrm{GF}(q^d)$, then

$$\mathrm{Tr}_q^{q^d}(\alpha) = \mathrm{Tr}_q^{q^k}\!\left(\mathrm{Tr}_{q^k}^{q^d}(\alpha)\right). \qquad (5\mathrm{A}.67)$$

Proof. Consider the nested trace-polynomial expression:

$$\mathrm{Tr}_q^{q^k}\!\left(\mathrm{Tr}_{q^k}^{q^d}(z)\right) = \sum_{j=0}^{k-1}\left(\sum_{i=0}^{\frac{d}{k}-1} z^{q^{ki}}\right)^{q^j} \qquad \text{(Definitions)}$$

$$= \sum_{j=0}^{k-1}\sum_{i=0}^{\frac{d}{k}-1} z^{q^{ki+j}} \qquad \text{(Theorem 5A.10)}$$

$$= \sum_{n=0}^{d-1} z^{q^n} \qquad (n = ki + j)$$

$$= \mathrm{Tr}_q^{q^d}(z) \qquad \text{(Definition)}$$

Evaluating the above polynomial relation at $z = \alpha$, α in $\mathrm{GF}(q^d)$, gives the final result. ∎

Property 6. Let $\mathrm{GF}(q^d)$ be the smallest field containing α, and let the minimum polynomial of $m_\alpha(z)$ be denoted by

$$m_\alpha(z) = \sum_{i=0}^{d} a_i z^{d-i}. \qquad (5\mathrm{A}.68)$$

Then

$$a_1 = -\mathrm{Tr}_q^{q^d}(\alpha). \qquad (5\mathrm{A}.69)$$

Proof. This property follows directly from Theorem 5A.12, whence

$$m_\alpha(z) = \prod_{i=0}^{d-1}\left(z - \alpha^{q^i}\right). \qquad (5\mathrm{A}.70)$$

∎

APPENDIX 5B. FACTORIZATIONS OF $2^n - 1$ AND SELECTED PRIMITIVE POLYNOMIALS

The table in this appendix contains prime factorizations of numbers of the form $2^n - 1$, and a listing of selected primitive polynomials of degree n over GF(2). For further information on factorizations of $2^n - 1$, see Reisel [1] and Brillhart et al. [9]. Useful tables of primitive polynomials have been published by Peterson and Weldon [2], Watson [3], Stahnke [4], Zierler and Brillhart [5, 6], Zierler [7], and Lidl and Niederreiter [8].

In certain listings the notation $F(n)$ will be used to represent the factorization of $2^n - 1$. For example, since

$$2^{40} - 1 = (2^{20} - 1)(2^{20} + 1),$$

the factorization of $2^{40} - 1$ reads $F(20) \times 17 \times 61681$, indicating that the factorization of $2^{20} - 1$ should be inserted. Clearly 17×61681 must be the factorization of $2^{20} + 1$.

The primitive polynomials in this table all have a small number of nonzero terms, and therefore will be represented by the set $\mathcal{I}_n$ of intermediate powers of z present in the polynomial. For example, the primitive polynomial $z^{24} + z^4 + z^3 + z + 1$ is represented by $\mathcal{I}_{24} = (4, 3, 1)$, i.e.,

$$z^{24} + z^4 + z^3 + z + 1 = z^{24} + \left(\sum_{i \in \mathcal{I}_{24}} z^i \right) + 1.$$

n	$F(n)$	$\mathcal{I}_n$
2	3	(1)
3	7	(1)
4	3×5	(1)
5	31	(2)
6	$3^2 \times 7$	(1)
7	127	(1), (3)
8	$3 \times 5 \times 17$	(4, 3, 2), (6, 5, 1)
9	7×73	(4)
10	$3 \times 11 \times 31$	(3)
11	23×89	(2)
12	$3^2 \times 5 \times 7 \times 13$	(6, 4, 1), (7, 4, 3)
13	8191	(4, 3, 1)
14	$3 \times 43 \times 127$	(5, 3, 1), (12, 11, 1)
15	$7 \times 31 \times 151$	(1), (4), (7)
16	$3 \times 5 \times 17 \times 257$	(5, 3, 2), (12, 3, 1)
17	131071	(3), (5), (6)
18	$3^3 \times 7 \times 19 \times 73$	(7), (9), (5, 2, 1)
19	524287	(5, 2, 1), (6, 5, 1)
20	$3 \times 5^2 \times 11 \times 31 \times 41$	(3)
21	$7^2 \times 127 \times 337$	(2)
22	$3 \times 23 \times 89 \times 683$	(1)

23	47×178481	(5), (9)
24	$3^2 \times 5 \times 7 \times 13 \times 17 \times 241$	(7, 2, 1), (4, 3, 1)
25	$31 \times 601 \times 1801$	(3), (7)
26	$3 \times 2731 \times 8191$	(6, 2, 1), (8, 7, 1)
27	$7 \times 73 \times 262657$	(5, 2, 1), (8, 7, 1)
28	$3 \times 5 \times 29 \times 43 \times 113 \times 127$	(3), (9), (13)
29	$233 \times 1103 \times 2089$	(2)
30	$3^2 \times 7 \times 11 \times 31 \times 151 \times 331$	(6, 4, 1), (16, 15, 1)
31	2147483647	(3), (6), (7), (13)
32	$3 \times 5 \times 17 \times 257 \times 65537$	(22, 2, 1), (28, 27, 1)
33	$7 \times 23 \times 89 \times 599479$	(13), (6, 4, 1)
34	$3 \times 43691 \times 131071$	(15, 14, 1), (27, 2, 1)
35	$31 \times 71 \times 127 \times 122921$	(2)
36	$F(18) \times 5 \times 13 \times 37 \times 109$	(11), (6, 5, 4, 2, 1)
37	223×616318177	(12, 10, 2), (6, 4, 1)
38	$F(19) \times 3 \times 174763$	(6, 5, 1)
39	$7 \times 79 \times 8191 \times 121369$	(4), (8), (14)
40	$F(20) \times 17 \times 61681$	(5, 4, 3), (21, 19, 2)
41	13367×164511353	(3), (20)
42	$F(21) \times 3^2 \times 43 \times 5419$	(23, 22, 1), (7, 4, 3)
43	$431 \times 9719 \times 2099863$	(6, 4, 3), (6, 5, 1)
44	$F(22) \times 5 \times 397 \times 2113$	(6, 5, 2), (27, 26, 1)
45	$7 \times 31 \times 73 \times 151 \times 631 \times 23311$	(4, 3, 1)
46	$F(23) \times 3 \times 2796203$	(21, 20, 1), (8, 5, 3, 2, 1)
47	$2351 \times 4513 \times 13264529$	(5), (14), (20), (21)
48	$F(24) \times 97 \times 257 \times 673$	(28, 27, 1), (7, 5, 4, 2, 1)
49	127×4432676798593	(9), (12), (15), (22)
50	$F(25) \times 3 \times 11 \times 251 \times 4051$	(4, 3, 2), (27, 26, 1)
51	$7 \times 103 \times 2143 \times 11119 \times 131071$	(6, 3, 1), (16, 15, 1)
52	$F(26) \times 5 \times 53 \times 157 \times 1613$	(3), (19), (21)
53	$6361 \times 69431 \times 20394401$	(6, 2, 1), (16, 15, 1)
54	$F(27) \times 3^4 \times 19 \times 87211$	(37, 36, 1), (6, 5, 4, 3, 2)
55	$23 \times 31 \times 89 \times 881 \times 3191 \times 201961$	(24), (6, 2, 1)
56	$F(28) \times 17 \times 15790321$	(7, 4, 2), (22, 21, 1)
57	$7 \times 32377 \times 524287 \times 1212847$	(7), (22), (5, 3, 2)
58	$F(29) \times 3 \times 59 \times 3033169$	(19), (6, 5, 1)
59	$179951 \times 3203431780337$	(22, 21, 1), (6, 5, 4, 3, 1)
60	$F(30) \times 5^2 \times 13 \times 41 \times 61 \times 1321$	(1), (11)
61	2305843009213693951	(5, 2, 1), (16, 15, 1)
62	$F(31) \times 3 \times 715827883$	(6, 5, 3), (57, 56, 1)
63	$F(21) \times 73 \times 92737 \times 649657$	(1), (5), (31)
64	$F(32) \times 641 \times 6700417$	(4, 3, 1)
65	$31 \times 8191 \times 145295143558111$	(18), (32), (4, 3, 1)
66	$F(33) \times 3^2 \times 67 \times 683 \times 20857$	(10, 9, 1), (8, 6, 5, 3, 2)
67	$193707721 \times 761838257287$	(5, 2, 1), (10, 9, 1)
68	$F(34) \times 5 \times 137 \times 953 \times 26317$	(9), (33), (7, 5, 1)
69	$F(23) \times 7 \times 10052678938039$	(6, 5, 2), (29, 27, 2)
70	$F(35) \times 3 \times 11 \times 43 \times 281 \times 86171$	(5, 3, 1), (16, 15, 1)

71	$228479 \times 48544121 \times 212885833$	(6), (9), (18), (20), (35)
72	$F(36) \times 17 \times 241 \times 433 \times 38737$	(53, 47, 6), (6, 4, 3, 2, 1)
73	$439 \times 2298041 \times 9361973132609$	(25), (28), (31), (4, 3, 2)
74	$F(37) \times 3 \times 1777 \times 25781083$	(7, 4, 3), (16, 15, 1)
75	$F(25) \times 7 \times 151 \times 100801 \times 10567201$	(6, 3, 1), (11, 10, 1)
76	$F(38) \times 5 \times 229 \times 457 \times 525313$	(5, 4, 2), (36, 35, 1)
77	$23 \times 89 \times 127 \times 581283643249112959$	(6, 5, 2), (31, 30, 1)
78	$F(39) \times 3^2 \times 2731 \times 22366891$	(7, 2, 1), (20, 19, 1)
79	$2687 \times 202029703 \times 1113491139767$	(9), (19), (4, 3, 2)
80	$F(40) \times 257 \times 4278255361$	(38, 37, 1), (7, 5, 3, 2, 1)
81	$F(27) \times 2593 \times 71119 \times 97685839$	(4), (16), (35)
82	$F(41) \times 3 \times 83 \times 8831418697$	(38, 35, 3), (8, 7, 6, 4, 1)
83	$167 \times 57912614113275649087721$	(7, 4, 2), (46, 45, 1)
84	$F(42) \times 5 \times 13 \times 29 \times 113 \times 1429 \times 14449$	(13), (8, 7, 5, 3, 1)
85	$31 \times 131071 \times 9520972806333758431$	(8, 2, 1), (28, 27, 1)
86	$F(43) \times 3 \times 2932031007403$	(6, 5, 2), (13, 12, 1)
87	$F(29) \times 7 \times 4177 \times 9857737155463$	(13), (7, 5, 1)
88	$F(44) \times 17 \times 353 \times 2931542417$	(72, 71, 1), (8, 5, 4, 3, 1)
89	$6189700196426901374449562111$	(38), (6, 5, 3)
90	$F(45) \times 3^3 \times 11 \times 19 \times 331 \times 18837001$	(5, 3, 2), (19, 18, 1)
91	$127 \times 911 \times 8191 \times 112901153 \times 23140471537$	(84, 83, 1), (7, 6, 5, 3, 2)
92	$F(46) \times 5 \times 277 \times 1013 \times 1657 \times 30269$	(6, 5, 2), (13, 12, 1)
93	$F(31) \times 7 \times 658812288653553079$	(2)
94	$F(47) \times 3 \times 283 \times 165768537521$	(21), (6, 5, 1)
95	$31 \times 191 \times 524287 \times 420778751 \times 30327152671$	(11), (17), (6, 5, 4, 2, 1)
96	$F(48) \times 193 \times 65537 \times 22253377$	(49, 47, 2), (7, 6, 4, 3, 2)
97	$11447 \times 13842607235828485645766393$	(6), (12), (33), (34)
98	$F(49) \times 3 \times 43 \times 4363953127297$	(11), (27), (7, 4, 3, 2, 1)
99	$F(33) \times 73 \times 199 \times 153649 \times 33057806959$	(7, 5, 4), (47, 45, 2)
100	$F(50) \times 5^3 \times 41 \times 101 \times 8101 \times 268501$	(37), (8, 7, 2)
101	$7432339208719 \times 341117531003194129$	(7, 6, 1)
102	$F(51) \times 3^2 \times 307 \times 2857 \times 6529 \times 43691$	(77, 76, 1)
103	$2550183799 \times 3976656429941438590393$	(9), (13), (30), (31)
104	$F(52) \times 17 \times 858001 \times 308761441$	(11, 10, 1)
105	$F(35) \times 7^2 \times 151 \times 337 \times 29191 \times 106681 \times 152041$	(16), (17), (37), (43), (52)
106	$F(53) \times 3 \times 107 \times 28059810762433$	(15)
107	$162259276829213363391578010288127$	(65, 63, 2), (7, 5, 3, 2, 1)
108	$F(54) \times 5 \times 13 \times 37 \times 109 \times 246241 \times 279073$	(31)
109	$745988807 \times 8700359860987209987332873$	(7, 6, 1)
110	$F(55) \times 3 \times 11^2 \times 683 \times 2971 \times 48912491$	(13, 12, 1)
111	$F(37) \times 7 \times 321679 \times 26295457 \times 319020217$	(10), (49)
112	$F(56) \times 257 \times 5153 \times 54410972897$	(45, 43, 2)
113	$3391 \times 23279 \times 65993 \times 1868569 \times 1066818132868207$	(9), (15), (30)
114	$F(57) \times 3^2 \times 571 \times 174763 \times 160465489$	(82, 81, 1)
115	$31 \times 47 \times 14951 \times 178481 \times 4036961 \times 2646507710984041$	(15, 14, 1)
116	$F(58) \times 5 \times 107367629 \times 536903681$	(71, 70, 1)
117	$F(39) \times 73 \times 937 \times 6553 \times 86113 \times 7830118297$	(20, 18, 2)
118	$F(59) \times 3 \times 2833 \times 37171 \times 1824726041$	(33), (45)
119	$127 \times 239 \times 20231 \times 131071 \times 62983048367 \times 131105292137$	(8), (38)
120	$F(60) \times 17 \times 241 \times 61681 \times 4562284561$	(118, 111, 7)

REFERENCES

[1] H. Reisel, *En bok om primtal*, Odense, Denmark: Studentlitteratur, 1968.

[2] W. Peterson and E. Weldon, *Error-Correcting Codes*. Cambridge, MA: MIT Press, 1972.

[3] E. Watson, "Primitive polynomials (mod 2)," *Math. Comp.*, vol. 16, 1962, pp. 368–369.

[4] W. Stahnke, "Primitive binary polynomials," *Math. Comp.*, vol. 27, Oct. 1973, pp 977–980.

[5] N. Zierler and J. Brillhart, "On primitive trinomials (mod 2)," *Inform. Contr.*, vol. 13, Dec. 1968, pp. 541–554.

[6] N. Zierler and J. Brillhart, "On primitive trinomials (mod 2), II," *Inform. Contr.*, vol. 14, June 1969, pp. 566–569.

[7] N. Zierler, "Primitive trinomials whose degree is a Mersenne exponent," *Inform. Contr.*, vol. 15, July 1969, pp. 67–69.

[8] R. Lidl and H. Niederreiter, *Finite Fields*, Reading, MA: Addison-Wesley, 1983.

[9] J. Brillhart, D. Lehmer, J. Selfridge, B. Tuckerman, and S. Wagstaff, Jr., *Factorizations of $b^n \pm 1$*. Providence, RI: American Mathematical Society, volume 22 in Contemporary Mathematics Series, 1983.

VOLUME I
INDEX